Information Sampling and Adaptive Cognition

A "sample" is not only a concept from statistics that has penetrated common sense but also a metaphor that has inspired much research and theorizing in current psychology. The sampling approach emphasizes the selectivity and biases inherent in the samples of information input with which judges and decision makers are fed. Because environmental samples are rarely random, or representative of the world as a whole, decision making calls for censorship and critical evaluation of the data given. However, even the most intelligent decision makers tend to behave like "naïve intuitive statisticians": They are quite sensitive to the data given but uncritical concerning the source of the data. Thus, the vicissitudes of sampling information in the environment together with the failure to monitor and control sampling effects adequately provide a key to reinterpreting findings obtained in the past two decades of research on judgment and decision making.

Klaus Fiedler is Professor of Psychology at the University of Heidelberg in Germany. Among his main research interests are cognitive social psychology, language and communication, social memory, inductive cognitive processes in judgment and decision making, and computer modeling of the human mind. Professor Fiedler is the winner of the 2000 Leibniz Award.

Peter Juslin is Professor of Psychology at Uppsala University in Sweden. His main research interests concern judgment and decision making, categorization, and computational modeling. He received the Brunswik New Scientist Award in 1994 and the Oscar's Award at Uppsala University in 1996 for young distinguished scientists. He has published numerous scientific papers in various journals, including many articles in the major APA journals such as *Psychological Review*.

Information Sampling and Adaptive Cognition

Edited by

KLAUS FIEDLER
University of Heidelberg

PETER JUSLIN
Uppsala University

CAMBRIDGE
UNIVERSITY PRESS

CAMBRIDGE
UNIVERSITY PRESS

32 Avenue of the Americas, New York NY 10013-2473, USA

Cambridge University Press is part of the University of Cambridge.

It furthers the University's mission by disseminating knowledge in the pursuit of education, learning and research at the highest international levels of excellence.

www.cambridge.org
Information on this title: www.cambridge.org/9780521831598

First published 2006

A catalogue record for this publication is available from the British Library

Library of Congress Cataloguing in Publication data

Information sampling and adaptive cognition / edited by Klaus Fiedler, Peter Juslin.
    p. cm.
Includes bibliographical references and index.
ISBN 0-521-83159-8 (casebound) – ISBN 0-521-53933-1 (pbk.)
1. Human information processing.   2. Cognition.   I. Fiedler, Klaus, 1951–
II. Juslin, Peter.   III. Title.
BF444.I55   2005
153.4 – dc22   2004025856

ISBN 978-0-521-83159-8 Hardback
ISBN 978-0-521-53933-3 Paperback

Contents

List of Contributors

Melissa Acevedo, Brown University

Greg Barron, Harvard Business School

David V. Budescu, University of Illinois

Nick Chater, University College, London

Robyn Dawes, Carnegie Mellon University

Ido Erev, Technion, Israel Institute of Technology

Klaus Fiedler, University of Heidelberg, Germany

Peter Freytag, University of Heidelberg, Germany

Gerd Gigerenzer, Max Planck Institute, Germany

Ralph Hertwig, University of Basel, Switzerland

Ulrich Hoffrage, Max Planck Institute, Germany

Robin M. Hogarth, Universitat Pompeu Fabra, Spain

Peter Juslin, University of Uppsala, Sweden

Yakoov Kareev, The Hebrew University of Jerusalem

Joshua Klayman, University of Chicago Graduate School of Business

Derek S. Koehler, University of Waterloo, Canada

Joachim Krueger, Brown University

Thorsten Meiser, University of Jena, Germany

Andreas Mojzisch, *Universität Göttingen,* Germany

Mike Oaksford, University of London

Fenna Poletiek, Leiden University, The Netherlands

Jordan M. Robbins, Brown University

Stefan Schulz-Hardt, Universität München, Germany

Peter Sedlmeier, Chemnitz University of Technology, Germany

Jack B. Soll, INSEAD Business School, France

Elke Weber, Columbia University

Chris M. White, Université de Lausanne, Switzerland

Anders Winman, Uppsala University, Sweden

PART I

INTRODUCTION

1

Taking the Interface between Mind and Environment Seriously

Klaus Fiedler and Peter Juslin

Metaphors mold science. Research on judgment and decision making (JDM) – for which psychologists and economists have gained Nobel prizes (Kahneman, Slovic & Tversky, 1982; Simon, 1956, 1990) and elevated political and public reputations (Swets, Dawes & Monahan, 2000) – is routinely characterized as having proceeded in several waves, each carried by its own metaphor. A first wave of JDM research in the 1960s compared people's behavior to normative models, a common conclusion being that, with some significant exceptions, human judgment is well approximated by normative models (Peterson & Beach, 1967): The mind was likened to an "intuitive statistician" (a metaphor borrowed from Egon Brunswik).

A second and more influential wave of JDM research, initiated by Tversky and Kahneman, emphasized the shortcomings, biases, and errors of human judgments and decisions (Kahneman et al., 1982). Because of the mind's constrained capacity to process information, people can only achieve *bounded rationality* (Simon, 1990), and – as inspired by research on problem solving (Newell & Simon, 1972) – they have to rely on *cognitive heuristics*. In contrast to problem-solving research, however, the focus was on heuristics leading to bias and error in judgment. The metaphor of an intuitive statistician was soon replaced by the notion that the mind operates according to principles other than those prescribed by decision theory, probability theory, and statistics. In the fond lurks Homo economicus, a rational optimizer with gargantuan appetite for knowledge, time, and computation defining the norm for rational behavior. In its shadow crouches Homo sapiens with its limited time, knowledge, and computational capacities, apparently destined for error and bias by the reliance on simple cognitive heuristics (see Gigerenzer, Todd, & the ABC Group, 1999, for a criticism and an alternative use of this metaphor).

This volume brings together research inspired by, or exemplifying, yet another developing metaphor that builds on and refines the previous metaphors, that of the mind as a *naïve intuitive statistician*. A sizeable

3

amount of recent JDM and social psychological research, in part covered by
the present volume, reevokes the metaphor of an intuitive statistician, but
a statistician who is naïve with respect to the origins and constraints of the
samples of information given. The basic idea is that the processes operating
on the given input information in general provide accurate descriptions of
the samples and, as such, are not violating normative principles of logic
and reasoning. Erroneous judgments rather arise from the naïvete with
which the mind takes the information input for granted, failing to correct for selectivity and constraints imposed on the input, reflecting both
environmental structures and strategies of sampling. A more comprehensive discussion of this conception – to be called a *sampling approach* in the
remainder of this book – can be found in Fiedler (2000).

Without making any claims for the superiority of this metaphor over its
predecessors, we provide in this volume an overview of research that can
be aligned with this sampling approach and we invite other researchers
to explore the possibilities afforded by this perspective. In the following,
we first try to characterize the metaphor of a naïve intuitive statistician
and discuss its potentials. Thereafter, we discuss the role of sampling –
the key construct with this approach – in behavioral science. We outline a
three-dimensional scheme, or taxonomy, of different sampling processes,
and we locate in this scheme the contributions to the present volume.

THE NAÏVE INTUITIVE STATISTICIAN

When judgment bias (defined in one way or another) is encountered, the
most common explanatory scheme in JDM and social psychological research is to take the information input as given truth and to postulate
cognitive algorithms that account for the deviation of the judgment output
from the input and its normatively permissible transformations. The input
itself is not assigned a systematic role in the diagnosis of judgment biases.
An obvious alternative approach is to assume that the cognitive algorithms
may be consistent with normative principles most of the time – as evident
in their successful use in many natural environments (Anderson, 1990;
Oaksford & Chater, 1998) – and to place the explanatory burden at the
other end: the information samples to which cognitive algorithms are
applied.

One way to explicate this alternative approach is by starting from the
assumption that the cognitive algorithms provide accurate descriptions
of the sample data and do not violate normative principles of probability theory, statistics, and deductive logic, provided they are applied under
appropriate task conditions and with sufficient sample size. Both normative and fallacious behavior may arise depending on the fit between the
algorithm and the sample input. From this point of view, the sampling constraints that produce accurate judgments should often be those that have

prevailed in our personal history or in the history of our species. Base-rate neglect may thus diminish when the process is fed with natural frequencies rather than single event probabilities (Gigerenzer & Hoffrage, 1995); overconfidence may diminish or disappear when general knowledge items are selected to be representative of the environment to which the knowledge refers (Gigerenzer, Hoffrage, & Kleinbölting, 1991; Juslin, 1994).

This formulation is reminiscent of the distinction between *cognitive processes* and *cognitive representations*. The proposal is that cognitive biases can be often explained not by cognitive processes that replace heuristic for normative algorithms, but by the biased samples that provide the representational input to the process. Both environmental and cognitive constraints may affect the samples to which the cognitive processes are applied, and the environmental feedback is often itself a function of the decisions that we make (Einhorn & Hogarth, 1978). At first, the distinction between bias in process or representation may seem of minor practical import; the outcome is a cognitive bias just the same. This conclusion is elusive however; the cure for erroneous judgment depends heavily on correct diagnosis of the reasons for the error.

Statistics tells us that there are often ways to adjust estimators and to correct for sampling biases. Under certain conditions at least, people should thus be able to learn the appropriate corrections. Specifically, both incentives are needed to detect errors in judgment and correct for them, because it matters to the organism, and unambiguous corrective feedback must be available. The latter condition may be violated by the absence of any sort of feedback whatsoever. More often, however, the sampling process is conditioned on unknown constraints and there exists no self-evident scheme for interpreting the observed feedback (Brehmer, 1980).

A fundamental source of ambiguity lies in the question of whether outcomes should be interpreted within a deterministic causal scheme or within a probabilistic one. If a deterministic scheme is assumed, any change can be traced to causal antecedents. A probabilistic scheme highlights statistical notions like regression to the mean. We might, for example, observe that a soccer player plays extremely well in a first game where she plays forward and then that she plays poorer as a libero in a second game. On a causal deterministic scheme, we might conclude that the difference is explained by a causal antecedent, such as playing as forward or as libero. With a probabilistic scheme, we might conclude that since the player was extremely good in the first game, there is little chance that everything turns out equally well in the second game, so there is inherently a large chance that performance is poorer in the second game. Moreover, if a probabilistic scheme is adopted, it matters whether the observed behavioral sample is a random sample or conditioned on some selective sampling strategy. The amount of mental computation required to assess the nature of the problem at hand is far from trivial.

The notion of a naïve intuitive statistician can indeed be regarded as an explication of the specific sense in which the mind *approximates* a statistician proper. Consistently with the heuristics and biases program, some of the failures to optimally correct sample estimates may derive from constrained and heuristic cognitive processing. Many of the phenomena that have traditionally been accounted for in terms of, say, the availability heuristic (Tversky & Kahneman, 1973) can with equal force be interpreted as the results of optimal processing by frequency counts that rely on biased samples. For example, misconceptions about the risk of various death causes may in part derive from biases in the newspaper coverage of accidents and diseases, that is, the environmental sample impinging on the judge, rather than biased processing of this input (Lichtenstein & Fischhoff, 1977). This recurrent working hypothesis, that cognitive algorithms per se are unbiased and that the explanatory burden is on the input end of the process, is central to the metaphor of a naïve intuitive statistician. We propose that this metaphor can complement previous approaches to judgment and decision making in several important ways.

One virtue is that it has the inherent potential to explain both achievement and folly with the same conceptual scheme, thereby reconciling the previous two research programs. Although fad and fashion have changed over the years, it seems fair to conclude that there exist numerous studies documenting both impressive judgment accuracy and serious error. In terms of the metaphor of a naïve intuitive statistician the issue of rationality is a matter of the fit between the cognitive processes and the input samples to which they are applied. Arguably, rather than having a continued discussion of the rationality of human judgment in a general and vacuous manner, it is more fruitful to investigate the sampling conditions that allow particular cognitive algorithms to perform at their best or at their worst.

In the light of the naïve intuitive statistician, the key explanatory concepts concern information sampling and the influence of different sampling schemes. In the next section, we discuss the manifold ways in which information sampling enters research in behavioral science. Thereafter, we suggest a taxonomy of different sampling processes leading to distinct cognitive phenomena. The goal of the taxonomy is to provide an organizing scheme within which the contributions to the present volume can then be located.

SAMPLING – A UNIVERSAL ASPECT OF PSYCHOLOGICAL THEORIZING

Although "sampling," a technical term borrowed from statistics, is quite narrow for a book devoted to the broad topic of adaptive cognition, it nevertheless sounds familiar and almost commonsensical, given the ubiquitous role played by statistics in everyday decisions and behavior. Sampling is actually an aspect that hardly any approach to psychology and behavioral science can miss or evade, though the role of sampling as a key theoretical term often remains implicit. Look what it takes to explain behavior in a nondeterministic, probabilistic world. Why does the same stimulus situation elicit different emotional reactions in different persons? Why does the time required to solve a problem vary with the aids or tools provided in different experimental conditions, or with personality traits of the problem solvers? What facilitates or inhibits perceptual discrimination? What helps or hinders interpersonal communication? How can we explain the preferences of decision makers and their preference reversals?

Theoretical answers to these questions share, as a common denominator, the principle of *subset selection*. To explain human behavior, it is crucial to understand which subset of the universe of all available stimuli and which subset of the universe of all possible responses are chosen. For instance, theories of emotion explain the conditions under which specific emotional reactions are selected from a broader repertoire of response alternatives. Problem solving entails selecting appropriate subsets of tools, talents, and strategic moves. Perceptual discrimination is a function of the stimulus features selected for the comparison of two or more percepts. Likewise, effective communication depends on whether the receiver finds the same subset of meaningful elements in the communication that the communicator has intended. Decision making has to rely on selected states of the world, which trigger subjective probabilities, and on desired end states, which trigger subjective utilities. Theoretical explanation means to impose subset constraints on psychological processes, which as a matter of rule entails sampling processes.

Sampling Constraints in the Internal and External Environments

Two metatheories of subset selection can be distinguished in behavioral science; one refers to sampling in the internal environment and the other to sampling in the external environment. Here we totally refrain from assumptions about cognitive processes but we are exclusively concerned with environmental information, which can however be represented internally or externally. A considerable part of environmental information encountered in the past is no longer out there as perceptual input but represented in memory, much like present environmental information may be represented in the sensory system. Both internally (memorized) as well as externally (perceived) represented information reflects environmental input, though, as clearly distinguished from cognitive processes.

It is interesting to note that the first metatheory, concerned with sampling in the internal environment stored in memory, has attracted much more research attention in twentieth-century psychology than has the second metatheory, which builds on information sampling in the external world. Countless theories thus assume that variation in performance reflects which subset of all potentially available information is mentally activated when an individual makes decisions in a given situation at a given point in time (Martin, 1992). Drawing on concepts such as accessibility, priming, selective knowledge, retrieval cues, domain specificity, schematic knowledge, or resource limitation, the key idea is that human memory is the site of subset selection, where constraints are imposed onto the stimulus input driving judgments and decisions. The domain of this popular metatheory extends well beyond cognitive psychology. Selective accessibility of knowledge structures or response programs also lies at the heart of the psychology of emotions, aggression, social interaction, development, and abnormal psychology.

The second metatheory, although similar in its structure and rationale, is much less prominent and continues to be conspicuously neglected in behavioral science (Gigerenzer & Fiedler, 2005). The same selectivity that holds for environments conserved in memory also applies to the perceptual environment originally encountered in the external world. The information provided by the social and physical environment can be highly selective as a function of spatial and temporal constraints, social distance, cultural restrictions, or variable density of stimulus events. People are typically exposed to richer and more detailed information about themselves than about others, just because they are closest to themselves. Similarly, they are exposed to denser information about in-groups than out-groups of their own culture than to other, exotic cultures. Not only does environmental input vary quantitatively in terms of the density or amount of detail; it is also biased toward cultural habits, conversational norms, redundancy, and specific physical stimulus properties.

Analysis of the stimulus environment impinging on the organism, and of its constraints on information processing, is of similar explanatory power as the analysis of memory constraints – as anticipated in the seminal work of Brunswik (1956), Gibson (1979), and Lewin (1951). In a hardly contestable sense, the perceptual input from the external world even has priority over internal representations in memory, because the former determines the latter. A precondition for organisms to acquire (selective) world knowledge and adaptive behavior is that appropriate stimuli, task affordances, or feedback loops are provided in the first place. Education can indeed be regarded as a matter of engineering the environment to provide the information samples most conducive of learning. Academic ecologies constrain education; professional ecologies constrain the development of specialized skills and expertise; television and media ecologies constrain the development of aggression and other social behaviors; marketing ecologies place constraints on the behavior of consumers, entrepreneurs, and bargainers; and the "friendliness" versus "wickedness" of learning environments can explain the occurrence of cognitive fallacies and illusions.

A nice example, borrowed from Einhorn and Hogarth (1978), is that personnel managers' self-confident belief that they have selected the most appropriate employees will hardly ever be challenged by negative feed-back, just because the applicants they have rejected are not available. Simi-larly, many police officers and lawyers continue to believe in the accuracy of polygraph lie detection, because validity studies selectively include those cases that confirm the polygraph (i.e., confessions after a positive test re-sult) but exclude those cases that could disconfirm the test (Fiedler, Schmid, & Stahl, 2002; Patrick & Iacono, 1991), for in the absence of a confession, the validity criterion is unknown.

Although the point seems clear and uncontestable, selective environ-mental input has rarely found a systematic place in psychological theo-ries, with a few notable exceptions (Chater, 1996; Oaksford & Chater, 1994; van der Helm & Leeuwenberg, 1996). One prominent exception is Eagly's (1987) conception of gender differences as a reflection of societal roles played by men and women. In most areas of research, however, envi-ronmental factors are either completely ignored or, at best, assigned the role of moderators, acting merely as boundary conditions to be counter-balanced in a well-conducted experiment. A conspicuous symptom of this state of affairs is that research methodology and peer reviewing put much weight on representative sampling of participants, but stimuli and tasks are often selected arbitrarily and one or two problems or task settings are usually considered sufficient (Wells & Windschitl, 1999). Contempo-rary psychology has not even developed appropriate concepts and tax-onomies for types of environmental constraints. As a consequence, many intrapsychic (i.e., cognitive, motivational, emotional) accounts of phenom-ena are susceptible to alternative explanations in terms of environmen-tal constraints (Fiedler & Walther, 2004; Gigerenzer & Fiedler, 2004; Juslin, Winman, & Olsson, 2000).

In spite of the uneven status attained by the two metatheories in the past, sampling processes in internal and external environments draw on the same principle of selectivity – the shared notion that behavioral variation can be explained in terms of subsets of information that are selected from a more inclusive universe. As this notion is equally applicable to internal and external sampling processes, it opens a modern perspective on adaptive cognition and behavior (cf. Gigerenzer, Todd, & the ABC Group, 1999), which takes the interface between mind and the environment seriously. However, given the traditional neglect of external ecologies in the study of internal cognitive processes (Gigerenzer & Fiedler, 2004), the present book emphasizes primarily sampling in the external world.

A TAXONOMY OF SAMPLING PROCESSES

The sampling metaphor is rich enough in implicational power to suggest an entire taxonomy, an elaborated conceptual framework, within which the various cognitive–ecological interactions described in the following

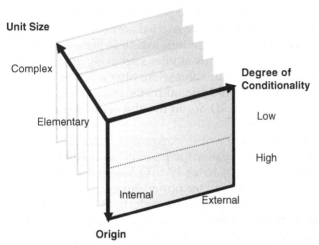

FIGURE 1.1. A three-dimensional taxonomy of sampling processes.

chapters can be located. For the present purpose (i.e., to provide a preview of this volume), we introduce only three dimensions underlying such a sampling taxonomy. (Further aspects will be discussed in Chapter 11). However, for the moment, the following three dimensions are required to structure the contents of the present volume. One dimension on which sampling processes can differ was already mentioned; we may distinguish samples by their *origin*, as either internally (cognitively) generated samples or externally (environmentally) provided. A second important dimension refers to *conditionality*; drawing samples can to varying degrees be conditional on a more of less constraining subset. On a third, multilevel dimension, the *unit* of information being sampled can vary from elementary to complex units. Within the resulting three-dimensional space, depicted in Figure 1.1, different types of sampling can be located.

As mentioned, other dimensions may also be revealing and theoretically fruitful, for example, *passive* sampling (being exposed to environmental stimuli) versus *active* sampling (a strategic information search in memory or external databases). Another intriguing dimension is whether samples are drawn *simultaneously*, in a single shot, or *sequentially* over time as in a continuous updating process (Hogarth & Einhorn, 1992). Last but not least, one may distinguish sampling processes going on in the *judges' minds* and in the *researchers' minds*. Note also that all three dimensions included in Figure 1.1 refer to antecedent properties of sampling processes, not to the outcome. When sampling outcomes are taken into account, we encounter further distinctions with important psychological implications, such as the *size* of samples (large versus small) or their *representativeness* (biased versus unbiased). The taxonomy also leaves the *content* unspecified, that is, whether the entities being sampled are persons, stimulus objects, task situations, or contextual conditions.

Origin: Internal versus External

Let us now explain and illustrate the three aspects of Figure 1.1, starting with the origin dimension. For a concrete illustration, consider the case of an ordinary consumer, who is concerned with prices, product quality, and personal need satisfaction. When facing the complex decision to purchase a car, for instance, the effective information input that determines or constrains the decision is subject to multifaceted sampling processes. With respect to the origin dimension, the consumer will consult both internal sources (i.e., memories of past purchase decisions; epistemic knowledge) and external sources (e.g., magazines, salespersons, fellow consumers, or Internet sites). It is crucial to realize that the information collected from both internal and external sources is merely a subset of the universal set of all information. Moreover, as a matter of principle, the amenable subset does not come with the guarantee of an unconstrained random sample. Given that sources of information differ in proximity, accessibility, familiarity, and reliability, as well as in the readiness to provide information selectively, being exposed to a true random sample with regard to all contributing sources would appear to be a highly unlikely exception. One may at best expect that – aggregating over many sources and subsamples – various biases will cancel each other out so that the resulting aggregate will produce quasi-random distributions on specific attribute dimensions. For example, the distributions of the price–service relation offered for the same category of cars by different providers may be quasi-random, providing a representative background or norm distribution (Kahneman & Miller, 1986) against which individual offers can be compared, even though all individual sources provide biased samples. However, granting that the resulting aggregate sample happens to give an unbiased impression of the price–service relations, the same sample could be highly biased and selective regarding other aspects such as, for instance, the frequency and intensity of worries to be expected with the car (in terms of breakdowns, overly expensive maintenance and repairs, etc.). These are clearly less likely to be encountered in advertising.

Degree of Conditionality

The latter example entails an allusion to the second dimension, conditionality, that is, to what extent samples are generated conditional on selective constraints. Certainly, all sampling is conditional to a particular position in time and space, to the availability of measurement tools, and to other limiting conditions. Nevertheless, the degree of conditionality can vary considerably, especially the degree to which sampling is output-bound, that is, conditional on the very dependent variable, or output, to be evaluated. In product marketing, for instance, the crucial dependent variable, or criterion, is positive versus negative evaluation of products, but information sampling in advertising is conditional on this very criterion, positivity. Thus, the information that is made available to consumers and customers is conditional on what producers and providers want to reveal, or what they are forced to reveal. Accident and deficit statistics are rarely available and much less visible in relevant magazines and information sources than assets and performance features of cars. To the extent that sampling is contingent on positive rather than negative valence or outcomes, variation in sample size will systematically reinforce the one aspect on which sampling is contingent. Given that product information sampling is contingent on positive valence, the information environment of advertising is subject to a positivity bias, with assets more likely to be advertised than deficits. Moreover, the apparent positivity of products should become more and more visible as sample size increases in an extended decision process.

Product information sampling is contingent not only on producers' and providers' intent but also on other informants' motives and constraints. These may further enhance the tendency to make product assets more visible than deficits. For example, when we ask our friends and acquaintances what their experience has been with a car of a certain brand, cognitive dissonance may prevent them from admitting that their own decision has been wrong or naïve. Radio and TV stations are obliged to broadcast commercials, whereas negative advertising is hardly possible for pragmatic reasons. Another reason for selective, conditional sampling is that insurance companies – repositories of statistics on accidents and many other negative outcomes, such as theft, vandalism, and material insufficiencies – are not interested in revealing such negative information.

Obviously, the selectivity of product information, as multiply conditionalized on positive valence, may be counteracted in part by opposite tendencies, such as a tendency for internal memory sampling to favor negative valence. However, again, there is no reason to hope that on aggregate multiple conditions will cancel each other out completely. The effective product of multiply conditional information sampling often entails a marked bias. The effective information input of consumers is strongly conditional on what they possess, on what they are interested in, on what is regularly covered in the media, in which country they are living, and on their age, gender, and level of education.

Statistically, the result of conditional sampling is a quota sample. If the quota used as conditionality criterion is explicitly known, the conclusions drawn from the resulting sample need not be mistaken or misleading. For instance, if Japanese cars make up 25% of the entire market but a quota of 50% Japanese cars is imposed on a consumer survey study, it is easy to correct for the oversampling by reducing the weight given to Japanese car data by the factor $25/50 = 1/2$. Cognitive processes might similarly correct for the influence of conditional information sampling. Even the consumer, who is fed selectively with positive product information and does not exactly know the precise quota (nor can the objective base rate of all assets and deficits ever be counted), should nevertheless understand, at a qualitative or ordinal level, that a correction in the negative direction is necessary.

However, as several chapters in this volume will show, even intelligent people are not very good at the metacognitive task of monitoring and correcting sampling processes. They act, in one word, as naïve intuitive statisticians. Thus, consumers normally take their information sample for granted and hardly ever reason about whether their sample has to be corrected upward or downward, for instance toward Japanese, American, or German cars, or toward externally provided data or internally generated feelings and scenarios. Similarly, voters do not correct for the fact that they mostly expose themselves to the election campaign of their favorite party and undersample data from rival parties. Likewise, scientists hardly ever correct their empirical and theoretical conclusions for the fact that published results are conditional on significance and some outcomes are more likely to be published than others.

As a consequence of this metacognitive myopia and unwillingness to engage in correction processes, resulting judgments and decisions are often severely biased (Fiedler et al., 2000). These biases can take incredible forms, such as in the famous social psychological finding that people infer someone's attitude from an essay written by this person, even when they know that the position advocated in that essay has been prescribed by the experimenter (Jones & Harris, 1967); that is, the information has been conditional on directive instructions. Likewise, people are typically insensitive to biases imposed by interviewers' leading questions on interviewees' answers (Semin, Rubini, & Fiedler, 1995; Swann, Guiliano, & Wegner, 1982), and learners are insensitive to the fact that the number of learning trials has an impact on their performance (Fiedler et al., 2002; Koriat, 1997) or that an increasing number of retrieval cues leads to richer samples retrieved from memory (Fiedler & Armbruster, 1994).

Unit Size: Elementary versus Complex

Let us finally consider the unit size dimension. All illustrative examples discussed so far have referred to sampling of natural, singular stimulus events: published assertions about cars, other people's utterances, experienced events related to cars, etc. However, these natural pieces of information can be decomposed into more elementary units, such as specific brands of cars or the sample of attributes or cue dimensions (e.g., performance, economy, design features, price, or comfort). Each of these specific brand names, or attribute dimensions, can in turn be decomposed into finer features, down to the molecular- or even neuronal-level features assumed in connectionist models (Roe, Busemeyer, & Townsend, 2001; Kashima, Woolcock, & Kashima, 2000). Sampling of these lower level features can have strong and significant effects on comparison processes (Tversky, 1977) and on the way in which products and brand categories are represented cognitively.

In any case, sampling processes at this elementary, subsymbolic level (Smolensky, 1988), below the natural level of singular stimuli, can be crucial to understanding the course of cognitive processes. Like the more familiar process of *extensional* sampling of stimulus events, this process of *intensional* sampling of stimulus aspects or meaning elements can originate in external stimulus sources as well as in internally generated associations or imaginations. Moreover, the generation of sampling of elementary stimulus features could in principle be conceived as a quasi-random process, as in a Thurstonian model (Juslin & Olsson, 1997; Thurstone, 1927), with little weight given to conditionality. The sampling of elementary features can also be strongly conditional on the context (Roe et al., 2001; Tversky & Simonson, 1991) or on the comparison stimulus (Tversky, 1977; Wänke, Schwarz, & Noelle-Neumann, 1995).

At the opposite end of the unit size dimension, consumers might sample complex knowledge structures, scripts, schemas, analogies, or decision-relevant networks that are superordinate to individual stimuli. For instance, our consumer may possess various scripts or decision-making routines that might all be applied to purchasing a car: one script for an impulse deal, one for systematic weighting of multiple attributes, one for imitating friends or role models, one for habitual adherence to the same brand one has always bought, etc. Which of these procedural options actually dominates the decision depends on knowledge-sampling processes at a level higher than individual stimulus events. Again, the sampling of superordinate units may be driven externally, through priming or social influence, or internally, through dominant memory functions or mental simulation, and the generation mode could be conditional, to varying degrees, on such factors as time pressure, financial constraints, or recency and on routine strength (Betsch et al., 2001).

APPLYING THE SAMPLING TAXONOMY

The different types of information sampling that can be located and distinguished in the three-dimensional space depicted in Figure 1.1 correspond to different theoretical approaches, with different implications. It is no coincidence, for instance, that theories based on outcome-conditional sampling rationales typically lead to empirical demonstrations of cognitive biases and judgment illusions. Priming effects on evaluative judgments (Musch & Klauer, 2002) reflect memory samples conditionalized on evaluative (positive or negative) primes. These primes can be elementary stimuli or complex, superordinate themes, schematic structures, or even diffuse mood states (Forgas, 1995). Errors in syllogistic reasoning can result from a belief bias, a selective tendency to conditionalize reasoning processes on endorsed assumptions or beliefs about the world (Klauer, Musch, & Naumer, 2000). Numerous manifestations of so-called confirmation biases or self-fulfilling prophecies can result from an information-search strategy called positive testing (Klayman & Ha, 1987), which consists in sampling information conditional on the task focus. The impression of a person solicited in an interview will be quite different depending on whether the interview focuses on extravert or introvert behavior (Snyder, 1984), honesty or dishonesty (Zuckerman et al., 1984), or overt or covert aggression (Fiedler & Walther, 2004). As already explained, conditional samples place a burden on metacognitive monitoring and correction. The quota samples that result from conditionalized information search require intelligent corrections of the obtained distributions, but this metacognitive function seems to be quite restricted (Fiedler, 2000).

In contrast, theoretical approaches within the Brunswikian tradition (Brunswik, 1956; Juslin, 1994; Gigerenzer et al., 1991) often attempt to realize representative sampling of persons, stimuli, and tasks that conserve the properties of the total ecological population. Within this approach, cognitive biases often disappear and give way to well-adapted, calibrated judgments. For example, when research participants rate the subjective confidence that their responses to knowledge questions are correct, they often are overconfident. That is, the results are miscalibrated in the sense that their average confidence expressed in percent is higher than the percent of actually correct responses. However, although overconfidence has been a research topic for decades (Adams & Adams, 1961; Lichtenstein, Fischhoff, & Phillips, 1982; Oskamp, 1965) and is often treated like a well-established finding, it could to a large degree be germane to conditional sampling. Because experimenters use to select experimental knowledge questions that are tricky or difficult – that is, conditionalize the sample on high difficulty – the so-called overconfidence bias may in part reflect that confidence is simply calibrated to an easier learning environment. Indeed, when representative samples of knowledge questions are used, the overconfidence bias is greatly reduced – though to varying degrees with different probability assessment formats (see Chapter 9) – and judgments appear much better calibrated (Juslin et al., 2000).

As an aside, the example shows that sampling processes in the researcher are as important a determinant of research findings as sampling of information impinging on research participants. Moreover, the pervasive overconfidence effect in arbitrary task samples, as contrasted with good calibration in representative samples, need not be ascribed to only the researcher's fault or only the participant's fault; crucially, it can be a consequence of the failure to correct for conditional samples on either side. Just as the researcher fails to correct the participants' ratings for increased difficulty, it can be claimed that judges fail to correct their experimental confidence scale for the changing environment or difficulty level. After all, the transfer of learning from one environment to another environment, with different properties, is a prominent aspect of adaptive intelligence. So there can be no absolution for participants either. Rather, the shared deficit of both researchers and research participants is that they both fail to apply the metacognitive corrections that would be necessary to utilize the tricky input stemming from conditionalized samples.

Keeping within the overconfidence phenomenon, Erev, Wallsten, & Budescu (1994) detailed an intriguing case of a research illusion through conditional sampling. These authors showed, provocatively, that within the same empirical data set one can find strong evidence for both over- and underconfidence, depending on the direction of conditionality. In traditional overconfidence research, objective accuracy rates have normally been conditionalized on subjective confidence. That is, the objective accuracy of judgments is analyzed given that judgments belong to a certain confidence level, say, 80%. Obtaining overconfidence (i.e., an objective accuracy rate less than 80%) under such conditions is a natural consequence of the regressiveness of all imperfect prediction. Because the correlation between objective accuracy and subjective confidence is less than perfect, the prediction of objective accuracy given a certain level of subjective confidence must be regressive. Conversely, Erev et al. (1994) were able to produce *underconfidence* when the very same data are analyzed the other way around, conditionalizing subjective confidence on objective accuracy. Thus, considering a subset of judgments that are actually correct at a rate of 80%, one will typically find average subjective confidence ratings within this subset to be less than 80%, this time pointing to underconfidence.

This "regression trap" (Rulon, 1941; Gigerenzer & Fiedler, 2005; Stelzl, 1982) contributes to many illusions and pitfalls resulting from conditional sampling. Regressiveness in part underlies the so-called hard–easy effect (Juslin et al., 2000), that is, the empirical fact that overconfidence is stronger for difficult problems and lower or even reversed for easy problems. In fact, when problems lead to extremely low confidence ratings, regression implies that the corresponding accuracy scores will not be as extremely low. Amazing fallacies induced by the regression trap have been documented in numerous other areas of everyday reasoning and decision making. For the stock market – a particularly regressive environment – it has been shown that conditionalizing the acquisition of shares each year on above-average performance can lead to lower profit across a longer period than conditionalizing share acquisition on below-average performance (Hubert, 1999).

In the educational domain, regression implies that desirable child behavior is likely to be followed by less desirable behavior and, vice versa, undesirable conduct will often be followed by less undesirable conduct – even in the absence of educational treatments. If in such an environment, characterized by a good deal of baseline variation or "noise," reward is conditionalized on highly desirable behavior and punishment on highly undesirable behavior, the illusion can be nourished that punishment is more effective as an educational technique than reward. The consequences of punishment and reward will be conditionalized on situations likely to be followed – through regression – by observable samples of improvement versus impairment, respectively.

Thus far, we have discussed at some length that Brunswikian sampling (unconditional samples of naturally occurring units in the external environment) often leads to optimistic appraisals of human intelligence, whereas strongly conditional sampling (whether internally from memory or from external ecologies) often produces serious misunderstandings and biases. Let us now turn to the notion of Thurstonian sampling, which highlights the fact that information samples are permanently generated in a random or quasi-random fashion within the organism itself. Thus, just as the nystagm of the eye produces a constant variation in the stimulation pattern on the retina, the central nervous system in general is permanently active in producing variation in the cognitive representation of stimulus and response patterns. The idea underlying Thurstone's (1927) law of comparative judgment is that the cognitive representation of two stimuli to be compared, and hence the position of stimuli on a metric attribute dimension, is not invariant but constantly changing, producing a (normal) distribution of values that a stimulus can take at different moments in time. For instance, the consumer's evaluation of two cars on the elegance-of-design dimension is not assumed to be constant; it varies as a function of associative processes and intrinsic system variation. The distributions of all momentary design evaluations that each car can produce can overlap more or less, and the consumer's degree of preference for one over the other car is a function of this overlap. With reference to our taxonomy, Thurstonian sampling processes can be typically characterized as internal, low in conditionality, and referring to elementary stimulus aspects (e.g., aspects of design, momentary associations, or other factors).

Theories inspired by Thurstonian sampling models, such as signal detection and statistical decision theories (cf. Swets et al., 2000), highlight the variable, oscillating, and permanently changing nature of cognitive representations. Although the underlying variation is typically conceived as a random, unconditional sample, it is but one step further to recognize that Thurstonian processes may reflect conditional sampling as well. If this happens, the resulting cognitive responses may again be biased – as is typical of conditional sampling. As an illustration, ordering effects in paired comparisons may be because features of the stimulus present first may largely determine the features sampled for the representation of the second stimulus. For instance, when comparing "tennis" and "football," the proportion of respondents who prefer tennis is higher than when comparing "football" and "tennis" (Wänke et al., 1995). This is, apparently, because tennis in the first position causes a different subset of football features being sampled than when football is considered first. Sampling of the second stimulus' features appears to be conditionalized on the first stimulus in paired comparison (Koriat, Fiedler, & Bjork, 2005). Conditional sampling may thus underlie a number of context influences on paired comparisons, such as the attraction effect, the similarity effect, and the compromise effect (see Roe et al., 2001). The common denominator of these context effects is that the internal sampling of features for two focal stimuli, A and B, is conditional on the features that characterize a task-irrelevant context stimulus, C.

ORGANIZATION OF THE PRESENT VOLUME

According to the present taxonomy, different types of sampling generate different theoretical approaches and different research findings. This taxonomy can be used to organize the contributions to the present volume. The clusters, or chapter sections, in the table of contents indeed reflect distinct approaches that can be located within the three-dimensional space.

Following this introduction, the next section has the headline or common denominator of *The Psychological Law of Large Numbers*. The research reported here typically refers to experimental tasks involving external sampling of singular stimulus events, at low levels of conditionality. However, crucially, the samples of information vary in size, and the impact of sample size on judgment and decision making will be examined from the standpoint of various novel theoretical positions. Although no serious bias or output-bound selectivity is involved in these approaches, sample size can have a systematic impact on decision making, for the size of a sample determines its reliability and perceived variability. As a consequence, inferences based on large numbers of observations are by default more accurate and stable and more likely to render even weak tendencies visible than inferences based on few observations. For instance, tests including many items are more reliable and to that extent promise to be more valid than tests including fewer items (Kuder & Richardson, 1937). There is also ample evidence in social psychology that prevailing tendencies in social behavior tend to be more visible in large groups than in small ones (Fiedler, 1996; Hamilton & Gifford, 1976) and that a greater number of attributes can be identified for larger groups, leading to more varied, less homogenous impressions of large groups than of small groups (Judd & Park, 1988).

However, the research covered in this section reveals that things may be slightly more complex and may sometimes counterintuitively diverge from Bernoulli's law of large numbers. In Chapter 2, Kareev points out various consequences of the simple and well-known fact that the variance of a sample shrinks with decreasing sample size. The loss of one degree of freedom in a sample of size N implies a variance reduction by $(N - 1)/N$, which affects small samples more than large ones. Kareev shows that when decisions have to rely on sampled information, decision makers take the variance of the sample for granted and do not correct for the shrinkage of variance in small samples. Kareev demonstrates, vividly, that this seemingly tiny failure to monitor and control for sample-size effects at the metacognitive level can have quite far-reaching consequences. The same metacognitive insensitivity for sampling effects – which is typical of the naïve intuitive statistician – will be encountered in many chapters even though in different forms.

Chapter 3 by Sedlmeier is concerned with the intuitive understanding of the law of large numbers, that is, whether people know that reliability increases with increasing sample size and that statistics of small samples are more variable than large-sample statistics. Judges are quite sensitive to the law of large numbers when their judgments are informed by frequency distributions of singular stimuli, but not when they have to deal with sampling distributions, that is, with samples of samples, or more complex units. Normatively, responsible judges and decision makers ought to understand that inferences become more and more unstable and error prone as the number of underlying observations decreases. Sedlmeier's chapter specifies the conditions under which this precondition of rational decision making is met, or is not met, and he presents a computer simulation model of associative learning that can account for the performance of human decision makers.

Whereas Chapters 2 and 3 revolve around the questions of how much information actually is, and how much information judges believe to be, in small and large samples, respectively, Chapter 4 addresses the relative attention or weight given to frequent and infrequent events as they compete for relative impact on decisions. The authors of Chapter 4, Hertwig, Barron, Weber and Erev, offer a compelling explanation for the empirical fact that rare events are sometimes overweighted and sometimes underweighted or even overlooked. When small frequencies or probabilities are received explicitly, stated numerically, they tend to be overestimated, but when frequencies are experienced and have to be extracted from a series of raw observations, they tend to be underestimated and to go unnoticed among the majority of events from more numerous classes. This kind of dependence of frequentistic information on presentation mode can be shown to have important implications for human and animal learning, and for the evolution of adaptive behavior in general (Weber, Shafir, & Blais, 2004).

Chapter 5 is a joint project of several authors – Juslin, Fiedler, and Chater – each advancing different views in a provocative controversy, which is still ongoing. This controversy tackles the common claim that the quality of judgments and decisions must always increase with increasing amount of information or sample size. The debate is inspired by a theoretical argument recently raised by Kareev (2000). Kareev's point was that small samples may actually render existing contingencies in the environment *more* visible than large samples, owing to the skewed sampling distribution of correlations, which tend to exaggerate existing population correlations in the majority of cases. Juslin and Olsson (2005) in turn, have provided a critique of Kareev's approach, arguing that the alleged advantage of small over large samples disappears when the expected values of both correct and incorrect decisions are taken into account. Nevertheless, the controversy exhibited in this chapter leaves the possibility open that, under specifiable boundary conditions, "less may actually be more." The reader is invited to actively participate and evaluate the arguments for and against the claim that small samples may – under specific conditions – inform better decisions, even when large samples generally result in more accurate estimates.

The second major section is about *Accurate Judgment of Biased Samples*. The theories and models that are grouped in this section are typically based on externally gathered samples of varying unit size, but samples are output-bound, conditionalized on the very attribute to be evaluated. As a consequence, a sampling bias is present in the first place, leading to biased judgments even when the judgment process itself is completely unbiased.

Chapter 6 by Freytag and Fiedler provides an overview of a suitable research paradigm that can be used to study the "biases without biased processes" topic, which characterizes the entire section. In this paradigm, participants are asked to evaluate the validity of scientific results, based on a sketch of the way in which researchers have composed their study samples. The question is to what extent judges are sensitive to biases built into the sampling or data-generation process. As it turns out, judges are particularly shortsighted at the metacognitive level for sampling biases resulting from output-bound selection of cases, that is, samples that overrepresent the feature to be estimated.

Chapter 7 by Dawes conveys a similar story, emphasizing the role of output-bound, conditionalized sampling on legal judgments, with particular reference to the claim that child abusers never stop on their own. Judging such a claim is conditionalized on the acquisition of a sample of observations about child abusers. Such a sample is clearly more likely to include information about cases of continued child abuse than about cases in which the abuse stopped – a negative event that is quite hard to access.

Dawes was clearly one of the first authors to point out "structural availability biases" since his earlier writings.

In Chapter 8, Klayman, Soll, Juslin, and Winman elaborate on the reasons for the divergent results when confidence is assessed by probability assessment versus assessment of confidence intervals: Overconfidence is much more profound with confidence intervals. The authors discuss both explanations that assume the information samples per se are unbiased but naïvely interpreted when used to infer population properties and explanations that emphasize that the information samples are biased by selective search of memory.

Chapter 9 by Meiser presents original research on a cognitive illusion that has not been recognized until very recently: the pseudocontingency illusion (Fiedler & Freytag, 2004). Pseudocontingencies originate in the failure to consider baseline constraints in the sampling process. For example, considering the contingency between two dichotomous attributes, say gender (male versus female) and achievement in science (high versus low), a pseudocontingency results when both marginal distributions are skewed, such that most students in a class are male, and a majority of all students show high achievement. Such an alignment of two skewed base rates results in the impression that males perform better than females, which is a category mistake. In Meiser's research, the pseudocontingency effect is applied to illusions in the social psychological domain of intergroup relations.

Chapter 10 closes this section with Chater and Oaksford's examination of mental mechanisms: speculations on human causal learning and reasoning. The major point conveyed in this chapter is that, very often, causal inferences are not driven by the statistics of large samples, but by the properties of very few observations. Even in the extreme case of only one or two observations – one pairing of a cause and an effect – the spatial–temporal contiguity or the intrinsic similarity of the cause and the effect may place quite strong constraints on the inference process. With reference to the prominent Monty Hall problem, it is demonstrated once more that the failure to understand the genesis of a sample – for instance, to distinguish between random and intentionally selected information – lies at the heart of problem-solving difficulties.

What information contents are sampled? This is the topic of the third major section, which is introduced in Chapter 11 by Gigerenzer, who provides a conceptual framework for the question: What's in a sample? Gigerenzer distinguishes among three aspects underlying this question: who does the sampling, why is the sampling done, and how does one sample.

The question of what is being sampled pertains to the partitioning, or grain size, of the categories used for sampling. Support theory (Tversky & Koehler, 1994) is the starting point for White and Koehler's

Chapter 12. When a category is unpacked or partitioned into its constituents or subcategories, the evidential support for this category will increase. In other words, support is a subadditive function. The sample of available information that supports the hypothesis that, say, the stock market will experience a crash this year is weaker than the aggregate support for the component hypotheses that a crash will occur in spring, in summer, in fall, or in winter. White and Koehler present empirical evidence and computer simulations for a model called ESAM that applies these ideas to the lowest level of information units, the sampling of various numbers of cues that describe decision options.

In Chapter 13, Mojzisch and Schulz-Hardt elucidate the social origins of sampled information in group interaction. The shared information effect, originally demonstrated by Stasser and Titus (1985), refers to the phenomenon that in group decision making, redundant arguments that are already shared by other group members are more likely to be raised, or sampled, than new, independent arguments. Thus, the samples generated in social interaction are likely to be biased toward consensual, socially shared information, and they will typically mimic more agreement than is actually there. Mojzisch and Schulz-Hardt provide an up-to-date picture both of their own research on this fascinating topic and of the work of other leading researchers in this area.

Budescu's Chapter 14 also speaks to social origins of information sampling. However, whereas the sampling unit in Chapter 12 is the argument, the unit of sampling in this chapter is the consultant. The question involves how decision makers aggregate information stemming from various sources or consultants that may vary in overlap and consensus. Variance turns out to be a major determinant of the confidence with which decisions are made. This chapter appears to be of particular interest for decision making in the economic context of principal–agent contracts.

Chapter 15 by Krueger, Acevedo, and Robbins addresses two different perspectives on the similarity between judgments of the self and judgments of reference groups. The self-stereotyping perspective assumes that self-referent judgments are derived from samples of group-referent information, whereas the social-projection perspective implies that group judgments are contingent on the sampling of self-referent information. Although the review reveals evidence for both perspectives, a social-projection model is presented that accounts for a remarkable number of empirical findings. In any case, the direction of inferences between groups and individuals constitutes another genuinely social approach to sampling mechanisms.

The final section, devoted to *Vicissitudes of Sampling in the Researcher's Mind and Method*, points out that sampling biases are by no means restricted to processes in experimental participants or decision makers but may be introduced by researchers themselves and by their methodologies. Hoffrage and Hertwig first address this intriguing issue in Chapter 16, with reference to Brunswik's (1956) notion of representative sampling. In particular, Hoffrage and Hertwig are concerned with the reluctance to sample stimuli in psychological experimentation. Whereas it is widely accepted that a large set of participants ought to be sampled and treated as a random variable in empirical research, one single task or one task context is usually considered sufficient to draw generalized conclusions from experimental findings.

Representative sampling of judgment tasks is also the focus of Winman and Juslin's Chapter 17, which provides a new sampling account of the familiar hindsight bias, that is, the tendency to misremember one's own prior judgments in the light of corrective feedback. Accordingly, judgments from the hindsight perspective reflect similar processes as confidence judgments from the foresight perspective, as evident in the confidence–hindsight mirror effect. The same factors that eliminate or reduce overconfidence – such as representative sampling of tasks – also reduce the hindsight effect, and the same factors that turn overconfidence into underconfidence can be shown to turn a normal hindsight effect into a negative hindsight effect.

Chapter 18 carries the notion of representative sampling over to a new area, artificial grammar learning, and to inductive learning of all kinds of complex structures. In traditional experiments on complex rule learning, participants have been presented with prediction tasks and feedback that speak to all kinds of rules that make up a grammar. However, the frequency distribution of the stimulus combinations generated by the grammar have not been controlled systematically. Poletiek shows that grammar learning can be improved considerably when the distribution of stimulus sequences or "strings" is representative of their occurrence rate in the language environment.

In Chapter 19, Hogarth applies Brunswik's experience-sampling method to the study of subjective confidence in decisions made by students and business executives. Combined with modern technology, experience sampling means to give participants a call at randomly selected points in time and to ask them to provide reports about ongoing decision processes. Hogarth demonstrates, among many other notable findings, that confidence in having made the right decision was higher on this online sample than in samples underlying retrospective reports. Confidence was generally quite high, even though no feedback was available to evaluate almost half of all decision problems.

CONCLUDING REMARKS

One theme of this volume is to reassign some of the explanatory burden from the use of biased cognitive heuristics per se to a refined analysis of the information samples to which the cognitive processes are applied. We tried to explicate this idea in terms of the metaphor of a naïve intuitive statistician equipped with cognitive algorithms consistent with normative principles (but not necessarily identical to them) with potential to produce both accurate and erroneous judgments depending on the sample input. The point, of course, is not to deny the importance of studying the cognitive mechanism, only to emphasize that this study entails a proper understanding of the informational input to which the cognitive algorithms apply – that one takes the interface between mind and environment seriously. Let us now end by pointing to some of the similarities and differences between this perspective and a number of other influential research programs in the behavioral sciences.

In this introduction we noted that this perspective refines on the metaphor of the intuitive statistician (Peterson & Beach, 1967) by being more theoretically oriented and by assigning a systematic theoretical status to the input samples to which the cognitive processes are applied in real environments. It extends beyond the heuristics and biases program (Kahneman et al., 1982) by paying more attention to the distinction between the cognitive algorithms and the information samples fed into the algorithms. It opens a more optimistic perspective on human judgment, allowing for both error and considerable competence, where the mediating factor is the fit between the cognitive processes and the input sample.

The research program on *fast and frugal heuristics* (Gigerenzer, Todd, & the ABC Group, 1999) reevoked Herbert Simon's notion of bounded rationality attempting to chart judgment heuristics that are successful adaptations to real environments. An important idea is that apparently simple heuristics may approximate more complex and allegedly normative models, once they are applied to the structure of real environments. This program instantiates the scheme with cognitive processes consistent with normative principles under appropriate sampling schemes and – because specialized judgment heuristics are more sensitive to environmental constraints than are computationally complex all-purpose algorithms – the way in which information is sampled indeed is crucial. However, thus far the research on fast and frugal heuristics has mainly been concerned with existence proofs of the proficiency of simple algorithms when applied to appropriate environmental structures, rather than on accounting for accurate and erroneous human judgments with the same heuristics. The perspective in the current volume puts more emphasis specifically on the effects of various sampling schemes and how these contribute to the explanation of both good and poor judgment.

The research program on *rational analysis* (Anderson, 1990; Oaksford & Chater, 1998) assumes that cognitive processes approximate optimal adaptations to the computational problems imposed by people's natural environments. Thus, if one knows (a) the goal or purpose of the computations performed by the cognitive system and (b) the structure of the environment to which the cognitive system has to adapt, one can derive the optimal behavior in this environment. To the extent that cognitive mechanisms are successful adaptations, observed behavior should approximate optimal behavior. Rational analysis obviously highlights the role of the environment over cognitive mechanisms. Yet – and despite an explicitly probabilistic scheme in terms of Bayesian optimization – issues about sampling are conspicuously absent in the implementation of rational analysis. Perhaps sampling issues would have been more salient if optimization had been framed more explicitly in frequentistic (rather than Bayesian) terms, suggesting sampling from a reference class. The perspective suggested by this volume complements rational analysis by providing a more detailed analysis of the effects of different sampling schemes on manifest behavior.

The metaphor of a naïve intuitive statistician with its concern with the fit between cognitive processes and the environmental input thus accords with the increased attention to environmental constraints in recent research (e.g., Anderson, 1990; Oaksford & Chater, 1998; Gigerenzer et al., 1999). Yet we believe that it complements previous approaches in crucial ways by making the sampling interface between mind and environment the explicit object of inquiry. We hope that the chapters that follow serve to substantiate this claim and that they can inspire other researchers to further explore the potentials afforded by this perspective.

References

Adams, J. K., & Adams, P. A. (1961). Realism of confidence judgments. *Psychological Review, 68,* 33–45.

Anderson, J. R. (1990). *The adaptive character of thought.* Hillsdale, NJ: Lawrence Erlbaum.

Betsch, T., Haberstroh, S., Glöckner, A., Haar, T., & Fiedler, K. (2001). The effects of routine strength on adaptation and information search in recurrent decision making. *Organizational Behavior and Human Decision Processes, 84,* 23–53.

Brehmer, B. (1980). In one word: Not from experience. *Acta Psychologica, 45,* 223–241.

Brunswik, E. (1956). *Perception and the representative design of psychological experiments.* Berkeley: University of California Press.

Chater, N. (1996). Reconciling simplicity and likelihood principles in perceptual organization. *Psychological Review, 103,* 566–581.

Eagly, A. H. (1987). *Sex differences in social behavior: A social-role interpretation.* Hillsdale, NJ: Lawrence Erlbaum.

Einhorn, H. J., & Hogarth, R. M. (1978). Confidence in judgment: Persistence of the illusion of validity. *Psychological Review, 85,* 395–416.

Erev, I., Wallsten, T. S., & Budescu, D. V. (1994). Simultaneous over- and underconfidence: The role of error in judgment processes. *Psychological Review, 101,* 519–527.

Fiedler, K. (1996). Explaining and simulating judgment biases as an aggregation phenomenon in probabilistic, multiple-cue environments. *Psychological Review, 103*, 193–214.

Fiedler, K. (2000). Beware of samples! A cognitive-ecological sampling approach to judgment biases. *Psychological Review, 107*, 659–676.

Fiedler, K., & Armbruster, T. (1994). Two halfs may be more than one whole: Category-split effects on frequency illusions. *Journal of Personality and Social Psychology, 66*, 633–645.

Fiedler, K., Brinkmann, B., Betsch, T., & Wild, B. (2000). A sampling approach to biases in conditional probability judgments: Beyond baserate neglect and statistical format. *Journal of Experimental Psychology: General, 129*, 399–418.

Fiedler, K., & Freytag, P. (2004). Pseudocontingencies. *Journal of Personality and Social Psychology, 87*, 453–467.

Fiedler, K., Schmid, J. & Stahl, T. (2002). What is the current truth about polygraph lie detection? *Basic and Applied Social Psychology, 24*, 331–324.

Fiedler, K., & Walther, E. (2004). *Stereotyping as social hypothesis testing*. New York: Psychology Press.

Fiedler, K., Walther, E., Freytag, P., & Plessner, H. (2002). Judgment biases in a simulated classroom – a cognitive-environmental approach. *Organizational Behavior and Human Decision Processes, 88*, 527–561.

Forgas, J. P. (1995). Mood and judgment: The affect infusion model (AIM). *Psychological Bulletin, 117*, 39–66.

Gibson, J. J. (1979). *The ecological approach to visual perception*. Boston: Houghton Mifflin.

Gigerenzer, G. (1991). From tools to theories: A heuristic of discovery in cognitive psychology. *Psychological Review, 98*, 254–267.

Gigerenzer, G., & Fiedler, F. (2005). *Minds in environments: The potential of an ecological approach to cognition*. Manuscript submitted for publication.

Gigerenzer, G., & Hoffrage, U. (1995). How to improve Bayesian reasoning without instructions: Frequency formats. *Psychological Review, 102*, 684–704.

Gigerenzer, G., Hoffrage, U., & Kleinbölting, H. (1991). Probabilistic mental models: A Brunswikian theory of confidence. *Psychological Review, 98*, 506–528.

Gigerenzer, G., Todd, P., & the ABC Group (1999) (Eds.). *Simple heuristics that make us smart*. Oxford: Oxford University Press.

Hamilton, D. L., & Gifford, R. K. (1976). Illusory correlation in interpersonal perception: A cognitive basis of stereotypic judgments. *Journal of Experimental Social Psychology, 12*, 392–407.

Hogarth, R. M., & Einhorn, H. J. (1992). Order effects in belief updating: The belief-adjustment model. *Cognitive Psychology, 24*, 1–55.

Hubert, M. (1999). The misuse of past-performance data. In L. E. Likson & R. A. Geist (Eds.), *The psychology of investing* (pp. 147–157). New York: Wiley.

Jones, E. E., & Harris, V. A. (1967). The attribution of attitudes. *Journal of Experimental Social Psychology, 3*, 1–24.

Judd, C. M., & Park, B. (1988). Out-group-homogeneity: Judgments of variability at the individual and the group levels. *Journal of Personality and Social Psychology, 54*, 778–788.

Juslin, P. (1994). The overconfidence phenomenon as a consequence of informal experimenter-guided selection of almanac items. *Organizational Behavior and Human Decision Processes, 57,* 226–246.

Juslin, P., & Olsson, H. (1997). Thurstonian and Brunswikian origins of uncertainty in judgment: A sampling model of confidence in sensory Discrimination. *Psychological Review, 104,* 344–366.

Juslin, P., & Olsson, H. (2005). Capacity limitations and the detection of correlations: Comment on Kareev (2000). *Psychological Review, 112,* 256–276.

Juslin, P., Winman A., & Olssen H. (2000). Naive empiricism and dogmatism in confidence research: A critical examination of the hard-easy effect. *Psychological Review, 107,* 384–396.

Kahneman, D., & Miller, D. T. (1986). Norm theory: Comparing reality to its alternatives. *Psychological Review, 93,* 136–153.

Kahneman, D., Slovic, P., & Tversky, A. (Eds.) (1982). *Judgment under uncertainty: Heuristics and biases.* Cambridge, UK: Cambridge University Press.

Kareev, Y. (2000). Seven (indeed, plus minus two) and the detection of correlations. *Psychological Review, 107,* 397–402.

Kashima, Y., Woolcock, J., & Kashima, E. S. (2000). Group impressions as dynamic configurations: The tensor product model of goup impression formation and change. *Psychological Review, 107,* 914–942.

Klauer, K. C., Musch, J., & Naumer, B. (2000). On belief bias in syllogistic reasoning. *Psychological Review, 107,* 852–884.

Klayman, J., & Ha, Y. (1987). Confirmation, disconfirmation, and information in hypothesis testing. *Psychological Review, 94,* 211–228.

Koriat, A. (1997). Monitoring one's own knowledge during study: A cue utilization approach to judgments of learning. *Journal of Experimental Psychology: General, 126,* 349–370.

Koriat, A., Fiedler, K., & Bjork, R. A. (2005). *The inflation of conditional predictions.* Manuscript submitted for publication.

Kuder, G. F., & Richardson, M. W. (1937). The theory of the estimation of test reliability. *Psychometrika, 2,* 151–160.

Lewin, K. (1951). *Field theory in social science.* New York: Harper & Row.

Lichtenstein, S., & Fischhoff, B. (1977). Do those who know more also know more about how much they know? The calibration of probability judgments. *Organizational Behavior and Human Performance, 20,* 159–183.

Lichtenstein, S., Fischhoff, B., & Phillips, L. D. (1982). Calibration of probabilities: The state of the art to 1980. In D. Kahneman, P. Slovic, & A. Tversky (Eds.), *Judgment under uncertainty: Heuristics and biases* (pp. 306–334). Cambridge, UK: Cambridge University Press.

Martin, L. L. (1992). Beyond accessibility: The role of processing objectives in judgment. In L. L. Martin, & A. Tesser (Eds.), *Construction of social judgment* (pp. 217–245). Hillsdale, NJ: Lawrence Erlbaum.

Musch, J., & Klauer, K. C. (Eds.) (2002). *The psychology of evaluation: Affective processes in cognition and evaluation.* Mahwah, NJ: Lawrence Erlbaum.

Newell, A., & Simon, H. A. (1972). *Human problem solving.* Englewood Cliffs, NJ: Prentice-Hall.

Oaksford, M., & Chater, N. (1994). A rational analysis of the selection task as optimal data selection. *Psychological Review, 101,* 608–631.

Oaksford, M., & Chater, N. (1998). *Rationality in an uncertain world: Essays on the cognitive science of human reasoning.* Hove, England: Psychology Press/Erlbaum (UK) Taylor & Francis.

Oskamp, S. (1965). Overconfidence in case-study judgments. *Journal of Consulting Psychology, 29,* 261–265.

Patrick, C. J., & Iacono, W. G. (1991). Validity of the control question polygraph test: The problem of sampling bias. *Journal of Applied Social Psychology, 76,* 229–238.

Peterson, C. R., & Beach, L. R. (1967). Man as an intuitive statistician. *Psychological Bulletin, 68,* 29–46.

Roe, R. M., Busemeyer, J. R., & Townsend, J. T. (2001). Multi-alternative decision field theory: A dynamic connectionist model of decision making. *Psychological Review, 108,* 370–392.

Rulon, P. J. (1941). Problems of regression. *Harvard Educational Review, 11,* 213–223.

Semin, G. R., Rubini, M., & Fiedler, K. (1995). The answer is in the question: The effect of verb causality on locus of explanation. *Personality and Social psychology Bulletin, 21,* 834–841.

Simon, H. A. (1956). Rational choice and the structure of environments. *Psychological Review, 63,* 129–138.

Simon, H. A. (1990). Invariants of human behavior. *Annual Review of Psychology, 41,* 1–19.

Smolensky, P. (1988). On the proper treatment of connectionism. *Behavioral and Brain Sciences, 11,* 1–74.

Snyder, M. (1984). When belief creates reality. In L. Berkowitz (Ed.), *Advances in experimental social psychology* (Vol. 18, pp. 247–305). New York: Academic Press.

Stasser, G., & Titus, W. (1985). Pooling of unshared information in group decision making: Biased information sampling during discussion. *Journal of Personality and Social Psychology, 48,* 1467–1478.

Stelzl, I. (1982). *Fehler und Fallen der Statistik* [Errors and traps in statistics] Bern: Huber.

Swann, W. B., Giuliano, T., & Wegner, D. M. (1982). Where leading questions can lead: The power of conjecture in social interaction. *Journal of Personality and Social Psychology, 42,* 1025–1035.

Swets, J., Dawes, R. M., & Monahan, J. (2000). Psychological science can improve diagnostic decisions. *Psychological Science in the Public Interest, 1,* Whole No. 1.

Thurstone, L. L. (1927). A law of comparative judgment. *Psychological Review, 34,* 273–286.

Tversky, A. (1977). Features of similarity. *Psychological Review, 84,* 327–352.

Tversky, A., & Kahneman, D. (1973). Availability: A heuristic for judging frequency and probability. *Cognitive Psychology, 4,* 207–232.

Tversky, A., & Koehler, D. J. (1994). Support theory: A nonextensional representation of subjective probability. *Psychological Review, 101,* 547–567.

Tversky, A., & Simonson, I. (1991). Context dependent preferences. *Management Science, 39,* 1179–1189.

van der Helm, P. A., & Leeuwenberg, E. L. J. (1996). Goodness of visual regularities: A nontransformational approach. *Psychological Review, 103,* 429–456.

Wänke, M., Schwarz, N., & Noelle-Neumann, E. (1995). Question wording in comparative judgments: Understanding and manipulating the dynamics of the direction of comparison. *Public Opinion Quarterly, 59,* 347–372.

Weber, E., Shafir, S., & Blais, A.-R. (2004). Predicting risk-sensitivity in humans and lower animals: Risk as variance or coefficient or variation. *Psychological Review, 111,* 430–445.

Wells, G. L., & Windschitl, P. D. (1999). Stimulus sampling and social psychological experimentation. *Personality and Social Psychology Bulletin, 25,* 1115–1125.

Zuckerman, M., Koestner, R., Colella, M. J., & Alton, A. O. (1984). Anchoring in the detection of deception and leakage. *Journal of Personality and Social Psychology, 47,* 301–311.

PART II

THE PSYCHOLOGICAL LAW OF LARGE NUMBERS

2

Good Sampling, Distorted Views

The Perception of Variability

Yaakov Kareev

Time pressure, structural limitations of the cognitive system, or paucity of available data often force people to make do with but a sample, when they try to learn of the characteristics of their environment. In such cases, sample statistics must be relied upon to infer the parameter values in the population. Because sample data provide, by definition, only a partial view of reality, the degree to which that view is a veridical reflection of reality is of great importance. It is well known that, in the quest for accurate estimates of population parameters, larger samples are to be preferred over smaller ones. This is so since the variance of the sampling distribution of any statistic is inversely related to sample size (typically to the square root of it). Still, often the size of the sample available is quite small. Consider, for example, one important limiting factor, working-memory capacity – a structural characteristic of the human cognitive system that determines the number of items that can be considered simultaneously; it is about seven items for an average adult (Miller, 1956) and could even be considerably lower than that (see Cowan, 2001, for an estimate as low as four). Nonetheless, even if the samples used by people are typically quite small in size, some solace may be obtained if one could assume their statistics to provide *unbiased* estimates of the population parameters in question. When such an assumption can be made, the use of even small-sample statistics, though not guaranteeing accuracy, would at least not result in a systematically distorted view of reality. For example, statistics involving measures of central tendency provide unbiased estimates of their respective population parameters. One could take people's unwarranted confidence in small-sample statistics (see Tversky and Kahneman, 1971) as an indication that, when unbiased estimates are involved, the mere potential lack of accuracy is not regarded as consequential as some statisticians would believe it to be.

In contrast with measures of central tendency, it is a statistical fact, one whose implications have hardly been considered by behavioral scientists, that for measures of variability sample statistics provide a biased estimate of their respective population parameters. As shown in the following, the bias is such that the use of small-sample data would result, more often than not, in the observation of variances that are smaller than their respective population values. Since variability indicates the degree of certain aspects of regularity in the environment, the statistical fact just pointed out implies that the use of sample data, as observed, could lead to viewing the world as more consistent than it actually is. It should be emphasized that the bias holds for randomly drawn samples; in other words, it is a necessary outcome of well-executed sampling (and accurate calculations).

Whereas the bias itself is a statistical fact, the relevant psychological question is whether or not people have learned to correct for the bias. There exist statistical procedures to correct for the bias, and there is no logical reason why the cognitive system could not have come up with – through evolution or learning – corrections designed to counteract the effects of the bias. The interesting question, then, would be to ascertain whether people actually correct for the bias inherent in the use of sample data. Many observations of human behavior could be taken to suggest that they do not. As long ago as 1620, Francis Bacon, describing the bad habits of mind that cause people to fall into error, wrote, "The human understanding is of its own nature prone to suppose the existence of more order and regularity than it finds" (1620/1905, p. 265). The fundamental attribution error – people seeing greater consistency (i.e., less variance) in other people's behavior than there actually is (Gilbert & Malone, 1995; Ross, 1977) – could be taken as another indication of misperceived regularity.

Explanations of biases in the perception of reality, whether motivational or cognitive, assume that the data available are unbiased. Accordingly, any distortions or systematic biases found in behavior must result from the operation of the mental system – from selective attention, from decisions (conscious or not) as to what to encode, from the way information is processed, or from the way information is retrieved. In contrast, the present thesis suggests that, with respect to the perception of regularity, the data available to us are themselves systematically biased. Consequently, the systematic biases often evident in people's perceptions of regularity may reflect the *accurate* processing of *systematically biased* data, rather than the *biased* processing of *unbiased* data.

In the remainder of this chapter the statistical argument concerning the sampling distributions of measures of variance is first laid out in some detail. Then empirical evidence is presented to show that, with respect to measures of variability, people tend to rely on sample data and do not correct for the biased values likely to be observed in small samples. People are thus shown to rely on biased estimates of variability.

THE STATISTICAL ARGUMENT

It is a statistical fact that the sampling distribution of variance is highly skewed and the more so, the smaller the sample size (e.g., Hays, 1963). The variance of a sample, drawn at random from a population whose values are normally distributed, provides a biased estimate of the population variance. Mainly because small samples are less likely to contain extreme values, the biasing factor is $(N − 1)/N$, for a sample of size N. As the correction implies, the attenuation decreases as N increases, but for $N = 7$, the sample size likely to be considered by an average adult, its value is 0.86, resulting in considerable attenuation. Another way to appreciate the bias is to note that the variance of a sample of seven items, randomly drawn from a normal distribution, is smaller than the variance of that distribution in about 0.65 of the cases. Thus, for sample sizes likely to be considered by people, sample statistics for variance are considerably attenuated. It should be noted that other measures of variability – measures that people may be more likely to use than variance itself (such as the range, the interquartile range, or the mean absolute deviation) – are all highly correlated with it, and their sample-based estimates all suffer from a similar attenuation. Moreover, earlier research has indicated that subjective estimates of variability are strongly correlated with variance (for reviews, see Peterson & Beach, 1967, and Pollard, 1984). At the same time, the measures employed in the studies mentioned were such that they could not be used to assess whether, and to what extent, the perception of variability is biased.

It follows that if people, when assessing variability, draw a random sample from the environment or from memory to calculate the statistic in question, more often than not the value in the sample would indicate greater consistency than that existing in the population from which the sample was drawn. Furthermore, if the size of the sample considered is close to people's working-memory capacity, the incidence of such misperceptions would be quite high. Unless some correction for sample size is applied, an accurate calculation of the sample statistic, and use of its value as the estimate of the parameter in the population, would most often result in the world being seen as more consistent than it is.

It may be noted that the explanation offered here has much affinity with that provided by Linville and her associates (Linville & Fischer, 1993, 1998; Linville, Fischer, & Salovey, 1989; Linville & Jones, 1980) to explain differences in the perception of variability and covariation among in-group and out-group members. There, differences in perception were ascribed to differences in the size of the available samples from in- and out-groups (and the implications that samples of different sizes have for the opportunity to observe a variety of cases).

The present argument rests, at least in part, on the assumption that initial samples strongly affect people's impressions. Surely people retrieve or are exposed to additional data at later stages. However, people are known to have much confidence in small-sample data (Tversky & Kahneman, 1971) and to exhibit strong primacy effects (e.g., Hogarth & Einhorn, 1992). Once an impression sets in, future data are encoded and interpreted in the light of that original impression. Thus, the data most likely to be encountered first point at what the subsequent (and the eventual) estimate might be.

With respect to the taxonomy presented in Chapter 1 of this book, the sampling effect discussed in the present chapter holds irrespective of origin. Moreover, it holds even when sampling is unconditional (and will most likely be even stronger when sampling is conditional). Finally, the units dealt with are natural, singular events, rather than features.

EXPERIMENTAL EVIDENCE

People surely have intuitive notions of variability, but there is no reason to expect that these notions match statistical measures developed to capture them. As a result there is no simple way to uncover a bias in perception, even if one exists. The discrepancy posed no problem for earlier studies of the perception of variability, since they aimed at the correspondence between "objective" values – as measured by variance – and its perception (i.e., subjective estimates; for reviews, see Peterson & Beach, 1967, and Pollard, 1984). Participants in such studies typically estimate variance on some arbitrary scale (e.g., a number between 0 and 100), with their responses then used to chart the correspondence between objective values and subjective perception. Such measures cannot capture, however, systematic shifts relative to the actual values and are therefore useless for the present purposes.

Instead of attempting to measure absolute levels of perceived variability, we have designed tasks in which participants compared or chose between populations. Typically these populations did not differ in their objective degree of variability, but they would be perceived as different if judged on the basis of samples differing in size without applying a correction. Since the degree of the bias is related to sample size, judgments of populations with identical characteristics – but viewed under conditions likely to result in samples of different size – can be used to test the present hypothesis. In all situations studied, if a sample-related bias did not exist, or was corrected for, behavior would be unrelated to the experimental manipulation. In contrast, if a sample-induced bias did operate, judgments in the different conditions were expected to differ in systematic ways. As described in greater detail in the following, sample sizes were manipulated in a number of ways: At times participants compared populations in front of them to populations that had been removed from view, or to populations from which only a sample was available. In other cases, we capitalized on individual differences and compared the behavior of people differing in their working-memory capacity. It should be emphasized that there could be numerous reasons why behavior would *not* differ in the conditions compared in the studies: If people are unaware of the parameters in question, or are unable to assess them, or have learned to correct for sample-related biases, one would expect no effect of the manipulations. In contrast, if people are sensitive to the values of the parameters in question, do assess their value on the basis of a small number of items, and do not correct for the biases, behavior in the compared conditions would differ.

The stimuli used in all the studies to be reported consisted of concrete everyday materials. The use of such tangible materials ruled out the possibility that the items presented were conditional on participants' earlier responses – a concern often expressed in computer-controlled experiments. In all experiments it was obvious that the experimenter could neither modify the makeup of the population considered nor affect the choice of items to be viewed or their order.

With respect to the perception of the variability of a population, one could start by asking whether people are at all sensitive to the variance of a distribution. Only if such sensitivity is present could one ask if that perception is veridical or biased and, if the latter is the case, whether the bias is related to the size of the sample likely to have been observed. The following experiments, described in greater detail in Kareev, Arnon, and Horwitz-Zeliger (2002), constitute an attempt to answer these questions.

In the following we report the results of five experiments in all. The first experiment was designed to find out whether people are at all sensitive to variance, and if they are, whether their memory for it is veridical or biased. The second experiment was designed to find out whether, when forced to rely on a sample when estimating variability, people employ a correction for sample size. The third experiment capitalized on individual differences in working-memory capacity (and hence, presumably, in the size of the sample employed) to find out whether people fully correct for the effect of sample size on observed variability. The fourth experiment was designed to find out whether variability is taken into account (as it normatively should) in determining the confidence with which choices are made, and, if it is, whether it is the observed sample value, or that value corrected for sample size, that is employed. Finally, the fifth experiment was conducted to find out which items of a sequentially presented group are most likely to be included in the sample used to infer population characteristics and to make choices.

Experiment 1 – Is Variability Noted and Remembered Veridically?

Experiment 1 was designed to answer two questions: The first was whether people note the variability of a population even when attention is not drawn to it. The second was whether variability, if noticed, is remembered veridically or is distorted in some manner. To answer these questions participants first viewed a group of items that were then removed. They then indicated which of two new groups, in full view at the time of decision, better resembled the original one. Participants were not aware that one of the two new groups was identical to the original, whereas the other was either less or more variable.

The study allowed for a stringent test of our question: If people perceive and remember the variance of a population and do so accurately – either because they use the whole population to estimate it, or because they use sample variability but correct it for the bias due to sample size – they should be as good at choosing the identical comparison group over both the less and the more variable ones. However, if people infer the variance of the out-of-view population on the basis of a sample retrieved from memory, and do not correct for sample size, their estimate of that variance would be downward attenuated. In that case, people would easily reject the more variable comparison group but would find it difficult to choose correctly between the identical population and the one of lower variability. Needless to say, if people are insensitive to variability, they would be unable to distinguish between the two comparison populations and would perform the task at chance level in both conditions.

As mentioned, one of the two comparison groups was identical to the original, item for item. The comparison group that was not identical to the original differed in variability such that its variance was either 6/7 or 7/6 of the variance of the original group. The values of the items comprising the group were either continuous or discrete (and binary). For the continuous values the original population consisted of 28 paper cylinders colored up to a certain height; the colored sections were normally distributed with a mean of 6.0 cm and a standard deviation of 1.955. The nonidentical comparison groups had the same mean but a standard deviation of either 1.811 or 2.112. For the discrete values the original population consisted of 19 small disks marked with the letter L and 9 disks marked with T. The nonidentical groups had either a 21–7 or a 15–13 division. The variability of the nonidentical comparison population was manipulated independently of the type of variable. Ninety-six students participated in the experiment for a monetary reward.

Participants first observed for 10 s one of the standard populations (of one type of variable) and then observed, for the same duration, the other standard population. They were then presented with a pair of populations for each of the original groups and were asked to indicate which of the two better resembled the corresponding original population. Performance during comparison was self-paced and typically very fast. Order of presentation, variability of the nonidentical comparison population (smaller or larger than that of the identical one), and position of the identical box were all counterbalanced. All boxes were shaken before being opened for inspection; this not only assured a different arrangement of the items but also implied that the spatial arrangement of the items was immaterial for the judgment.

Choices were scored +1 when the comparison box with smaller variability was judged as being more similar to the original, and −1 when that with larger variability was so judged. Thus, the scoring of choices reflected not their accuracy but rather the tendency to judge the less variable one as being more similar to the original. This, in turn, reflects the degree of support for the prediction that people would perceive variability derived from memory as smaller than that of a population in full view. The mean score of .33 differed significantly from zero ($t(95) = 2.364$, $p = .020$). Importantly, type of variable did not have a significant effect on the choices ($F(1, 95) < 1$). This is of interest since, for the discrete variable, the change in variability also involved a change in the central value (the proportion). The similar effect observed for both types of variable rules out an explanation based on sensitivity to a change in proportion, rather than to a change in variability. A breakdown of the participants' choices revealed that the identical comparison box was judged as more similar to the original box in only .49 of the cases when compared with that of the smaller variability, but in fully .66 of the cases when the comparison box had the larger variability. As reasoned in the foregoing, such a result could not be obtained if people either assess variability correctly or do not notice it at all. These findings are compatible, on the contrary, with the claim that people, when retrieving a population from memory, indeed conceive of its variability as smaller than it actually is.

Experiment 2 – Perception of Variability: Sample-Based versus Population-Based

Experiment 1 established that, although people are sensitive to the variability of a distribution, they perceive it as smaller when derived from memory. Use of a sample, retrieved from memory without correction for sample size, was implicated as the mechanism bringing about that attenuation. Experiment 2 was designed to further explore the role of sampling in the perception of variability. In the experiment participants judged two pairs of populations. Unbeknown to the participants, the two populations within a pair were identical. Every participant performed two judgment tasks, each involving a pair of populations. In the first task, participants responded to a direct question as to which of the pair of populations was more (or less) homogenous. In the second task, participants chose one of two populations from which two items would later be drawn. Reward was to be provided if the two items drawn were identical in color or similar in the size of their colored sections (see the description of materials to follow).

Thus, the chances for a reward were inversely related to the variability of the population.

The main manipulation in the study was viewing condition: For each pair of populations, one of them was in full view at the time a judgment was made, whereas the other was not. Information about that population of the pair that was not in full view at the time of judgment was gleaned in one of two ways: In the out-of-sight condition, the population had been seen in its entirety but was removed at the time of judgment. In the sample-only condition, what participants knew of the population was based on a sample of seven items drawn, by the participant, out of the twenty-eight items comprising the population. The sample itself was in view at the time of judgment. Viewing condition was manipulated between participants. Type of task (direct question or choice) and type of variable (binary or continuous) were manipulated within participants, with each participant performing two of the four combinations of the two within-participant variables.

Each population consisted of twenty-eight items, a number far exceeding working-memory capacity. We assumed that, for the population that was out of sight, participants would have to rely on a sample drawn from their memory. As such, it should seem less variable than the population in full view if no correction were applied. For the sample-only condition, there the availability of only a sample was highly prominent. If people had any inkling of the effects of samples, their attempts to correct for them should be evident there. Here too, if people are unaware of variability, incapable of assessing its value, or aware of the consequences of having to rely on sample data, there would be no preference of one population over the other. In contrast, if people are sensitive to variability, but unaware of the consequences of relying on but a sample, then the population out of view, or that from which but a sample had been seen, would be judged as the less variable of the two.

One hundred and forty-four students participated in the study for payment. For the continuous-variable items, both populations consisted of twenty-eight paper cylinders, identical to those employed in the original distribution of Experiment 1. For the binary-variable items, the populations consisted of twenty-eight uniformly colored pieces of wood, eighteen of which were of one color, and the remaining ten of another.

For the direct-question task, participants were asked either in which of the two populations the distribution of values was more homogeneous (continuous variable) or which population had a split of colors closer to .5/.5 (binary variable). Note that, for the direct-question response task, a response bias to choose always the population in one viewing condition over the other would have the opposite effect for the two types of variables. For the choice task, the participants were rewarded if the pair of items drawn from the chosen population fell within a certain range (for

TABLE 2.1. *Proportion of Answers in Which Populations Seen through Sample Only, or Out of Sight, Rather than That in Full View, Were Judged as Having the Smaller Variability*

Task	Direct Evaluation		Choice		
Condition	Binary	Continuous	Binary	Continuous	Mean
Sample only	.667	.555	.528	.555	.577
Out of sight	.611	.639	.528	.528	.577
Mean	.639	.597	.528	.542	.577

the continuous variable) or were identical in color (for the binary variable). Because the chance of winning the reward is greater, the smaller the variability of the population from which items are to be drawn, we took the participants' choice as an indication of which population they believed had smaller variability. All combinations of task and type of variable were equally frequent, the order of presentation being fully counterbalanced within each viewing condition.

A judgment was scored 1 if it was in line with our hypothesis, that is, if the out-of-sight or the sample-only population was considered less variable. A judgment was scored 0 if the population in full view was considered the less variable. The mean results for each condition are presented in Table 2.1. An analysis of variance of these data revealed that the only significant effect was the deviation of the overall mean from the value of .5 expected by chance ($F (1, 140) = 6.34$, $MSE = 1.06$, $p = .013$). Thus, the results indicate that, when a population is out of sight, or when only a sample of it is available, its variability is perceived to be smaller than that of an identical population but in full view at the time of judgment. Since in real-life situations a complete population is rarely available, these results imply that variability is usually attenuated.

It is worth noting that the overall effect was identical under both viewing conditions. This further suggests that, when a population is out of sight, its parameters are estimated by recalling a sample. Note also that the overall proportion of cases in which the variability of the population in full view was judged as larger was .577. Although this value is significantly greater than chance, it is still lower than the .65 or so expected if people correctly judge the variability of a population in full view and compare it to a population from which only a sample of seven is available (whether by the experimental manipulation or by retrieving samples of that size from memory). One possible explanation of the difference is that people partially correct for the bias induced by small-sample data. Another possibility is that, even for populations in full view, variability is assessed on the basis of sampling (even if more than one sample), resulting in slightly biased perception and hence in a smaller difference between it and the other population than that predicted by theory. As we shall see, the results of Experiment 3 favor the latter explanation.

Experiment 3 – Perception of Variability: Relationship to Working-Memory Capacity

Experiment 3 was designed to explore the relationship between individual differences in working-memory capacity and the perception of variability. If people indeed sample data to infer population characteristics when the amount of data is large, working-memory capacity could be a factor limiting the size of such samples. If this is the case, and no correction for sample size is applied, people with smaller working-memory capacity would be expected to perceive variability as lower than would people with larger capacity. This hypothesis was tested in an experiment in which participants first observed a population consisting of twenty-eight items and then predicted the makeup of a sample of seven items to be drawn, without replacement, from that population. Participants then drew a sample and were rewarded according to the degree of correspondence between their prediction and the actually drawn sample. We compared the predictions of participants differing in their working-memory capacity, expecting predictions made by participants with lower capacity to be less variable than those made by participants with larger capacity.

Each participant performed the task four times: twice for populations that were fully in view at the time of prediction, and twice for populations seen shortly before prediction but out of sight at the time of prediction. Type of variable was manipulated within each viewing condition, such that one population involved a continuous variable and the other a binary-valued variable. Participants' working-memory capacity was estimated by their performance in a standard digit-span task. The participants were fifty-nine Hebrew University undergraduate students, participating for course credit (plus the monetary reward, if they earned any). The materials employed in the study were the same paper cylinders and pieces of wood used in Experiment 2. Different colors were used for the two continuous and the two binary populations.

The items were stored in opaque boxes that were vigorously shaken, once before exposure and again before drawing. In the out-of-sight condition, the lid of the box was raised for 10 s and then closed; for the full-view condition, the lid was raised and remained open. The out-of-sight boxes were presented before the full-view boxes, with order of presentation of different types of variables counterbalanced. Predictions for all four boxes were made after all of them had been seen, but before any drawing began. For the binary-valued box, the participants predicted the number of items of each color out of the seven items of the sample to be drawn. For the continuous variable, they predicted the number of items that would be colored above or below a reference height that was shown to them. That height created the same 18–10 split of the 28 items as in the case of the binary variable. After all predictions had been made, the open boxes were closed and the participants drew a sample of seven items without replacement from each of the four boxes. The participants were rewarded 4 ILS (about $1 at the time) when their prediction exactly matched the makeup of the sample, and 2 ILS when it deviated by 1. The digit-span task was administered last.

An extreme group design, comparing participants who had a digit span of less than 6 ($N = 21$) with those whose digit span exceeded 6 ($N = 24$), revealed that, as expected, sample makeups predicted by participants with smaller capacity were significantly less variable than those made by participants with larger capacity ($F(1, 43) = 4.45, MSE = .32, p = .041$). Capacity also interacted with viewing condition ($F(1, 43) = 6.29, MSE = .34, p = .016$): The two groups did not differ in the out-of-sight condition, but they did in the full-view condition. Finally, predicted sample make-ups were less variable for the binary than for the continuous variable ($F(1, 43) = 6.61, MSE = .34, p = .014$). The main effect of capacity indicates that people with smaller working-memory capacity exhibit smaller variability in their products than do people with larger capacity. As reasoned previously, such a difference could occur if, even when more data were available, the participants relied on samples whose size corresponded to their working-memory capacity, without correcting their estimates of variability for sample size.

Although the main effect of capacity was in line with our predictions, the significant interaction between capacity and viewing condition was unexpected, as the difference between the two groups was expected to be larger in the out-of-sight rather than in the full-view condition. Furthermore, the results of Experiment 2 led us to expect a main effect of viewing condition, but none was obtained. In light of the unexpected interaction between capacity and viewing condition, we looked for another way to test the hypotheses that people, in making predictions, rely on limited samples, commensurate with their working-memory capacity, and more so in the out-of-sight than in the full-view condition. The setup of the experiment provided us with this other way of testing both hypotheses. Since the variability of a sampling distribution is larger, the smaller the sample on which it is based, if people with smaller capacity use smaller samples to estimate variability, the overall distribution of their answers should be more variable. Similarly, if the use of sample data is more prevalent in the out-of-sight condition, one expects larger variability in answers there than in the full-view condition. A comparison of the within-group variance of the predictions made by members of the low-capacity group to that of members in the high-capacity group revealed that, across all four tasks, the variance of the former was significantly larger than that of the latter ($F(83, 95) = 1.365, p = .036$, one-tailed) – another indication that people use sample data in their assessment of variability.

The same kind of analysis was also employed to compare performance in the two viewing conditions. If people use sample data, sample size (or number of samples considered) would be smaller when the population is out of sight than when it is in full view. Indeed, the variance of the predictions made in the out-of-sight condition was significantly larger than that in the full-view condition ($F(83, 89) = 1.365$, $p = .048$, one-tailed).

Taken together with the main effect of capacity reported in the forego-ing, these two findings support the claim that people use samples to infer population variability, with people with smaller capacity apparently using smaller samples than do people with larger capacity.

The results of Experiment 3 also help address an issue left unsettled in Experiment 2. There the overall proportion of cases in which the out-of-sight or the sample-only population was judged to be of smaller variability was significantly greater than chance but smaller than that expected if no sampling had taken place for the population in full view (while samples of size seven were used for the others). It was argued there that such a discrepancy could have resulted either from partial correction for sample size or from the use of samples even for the population in full view (in which case the number of samples or the size of the sample was larger). The data observed in Experiment 3 favor the latter explanation: The difference in the overall variance of the answers provided by the low-capacity and high-capacity groups persisted, even when only cases in full view were considered ($F(41, 47) = 2.100$, $p = .0037$, one-tailed). Thus, it seems that the use of small-sample data to infer population statistics is prevalent, when amount of available data exceeds working-memory capacity.

Experiment 4 – Variability as Observed, or Corrected for Sample Size? Evidence from Choices and Confidence in Them

Variability should affect the confidence with which one can choose one of two populations on the basis of the difference between their means. Obvi-ously, the population whose mean is estimated to have the more desirable value should be preferred over the other; at the same time, the likely ac-curacy of the estimate of the mean, and hence the likelihood of an error, is indicated by the variability within each population. In other words, norma-tively, in making such a choice, variability should be noted and taken into account, with larger variability reducing one's confidence in the decision. Experiment 4 was designed to find out (a) whether people take variabil-ity into account when choosing between two populations with different means and, if they do, (b) whether it is sample variability, as observed, or sample variability corrected for sample size that better predicts choices and confidence in them.

The task was performed with a pair of distinct populations, stored together in a big box. Each population consisted of fifty matchboxes containing a few matches. The two populations differed in their mean number of matches but had either the same or very similar variance. As explained in more detail in the following, participants drew several match-boxes of each type and noted the number of matches in each. They then chose the type of matchbox that they believed had a higher number of matches, and they placed a monetary bid that indicated their confidence in their choice. Characteristics of the samples viewed by the participants were then used in a multiple regression analysis to find out what statistics best predicted choices and bids. In particular we were interested in finding out whether variability was at all taken into account, and if so, whether it was sample variability, as observed, or sample variability corrected for sample size that better predicted the decisions made.

Because our analysis was to relate participants' performance to characteristics of the samples they actually drew from the pair of populations, it was necessary to ensure a wide range of sample values. We therefore used eight pairs of populations in all. One of the two populations in every big box – the "standard population" – always had the same characteristics: It had a mean of 5.0 and a standard deviation of 1.92. The mean number of matches of the other population in a pair was larger than that of the standard by 1.5, 2, 2.5, or 3 matches. Of the eight pairs of populations, there were two pairs for each of the four possible differences in means. In addition we manipulated the size of the sample participants drew from each population; sample sizes were always different for the two pairs of populations with the same difference in means. Note that changes in sample size affect the expected values of inferential statistics and of their significance but not estimates of differences between means or of effect sizes. The characteristics of the various populations are described in detail in Kareev et al. (2002).

In performing the task, the participants first sampled the specified number of boxes with replacement, opening and noting the number of matches in each of the sampled boxes. Sampling was self-paced and the order in which boxes were sampled was at the participant's discretion. When the predetermined sample size was reached for the matchboxes of both types, the participant was told that two additional boxes, one of each color, would be drawn later; he or she was then asked to bet on which type would contain more matches, placing a bid on the difference in number of matches between them. To calculate the reward (positive or negative), the bid was multiplied by the difference in the number of matches between the two boxes actually drawn. The task was performed three times by each participant, with a different pair of populations used each time. The participants were eighty students; each was paid for participation and was rewarded according to the aforementioned calculation.

Correlations were calculated to determine which statistics of the actually observed samples best accounted for participants' choices and bids.

Choices were scored 1 when the population with the higher mean was chosen, and 0 when the other was.[1] The following statistics were derived from the samples actually observed and correlated with choices and bids: (a) difference between sample means; (b) variances (actual sample variance and estimated population variance based on sample variance, corrected for sample size); (c) ratio between the variances of the samples; (d) sample sizes; (e) point-biserial correlation and point-biserial correlation squared, both serving as estimates of the effect size of the difference between the two populations (each of these being calculated twice, once with the actual sample variance and once with sample variance corrected for sample size); (f) Z and t inferential statistics for the difference between the two means; and (g) significance of the Z and t statistics. (Recall that the value for Z uses actual sample variances divided by sample sizes, and that for t uses estimates, employing a correction for sample sizes.) Thus, the analysis included measures that do have variance and measures that do not as one of their components, juxtaposing the predictive power of measures that use actual sample variance and measures that use estimates of population variance (derived from sample variance corrected for sample size). Our main objective was to determine which of the sample statistics was the best predictor of behavior and to ascertain (a) whether it was one having a measure of variance as one of its terms and, if it did, (b) whether that term reflected sample variance, as observed, or sample variance corrected for sample size. Since many of the predictive measures compared had common components, the correlations between them were very high, and there was no point in testing for the significance of the difference between correlation coefficients. Instead, the measure of interest was the rank order of those correlations. The best single predictor of the participants' choices turned out to be the significance of the Z test– a measure affected not only by the difference between the two means but also by their (actual) variances and by the sample size: Its correlation with choice was $r = .522$ ($F(1, 238) = 89.31$, $p < .001$). The best single predictor of the bids was the uncorrected estimate of effect size – a measure that takes into account the difference between the two means and the variance in the combined samples, without correcting it for sample size. Its correlation with bid size was $r = .302$ ($F(1, 238) = 24.45$, $p < .001$). When individual differences in bid size were taken into account, by subtracting each individual's mean bid from each of his or her bids, that same measure remained the best single predictor, with prediction improving to $r = .387$ ($F(1, 238) = 44.50$, $p < .001$).

[1] Another possible scoring of choices would be to score a choice as 1 when that population having the higher sample mean was preferred over the other. Owing to the large differences between population means, the difference between sample means was in the same direction as that between population means in most cases (229 out of 240, with 3 other cases of identical sample means), rendering the distinction unimportant. Indeed, for both choices and bids, the same measures turned out to be the best predictors for either scoring systems.

The most important findings from the present perspective are that, for choices and bids alike, (a) variability was a component of the best predictors of behavior (i.e., people were taking not only the means but also variability into account), and (b) the best predictors included a measure of variability that was based on actual sample variance, not on estimates of variance corrected for sample size. It is also of interest that the best predictor of choices (the significance of Z) also reflects the sizes of the samples considered. We do not contend, of course, that in making choices or bids people calculate the significance of a Z test or effect size. Rather, we see the current findings as an indication that people, in making decisions, use some composite measure that takes variability into account, but in which the variability of the sample, as observed, rather than an estimate of population variability corrected for sample size is used. As pointed out in the introduction, the sample variability is, on average, downward attenuated, and the more so, the smaller the sample. Such attenuation does not affect the choice itself, which depends on the difference between the means, but increases the confidence with which the choice is made.

Experiment 5 – Sample Composition: A Primacy Effect

With the former experiments already establishing people's use of sample statistics, without correcting for sample size in tasks involving variability, Experiment 5 was designed to explore what part of the data available constitutes the sample considered by the participants. The participants first drew samples from two populations, then chose one of them to be used in a subsequent prediction task, and finally predicted – for a reward – the value of twenty items, drawn one at a time, with replacement, from the population of their choice. With the items in this experiment presented sequentially, and their total number exceeding working-memory capacity, the study enabled us to check for primacy and recency effects, as well as to estimate the size of the sample considered. The experiment was also used to find out, once more, if people take note of variability, in cases in which choosing the more homogenous population would increase their chances for reward.

Type of variable and total number of sampling trials were manipulated between participants. First, the items constituting each pair of populations had either one of two values (binary populations), or one of five values (multivalued populations). Second, the total number of sampling trials (for both populations combined) was either 13 or 19. These numbers ensured that samples drawn from the two populations were of unequal size. The total sample size was announced in advance, rendering the eventual difference in sample size highly prominent. Tall, opaque urns with an opening wide enough to insert a hand but not to see the contents were used to store the 36 items comprising each population. For the binary-valued populations, the items were beads – 24 of one color and 12 of another. The beads in one urn were red and green; those in the other were blue and yellow. For the multivalued populations, the items were playing cards with face values ranging from 6 to 10, with frequencies of 4, 8, 12, 8, and 4, for the five values, respectively. The cards in one urn were from the red suits; those in the other were from the black suits. Participants were 160 students who participated to fulfill a course requirement or for payment. They also received a reward reflecting their success in the prediction phase.

As items were drawn sequentially, and with the samples drawn from each urn in a pair always differing in size and almost always in variability, we could use multiple regression analysis to assess the degree to which characteristics of the actually drawn samples (or parts of them) accounted for the participants' choices.

Following the sampling stage, the participants chose which urn to use in the subsequent prediction phase, in which they predicted, for each of twenty trials, the value of an item to be drawn from the chosen urn. For the beads, a correct prediction was rewarded 0.5 ILS (about $0.12). For the cards, a correct prediction was rewarded 0.6 ILS, and a prediction deviating by 1 from the value drawn was rewarded 0.3 ILS. This scheme resulted in an expected reward of 1/3 ILS at each trial to participants who predicted the more frequent value of the beads or the mean value of the cards. During prediction, items were also drawn with replacement.

Our analyses revealed that people are apparently aware that predictions based on larger samples are more reliable than those based on smaller samples and that they were sensitive to differences in variability. On average, the urn from which a larger sample had been drawn was chosen in .63 of the cases, and the urn with the smaller sample variance was chosen in .54 of the cases; when one of the urns was both the one from which a larger sample had been drawn and the one with the smaller sample variance, the proportion with which it was chosen was .69. A multiple regression analysis using the stepwise method revealed that for all four conditions combined (with sample sizes and variances standardized within condition), the best combination of predictors involved the difference between sample sizes (whose correlation with choice was $r = .341$, $p < .001$) and the difference between the two standard deviations ($r = -.169$, $p < .003$); with these, the multiple regression reached a value of $R = .418$ ($F(2, 157) = 16.65$, $p < .001$).

More important to the specific question of the experiment, the data were analyzed to determine if all the available data or only a subsample of them best predicted people's choices. Given the large body of literature on primacy and recency effects (e.g., Hogarth & Einhorn, 1992), we analyzed the quality of prediction using subsamples taken either from the beginning or from the end of the actual sample. The size of the subsamples varied from six to thirteen, in steps of one. For each of the four conditions, it was possible to identify a subset whose data better predicted the participants' choices than did the sample as a whole. That subset tended to include the first items and had a median size of ten. Across all four conditions combined, not only did the use of the first ten items to predict choices not impair our ability to predict participants' choices correctly, but in fact it slightly improved it. The value of R was .434, and the proportion of correct predictions rose from .675, when all data were used, to .688, when data from only the first ten items were used. The results of this analysis indicate that people are indeed restricted in the size of the sample they consider when the total number of available items exceeds their capacity. With a total of ten items considered, on average five items from each population were used – a value commensurate with estimates of working-memory capacity when data can be assigned to different categories (Mandler, 1967). The analysis also reveals that the first items encountered are the ones more likely to be taken into account (i.e., a primacy effect), a finding in line with results reported in other studies comparing the strength of primacy and recency effects (see Hogarth & Einhorn, 1992).

GENERAL DISCUSSION

The five experiments dealing with the perception of variability lead to a number of conclusions: First, people are sensitive to variability and take it into account (as they should) in tasks in which it is of consequence, such as predicting a value from a single population or estimating the difference between two population means. Second, people use sample data to assess variability, without correcting for the bias likely to be brought about by the use of such data. As a result, perceived variability is downward attenuated: People regard the world as less variable than it really is.

As noted in the introduction, observers of human behavior have long suspected that the perception of consistency is persistently biased in such a way that the world is perceived as less variable and more regular than it actually is. This chapter offers an explanation of this misperception. Of necessity – owing to paucity of data, time pressure, or working-memory limitations – people have to infer characteristics of the environment from sample data. In other words, they see the world around them through a narrow window. Given that people indeed use sample data to assess the degree of variability, the statistical nature of the sampling distributions of that variable would lead to such misperception: Since sample variance tends to be smaller than its corresponding population parameter, people would see a world that is more consistent than it really is. It follows that the fault for the bias in the perception of consistency need not be faulty information processing, as hitherto assumed. Rather than the *inaccurate* processing of *unbiased* data, the reason for the misperception of consistency could be the *accurate* processing of *biased* data. No biased processing need be postulated for the bias to emerge: Accurate processing of well-sampled data leads to such a distorted view.

The present set of studies tested whether people use all the data when many data points are available, and whether they do or do not correct for the bias likely to result, when only a small number of data points are used. The results indicate that although people are sensitive to variability, they use only a small number of data points even when more are available. Furthermore, people do not (at least do not fully) correct for the bias resulting from such use. Working-memory capacity was implicated as a major factor limiting the size of the sample considered, and hence of the bias resulting from the use of small-sample data. The results reported here complement those observed in another set of studies, involving the perception of correlations (Kareev, Lieberman, & Lev, 1997). Starting with a similar statistical observation, that the sampling distribution of measures of correlation is highly skewed, and the more so the smaller the sample on which it is based (Kareev, 1995), we compared the perception of correlation by groups of people likely to differ in the size of the sample they used in assessing the degree of relationship between the two variables. There, too, people likely to have inspected smaller samples – either because of differences in working-memory capacity or because of experimental manipulation of sample size – perceived the correlations to be stronger than people likely to have inspected larger samples. Further discussion of the effects of sample size on the perception of relationships and of evaluating the potential benefits and costs of using small samples to infer population correlations appears in Chapter 5 of this book.

Seeing the world as more consistent than it actually is has some obvious implications: If variances around a central value are small, confidence in one's ability to make accurate predictions and good choices between alternatives is bound to increase. Such an increase in confidence will increase optimism and might even affect one's sense of well-being.

Our results also suggest that, although everyone views the world through a narrow window, people viewing the world through an even narrower window – children (whose working-memory capacity is smaller than that of adults, e.g., Huttenlocer & Burke, 1976), people under stress (who have less available cognitive capacity, e.g., Gilbert, Pelham, & Krull, 1988), and less intelligent people (who have less working memory capacity, e.g., de-Jong & Das-Smaal, 1995, Engle, 2002, Jurden, 1995) – will see the world around them as even more consistent than others. For them, there will be even fewer shades of gray, and wider areas of black and white; though their estimates of central values and differences between them will be, on average, less accurate, their confidence in their estimates and decisions will typically be higher.

Francis Bacon pointed out people's tendency to see too much regularity in the world around them and lamented it. The present findings prove his observation to be correct. The prevalence of the misperception could indicate, however, that it is not as dysfunctional as his lament implies.

References

Bacon, F. (1620/1905). *Novum Organum* (transl. R. L. Ellis & J. Spedding, edited by J. M. Robertson). London: George Routledge and Sons.

Cowan, N. (2001). The magical number 4 in short-term memory: A consideration of mental storage capacity. *Behavioral and Brain Sciences, 24*, 87–185.

de-Jong, P. F., & Das-Smaal, E. A. (1995). Attention and intelligence: The validity of the Star Counting Test. *Journal of Educational Psychology, 87*, 80–92.

Engle, R. W. (2002). Working memory capacity as executive attention. *Current Directions in Psychological Science, 11*, 19–23.

Gilbert, D. T., & Malone, P. S. (1995). The correspondence bias. *Psychological Bulletin, 117*, 21–38.

Gilbert, D. T., Pelham, B. W., & Krull, D. S. (1988). On cognitive busyness: When person perceivers meet persons perceived. *Journal of Personality and Social Psychology, 54*, 733–740.

Hays, W. L. (1963). *Statistics for psychologists.* New York: Holt, Rinehart, & Winston.

Hogarth, R. M., & Einhorn, H. J. (1992). Order effects in belief updating: The belief-adjustment model. *Cognitive Psychology, 24*, 1–55.

Huttenlocher, J., & Burke, D. (1976). Why does memory span increase with age? *Cognitive Psychology, 8*, 1–31.

Jurden, F. H. (1995). Individual differences in working memory and complex cognition. *Journal of Educational Psychology, 87*, 93–102.

Kareev, Y. (1995). Through a narrow window: Working memory capacity and the detection of covariation. *Cognition, 56*, 263–269.

Kareev, Y., Arnon, S., & Horwitz-Zeliger, R. (2002). On the misperception of variability. *Journal of Experimental Psychology: General, 131*, 287–297.

Kareev, Y., Lieberman, I., & Lev, M. (1997). Through a narrow window: Sample size and the perception of correlation. *Journal of Experimental Psychology: General, 126*, 278–287.

Linville, P. W., & Fischer, G. W. (1993). Exemplar and abstraction models of perceived group variability and stereotypicality. *Social Cognition, 11*, 92–125.

Linville, P. W., & Fischer, G. W. (1998). Group variability and covariation: Effects of intergroup judgment and behavior. In C. Sedikides, J. Schopler, et al. (Eds.), *Intergroup cognition and intergroup behavior* (pp. 123–150). Mahwah, NJ: Lawrence Erlbaum.

Linville, P. W., Fischer, G. W., & Salovey, P. (1989). Perceived distributions of the characteristics of in-group and out-group members: Empirical evidence and a computer simulation. *Journal of Personality and Social Psychology, 57*, 165–188.

Linville, P. W., & Jones, E. E. (1980). Polarized appraisals of out-group members. *Journal of Personality and Social Psychology, 38*, 689–703.

Mandler, G. (1967). Organization and Memory. In K. W. Spence & J. T. Spence (Eds.), *The psychology of learning and motivation*, (Vol. 1, pp. 328–372). New York: Academic Press.

Miller, G. A. (1956). The magical number seven, plus or minus two: Some limits on our capacity for processing information. *Psychological Review, 63*, 81–97.

Peterson, C. R., & Beach, L. F. (1967). Man as an intuitive statistician. *Psychological Bulletin, 68*, 29–46.

Pollard, P. (1984). Intuitive judgments of proportions, means, and variances: A review. *Current Psychological Research and Review, 3*, 5–18.

Ross, L. (1977). The intuitive psychologist and his shortcomings. In L. Berkowitz (Ed.), *Advances in experimental social psychology* (Vol. 10, pp. 173–220). San Diego: Academic Press.

Tversky, A., & Kahneman, D. (1971). Belief in the law of small numbers. *Psychological Bulletin, 76*, 105–110.

3

Intuitive Judgments about Sample Size

Peter Sedlmeier

Suppose a state election is coming up in a certain European country and somebody asks you if you would bet on whether the Social Democrats will beat the Christian Democrats again as they did in the last election. Shortly before Election Day, you are presented the results of two opinion polls, based on random samples of either 100 or 1,000 voters. In the smaller sample, the Social Democrats have an advantage of 4 percentage points whereas in the larger sample, the Christian Democrats lead by the same amount. Would you take the bet? You probably would not and I assume that this would be the choice of most people asked this question. As we will see later, your choice would be justified on both empirical and theoretical grounds.

This chapter offers an explanation for how and when the "naïve intuitive statistician" (Fiedler & Juslin, this volume) understands the impact that the size of a sample should have on estimates about population parameters such as proportions or means. I will argue that, in general, people know intuitively that larger (random) samples yield more exact estimates of population means and proportions than smaller samples and that one should have more confidence in these estimates the larger the size of the sample. My main claim is that the *size-confidence intuition* – an intuition that properly captures the relationship between sample size and accuracy of estimates – governs people's judgments about sample size. This intuition can be regarded as the result of associative learning. At first glance, the results of studies that explored whether people take sample size into account in their estimates speak against this explanation, because the rates of correct solutions varied widely in those studies. I will suggest an explanation for this variance in solution rates: It can be mainly accounted for by differences in task structure. Basically two kinds of tasks have been examined in this area of research: One type can be solved by applying the size-confidence intuition whereas the other cannot. I will discuss these different kinds of tasks and review the pertinent empirical evidence. Then I will suggest the PASS (probability associator) model, an associative learning model, as a possible explanation for the size-confidence intuition and other kinds of intuitive sample-based judgments and will explore the implications of this explanation. Finally, I will discuss cases in which intuitive sample-size judgments may lead to inadequate results.

DIFFERENT KINDS OF SAMPLE-SIZE TASKS

The literature on whether people take sample size properly into account in their judgments contains contradictory conclusions. Evans and Pollard (1985, pp. 68–69), for instance, concluded after their studies that "overall, subjects did quite well as intuitive statisticians in that their judgements tended, over the experiments as a whole, to move in the direction required by statistical theory as the levels of Mean Difference, Sample size and Variability were varied." Reagan (1989, p. 57), in contrast, came to a different conclusion: "The lesson from 'sample size research' is that people are poorly disposed to appreciate the effect of sample size on sample statistics." How can such different conclusions coexist after a substantial amount of research? To us, the most plausible explanation is that the researchers used different kinds of tasks: Those who used what we termed *frequency distributions tasks* usually arrived at positive conclusions and those who studied *sampling distribution tasks* did not (Sedlmeier & Gigerenzer, 1997).

Frequency Distribution Tasks versus Sampling Distribution Tasks

I will illustrate the difference between the two kinds of tasks with the original version (sampling distribution task) and a variant (frequency distribution task) of the maternity ward task (Kahneman and Tversky, 1972; headings are inserted for discussion only):

Introductory text (both versions):
A certain town is served by two hospitals. In the larger hospital about 45 babies are born each day, and in the smaller hospital about 15 babies are born each day. As you know, about 50% of all babies are boys. The exact percentage of baby boys, however, varies from day to day. Sometimes it may be higher than 50%, sometimes lower.

Sampling distribution version:
For a period of 1 year, each hospital recorded the days on which more than 60% of the babies born were boys. Which hospital do you think recorded more such days?

Frequency distribution version:
Which hospital do you think is more likely to find on a given day that more than 60% of the babies born were boys?

Answer alternatives (both versions):
   A) The larger hospital
   B) The smaller hospital
   C) About the same

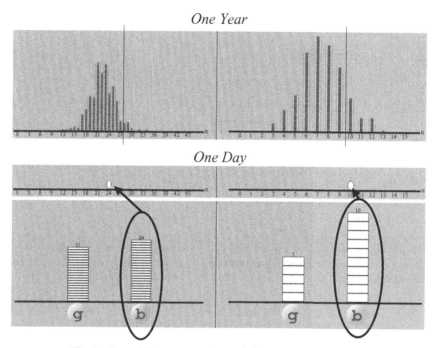

FIGURE 3.1. Illustration of the two versions of the maternity ward task. The frequency distribution version of the task applies to one day (bottom, "g" refers to "girl," and "b" to "boy") and the sampling distribution task to one year (top). The simulation results (relying on Bernoulli trials with $p = .5$) were obtained by using the software that accompanies a textbook on elementary probability theory (Sedlmeier & Köhlers, 2001).

In the frequency distribution version of the task, two frequency distributions (frequencies of baby boys and baby girls in the two hospitals on that given day) have to be compared with respect to how extreme their proportion of baby boys is. The lower portion of Figure 3.1 ("One Day") shows a possible result, obtained in a simulation, in which the "birth" of a "boy" and a "girl" were equally likely (Bernoulli trials with $p = .5$): on a given "day," in the larger hospital 24 out of 45 (53%) newborns are boys whereas in the smaller hospital, 10 out of 15 (67%) are boys. This result conforms to the intuition of a strong majority of respondents as shown in a review of six studies that used similar problems (Sedlmeier & Gigerenzer, 1997): Larger deviations from the expected proportion (e.g., from 50% in the maternity ward task) were judged to be more likely in the smaller sample. The median percentage of correct solutions in these frequency distribution tasks was 76% (Figure 3.2, left box plot; for details see Sedlmeier & Gigerenzer, 1997).[1]

---

[1] Very high rates of correct solutions were also found in 31 out of 35 studies that examined frequency distribution tasks that did not offer three choices but instead required confidence or preference ratings (Sedlmeier & Gigerenzer, 1997).

Peter Sedlmeier

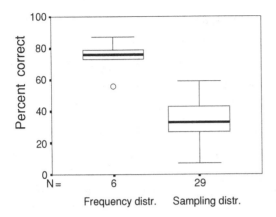

Tasks

FIGURE 3.2. Percentages of participants that took sample size into account in frequency distribution tasks (left) and sampling distribution tasks (right). Results are taken from the stem-and-leaf display in Exhibit 1 of Sedlmeier & Gigerenzer (1997).

A possible result for the sampling distribution version of the task – the distributions of proportions of baby boys in the 365 days of a year for the large and the small hospitals – is shown in the upper part of Figure 3.1 ("One Year"): The procedure to arrive at this result is depicted in the lower part of Figure 3.1. For every day, the number of boys is recorded as a "point" on an empirical sampling distribution (e.g., the 24 boys for the larger hospital and the 10 boys for the smaller hospital in the lower part of Figure 3.1). This procedure has been repeated 364 times and the resulting 365 proportions per hospital – depicted as stacked "disks" in the upper part of Figure 3.1 (with one disk corresponding to one day) – constitute an (empirical) sampling distribution. "More than 60% boys" means more than 27 boys out of 45 newborns (larger hospital) and more than 9 boys out of 15 (smaller hospital). The question in the sampling distribution task concerns the variance of the sampling distribution or, to be more specific, the number of proportions (days) that lie to the right of the two vertical lines in the top part of Figure 3.1. Clearly, in the smaller hospital, there are more such days. This result also can be theoretically derived by comparing the standard errors of binomial distributions with $p = .5$ and $n = 45$ versus $n = 15$. The solution of the sampling distribution tasks apparently defies intuition, as Kahneman and Tversky (1972) have already noted. In 29 studies that examined such sampling distribution tasks the solution rates were quite low (Figure 3.2, right box plot) with a median solution rate of 33%, which is just the result that can be expected by chance, given the three possible choices. [For details see Sedlmeier & Gigerenzer (1997).]

The difference in solution rates found in published research was later independently confirmed in specifically designed studies (Sedlmeier, 1998). Why is there this difference in solution rates? Gigerenzer and I argued that it might be because people are relying on an intuition, the *size-confidence intuition*, that conforms to the *empirical law of large numbers*. The empirical law of large numbers is not a mathematical law and can be experienced by anybody without mathematical sophistication: As (random) samples become larger, means or proportions calculated from these samples tend to become more accurate estimates of population means or proportions. The increasing accuracy is just a tendency and does not necessarily hold for every sample; it holds, however, on average (Sedlmeier & Gigerenzer, 2000). The size-confidence intuition makes it easy to solve frequency distribution tasks. For instance, in the frequency distribution version of the maternity ward task, the results from the larger hospital can be expected to be closer to the "true" 50% than the results from the smaller hospital, or – in other words – a large deviation of the proportion of baby boys from 50% is more likely in the smaller hospital. However, the size-confidence intuition is not directly applicable to sampling distribution tasks, which explains the difference in solution rates found in a review of relevant studies (Sedlmeier & Gigerenzer, 1997; see also Figure 3.2). One could argue that what we called a sampling distribution task can be solved by repeatedly applying the empirical law of large numbers. This is immediately evident to persons trained in statistics, but the empirical evidence indicates that lay people are unable to do this spontaneously.

In addition to our explanation, there have been two alternative attempts to explain the large variance in the solution rates for sample-size tasks. The first, advanced by Evans and Dusoir (1977), says that the descriptions of what we called sampling distribution tasks may just be more complex and so more difficult to understand than descriptions of what we called frequency distribution tasks. (Evans & Dusoir did not make an explicit distinction between the two types of tasks.) This means that if both types of tasks are equally well understood by participants, one should expect no difference or at least a smaller difference in solution rates. To examine this explanation, Sedlmeier (1998, Studies 1 and 2) made sure that all tasks were thoroughly understood by using computer animations and by allowing participants to perform simulations themselves. Despite this manipulation, the large difference in solution rates between the two types of tasks remained unchanged. However, the solution rates for both types of tasks were higher than in comparable studies, indicating that the solution rates in the literature, which were obtained by presenting tasks in the usual text form, may be underestimates of participants' true knowledge.

The second possible explanation for the variance in solution rates, which also does not refer to different types of tasks, was advanced by Nisbett et al. (1983). They hypothesized what determines whether people take sample size into account is mainly the salience of three "chance factors": clarity of the sample space and the sampling process, recognition of the role of chance in producing an event, and cultural prescriptions to think statistically in a particular domain. Because Nisbett et al. (1983) did not differentiate these different types of tasks, a sampling distribution task that fulfills these criteria should also yield high correct solution rates. A study by Sedlmeier (1998, Study 3) tested this prediction. In this study, assurance was first made that participants thought statistically about the process involved in coin tossing. Then participants were handed an urn containing 50 coins and were asked to take a handful, throw them onto a large tray on the floor, and count the number of heads and tails. This procedure was repeated once to ensure that the sample space and the sampling process were clear and that the role of chance in producing a head or tail was salient to the participants. After these preparations, participants solved a frequency distribution task (about the proportion of heads to be expected in one large sample versus one small sample of coins) and a sampling distribution task (about a given part of the distribution of proportions of heads in 100 large versus 100 small samples). If the salience of chance factors alone accounts for whether sample-size tasks can be solved correctly, there should be no difference in solution rates between the frequency and the sampling distribution tasks. However, whereas 75% of participants solved the frequency distribution task correctly, only 32% did so for the sampling distribution task. Thus, it seems that Nisbett et al.'s (1983) criteria may account for some differences in solution rates for frequency distribution tasks (which they exclusively used in their studies) but they cannot account for the contradictory conclusions in the literature, which seem to be mostly due to the difference in solution rates for sampling and frequency distribution tasks.

In sum, the size-confidence intuition seems to be the best explanation for the diverging results in sample-size tasks, so far: Frequency distribution tasks can be solved by applying this intuition wheras sampling distribution tasks cannot (see the following discussion). Where does the size-confidence intuition come from? Here is a suggestion: It is the result of simple associative learning.

ASSOCIATIVE LEARNING AND SAMPLE-BASED JUDGMENTS

The specific associative learning model I am proposing is the PASS model (Sedlmeier, 1999, 2002). This model has been used to predict estimates of relative frequency and probability. I will first describe how PASS works in principle and then address its predictions about increasing accuracy of estimates. After that, I will describe how PASS may account for the size-confidence intuition, and finally, I will address the important role of representational format in the model.

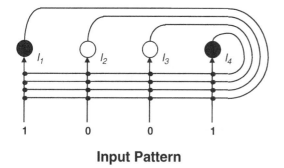

Input Pattern

FIGURE 3.3. Simplified variant of the core part of PASS, a recurrent neural network that encodes objects and events by their featural descriptions.

PASS: Basic Mechanisms

PASS encodes events consecutively, either in reality or in imagination. An event (hereafter standing for both "events" and "objects") is represented by its features, to which the associative learning mechanism applies. The core of PASS is a neural network that encodes features by its input nodes, modifies the associations among features by changing the weights among nodes, and elicits reactions by producing activations at its output nodes. Figure 3.3 shows a simplified neural network as used in the PASS framework. PASS has just encountered an event consisting of the feature array "1 0 0 1"; that is, the first and the fourth features are present and the second and third features are absent. Learning consists in modifying the associations among the features in memory. (In this simplified example, the memory consists of only four features and their associations, depicted in the four by four matrix shown as little circles in Figure 3.3.)

PASS operates in discrete time steps. At each time step, the events (including complex events that consist of several simple ones) toward which the attention is directed are encoded. At every time step, learning takes place. If two features co-occur in an event (such as the first and the fourth in Figure 3.3) the association between them is strengthened; if this is not the case, the association between the respective features becomes weaker. This learning process results in a memory matrix that is updated after every time step and that contains the association strengths among all the features in memory. (Note that in all prior simulations a time step was equivalent to a newly encoded event, but this need not necessarily be so.) When PASS is prompted with an event, that is, the features that define this event, activations (ranging from 0 to 1) occur at all output nodes.[2] The

---

[2] In this kind of recurrent neural network model, input and output nodes are identical. PASS does not, however, depend on a specific kind of neural network. It only requires that the network learn without feedback (Sedlmeier, 1999).

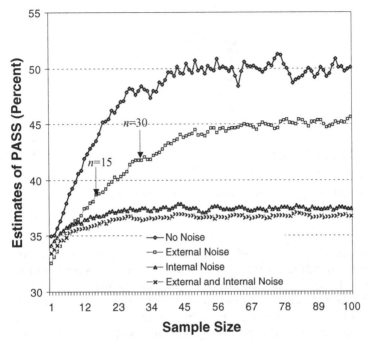

FIGURE 3.4. Typical PASS predictions for how the accuracy of relative frequency estimates increases with sample size. Shown are results for a simple pattern that was presented randomly in 50 out of 100 presentations in four different conditions. External and internal noise at encoding act additively in regressing the relative frequency estimate toward the relative frequency obtained by 1 divided by the number of different patterns (with three patterns used in this simulation). Results are averaged over 100 runs, that is, 100 sequences of 100 patterns each. See text for further explanation.

sum of these activations is taken as PASS's response, which in turn is the basis for estimates about the event's relative frequency or probability (see Sedlmeier, 2002, for details).

Accuracy and Sample Size

In general, PASS's estimates become more and more accurate as sample size increases. In one simulation (for which results are partially shown in Figure 3.4), three different events, consisting of four features each (such as those in Figure 3.3) were presented to PASS 20, 30, and 50 times each, in random order, leading to a final sample size of $n = 100$. After every new presentation, PASS made relative frequency estimates for every event. The whole procedure was repeated 100 times; that is, the simulation consisted of 100 "runs." Figure 3.4 shows PASS's estimates – averaged over the 100 runs – only for the event that was presented 50% of the time (50 out of 100 presentations), for all sample sizes from 1 to 100. Under ideal circumstances, PASS's relative frequency estimates for the 50% event asymptotically reach 50% (at a sample size of about $n = 40$ in this simulation; see curve for "No Noise" in Figure 3.4). However, under realistic conditions, events are encoded under external noise (since events have features in common) and internal noise (since previous events interfere with new events to a certain degree). PASS predicts regressed estimates (underestimation of high and overestimation of low relative frequencies) under these conditions, a result also commonly observed in empirical studies (see Sedlmeier, 1999, 2002; Sedlmeier, Hertwig, & Gigerenzer, 1998). Figure 3.4 shows that external and internal noise will determine the amount of regression in an additive way. For demonstration purposes, the amount of regression in Figure 3.4 is unrealistically strong owing to the simplicity of the patterns used; nonetheless, estimates become more accurate (come closer to 50% in the example) as sample size increases until they reach an asymptote value that is determined by the amount of regression.[3]

That judgment of means or proportions becomes more exact with increasing sample size can also be found in human estimates (e.g., Erlick, 1964; Shanks, 1995). In one of my own experiments (Sedlmeier, 2003) I repeatedly presented participants one of three different symbols, one at a time (with symbols completely counterbalanced across participants). In each subset of twelve presentations, the first symbol was presented six times (50%), the second four times (33%), and the third twice (17%). Participants had to make relative frequency judgments after every twelve presentations, up to a total sample size of sixty. Figure 3.5 shows that with increasing sample size, the estimates tend to become more exact, as predicted by PASS. The figure also shows typical regression effects for the 50% and 17% patterns.

Confidence and Sample Size

But where is the size-confidence intuition in the model? In PASS it is assumed that confidence judgments covary with how well the model can differentiate among different events. Before learning, that is, with a sample size of 0, the associations between the features are either identical or randomly distributed (see Sedlmeier, 1999). If at this stage a prompt, that is, the features that make up an event, were to be given to the model, the

---

[3] A restriction applies to this statement, however: Empirical estimates for very small quantities are often quite exact. This finding is usually explained by reference to a mechanism called *subitizing* (e.g., Wender & Rothkegel, 2000).

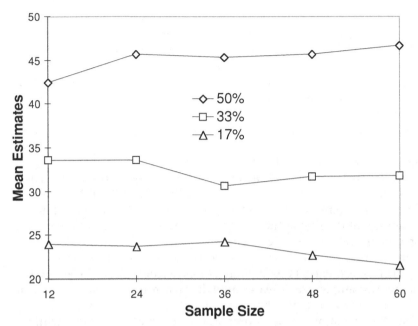

FIGURE 3.5. Participants' relative frequency estimates in a study of Sedlmeier (2003) for three patterns that were presented 30 times, 20 times, and 10 times in a total of 60 presentations. Sequences were created randomly for every participant. Data are averages.

model's average response would be nearly the same for different events: PASS would not be able to "keep one event apart from another." The more variation there is in the output node activations after being prompted with a given event, the better PASS "knows" this event. Therefore, the variation in the output units' activations is taken as a measure of confidence: The higher the variance of the output units' activations, the higher the confidence. Figure 3.6 shows an example of how PASS's confidence increases as sample size increases. Confidence values (variances of output unit activations) are averaged over 100 runs with three simple patterns presented in 50%, 30%, and 20% of all cases, respectively. It is assumed that PASS's confidence judgments refer to both the estimate in the sample as well as the estimate of a population value based on that sample value.

Figure 3.6 indicates that confidence should not increase linearly but should have a monotonically decreasing slope. An equal increase in sample size should lead to a more pronounced increase in confidence when compared to a small sample size than when compared to a large sample size. The overall pattern in Figure 3.6 does not depend on PASS's specific parameter values or kinds of feature patterns, although the specific form, that is, the relationship between sample size and slope, does, of course.

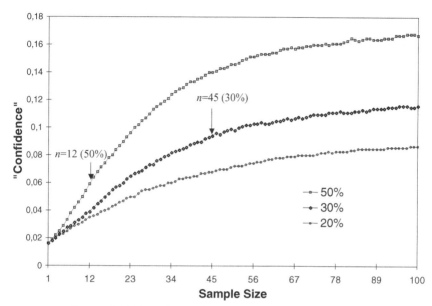

FIGURE 3.6. A typical PASS prediction for how confidence increases with increasing sample size. Shown are results for three simple patterns that were presented in 50, 30, and 20 out of 100 presentations, respectively, in random order. Results are averaged over 100 runs. See text for further explanation.

Again, empirical evidence can be found for PASS's prediction of a non-linear increase in confidence. For instance, Evans and Pollard (1985) had their participants judge the odds that the mean of IQ values, which these participants saw on a computer screen, was above or below 100. Figure 3.7 shows these odds transformed into confidence values for three experiments. In all three experiments the sample sizes (number of IQ values presented) were 64, 144, and 324. Consistent with PASS's prediction, the increase in confidence between sample sizes of 144 and 324, although expressing a difference of 180, is smaller than the increase in confidence between sample sizes of 64 and 144, with a difference of only 80.

Implicit Sample-Size Information: The Role of Representational Format

What happens when sample-size information is not encoded serially? (Here, serial encoding may also include the serial encoding of frequency information from pictograms or similar representational formats.) It seems that the crucial point for determining whether the PASS model is applicable is whether the information is transformed in a way that enables the imaginableness of discrete events. (For recent evidence that memories

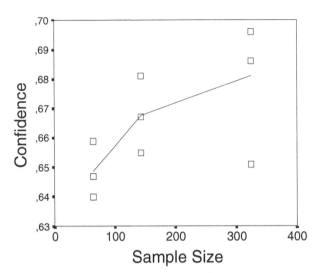

FIGURE 3.7. A visualization of data reported by Evans and Pollard (1985) from their tables 1, 2, and 3. The line shown is a *Lowess* line (e.g., Cleveland, 1979) with a tension of .9 (*Lowess* is short for a graphical method called *robust locally weighted regression* that connects successive smoothed values. The Lowess curve summarizes the *middle* of the distribution of *y* for each value of *x*). See text for further explanation.

develop with imagination alone see Mazzoni & Memon, 2003.)[4] Whether people can imagine numerical information as discrete events may also be the decisive factor in whether training programs for some aspects of statistical reasoning will work (Sedlmeier & Gigerenzer, 2001). So whenever one can expect that implicit sample-size information elicits an imagining of discrete events, either by giving prompts or as the result of specific training, one may expect intuitive judgments according to the PASS model. When imagination is not possible, biased judgments may be the consequence.

WHEN TO EXPECT BIASED SAMPLE-BASED JUDGMENTS

When people properly use the size-confidence intuition this does not mean that their judgments will always reflect the true state of affairs. For instance, judgments may be biased when people use biased input. However, even if samples are not biased, the size-confidence intuition may have a biasing impact on judgments. Also, encoding processes might influence confidence

---

[4] To stress the importance of imagination in judgmental processes is not new: Whether other events can be readily imagined seems to play a crucial role in how much pleasure or frustration people feel after a given event has occurred (Kahneman & Tversky, 1982); and imagination also helps to develop valid intuitions by learning "from what we do not see" (Hogarth, 2001, p. 276).

judgments. Moreover, biased judgments may arise when people's valid intuitions are overridden by deliberative processes, and finally, judgments may be incorrect when the size-confidence intuition does not apply, as in the case of sampling distribution tasks. These cases will now be discussed in turn.

Biased Input and Biased Responses

Fiedler (2000) gives an illuminating account of how biased samples may arise and how they can influence all kinds of judgments. He notes that "inductive judgments are often remarkably accurate when compared with actually given samples" (p. 661). This is just what the PASS model and its version of the size-confidence intuition would predict.[5] If, however, the sample is selectively drawn, judgments may be biased. For instance, when judges are to make judgments about the conditional probability of a disease given a symptom (e.g., breast cancer, given a positive mammogram), and they are not restricted in their sampling process, they may focus on cases *with* the disease rather than on those without the disease and therefore oversample cases for which the symptom correctly predicts the disease (Fiedler et al., 2000). Similar results were obtained in simulated classrooms where "teachers" were found to be quite selective in sampling from students' behavior (Fiedler et al., 2002). This kind of selective sampling may frequently occur when people seek to confirm hypotheses they are already holding (Klayman & Ha, 1989).

According to Fiedler (2000) people lack the metacognitive skills to correct for the sampling biases. Such metacognitive skills are also missing from the PASS model. However, there is increasing evidence that metacognitive skills conforming to Bayes's rule that can correct for biases in the selection process in the aforementioned examples can be effectively taught (Ruscio, 2003; Sedlmeier & Gigerenzer, 2001; Wassner, Martignon, & Sedlmeier, 2002). Another way to avoid such biased judgments may be to educate intuitions by exposing "learners" to a representative environment (Hogarth, 2001).

Unbiased Input and Biased Responses

As Fiedler (2000) argues, sample-based judgments may even be biased when representative samples are drawn that differ only in size. A prominent bias of this sort is the illusory correlation between group size (majority versus minority) and behavior (e.g., Hamilton & Gifford, 1976). If, for instance, some kind of positive behavior is more common overall than a

---

[5] In his examples, Fiedler usually uses predictor criterion pairs, such as symptom and disease, or sex and aggression. This combination of events is treated in PASS as one complex event (for details see Sedlmeier, 2005).

corresponding negative behavior, and if the relative frequency of the positive behavior is equal in the majority and the minority groups, relative frequency estimates for the positive behavior are nonetheless found to be typically higher for the majority, and so are evaluative ratings. Fiedler (1991) proposed that this illusory correlation may be solely due to the difference in sample size.

PASS offers two possible kinds of explanations for illusory correlations. The first kind is about relative frequency estimates and seems to be consistent with Fiedler's (1991, 1996) explanation. Consider two samples of behaviors, a large one drawn from the native inhabitants of a city and a small one drawn from recent immigrants. Now assume that both samples exhibit the same relative frequency of some kind of positive behavior. An illusory correlation effect would mean that the proportion of positive behavior is judged to be higher in the larger sample (majority) than in the smaller one (minority). This is what PASS would predict. PASS's average relative frequency estimates for small samples are close to a state of "indifference," that is, close to $1/k$, where k is the number of different events or "behaviors." For illustration consider the results shown in Figure 3.4: There are three events and PASS's initial relative frequency estimates are close to 1/3. As sample size increases, PASS's estimates tend to move toward the actual relative frequencies and eventually reach asymptotically a relative frequency estimate determined by the amount of regression induced by external and internal noise. As long as the asymptote is not reached, the relative frequency of a given kind of "behavior" in a smaller sample (minority) is judged less extreme than that in the larger sample, even if the true relative frequency does not change (compare, for illustration, the relative frequency estimates for $n = 15$ and $n = 30$, indicated by the two arrows in Figure 3.4).

The second explanation can be applied to evaluative ratings and involves the size-confidence intuition. As sample size increases, PASS's confidence in its estimates increases (see Figure 3.6). Confidence, in turn, might have a direct impact on the strength or extremity of evaluative ratings. PASS also makes an interesting prediction for the case when the behavior in question has a higher relative frequency in the minority group than in the majority group. Even in this case it should be possible to elicit both higher relative frequency estimates and higher evaluative ratings for the majority group. The latter is again illustrated in Figure 3.6 (compare the confidence judgments for $n = 45$ for a relative frequency of 30% – right arrow – to that for $n = 12$ and a relative frequency of 50% – left arrow). Of course, the explanations sketched out here and illustrated with the simulation results in Figures 3.4 and 3.6 do not model any specific empirical results, but they show how an illusory correlation between membership in groups that differ in sample size and some kind of rating can arise solely as the result of unequal group sizes.

The Impact of Encoding Processes

PASS predicts an even more subtle effect on sample-based judgments that depends on the conditions at the encoding stage. Samples of identical size and of the same events may elicit different estimates and confidence ratings, depending on attentional processes and memory contents. This prediction hinges on the assumption that during associative learning, the amount of change in the associative strengths among the features of events is influenced by the amount of attention focused on a given event. If a high amount of attention is given to an event, the associations among its features increase more than if the event is encoded with low attention. PASS predicts that frequency judgments should be higher for events that were encoded with high attention and also that confidence ratings should be higher for these events than for the same events when encoded with low attention. The first of these predictions has already been confirmed in two experiments (Sedlmeier & Renkewitz, 2004).

Deliberative versus Intuitive Judgments

As discussed in the foregoing, intuitive sample-based judgments conform to true states of affairs when the sample comprising the basis for the judgments is representatively sampled. However, even when the sample is representative of the environment and intuitive judgments would express this, biased judgments can arise when deliberations contradict intuitions. If, for instance, the predictions of a polling firm have been found to be way off the mark at the previous election, one may doubt its prediction even if the sample size it uses may be substantially larger than the ones used by its competitors (and even if its prediction would be the best this time). Mistrust in intuitions, especially for personally important decisions, may in part be a result of school education, with its emphasis on a deterministic world with deliberative explanations for everything (Fischbein, 1975; Engel & Sedlmeier, 2004). With representative samples, deliberative judgments seem to be more prone to errors than intuitive ones (e.g., Haberstroh & Betsch, 2002).

Sampling Distribution Tasks

Sampling distribution tasks are abstractions of naturally occurring events, because they deal with distributions of *aggregated* values, such as proportions or means. Although frequency distribution tasks can be seen as sampling distribution tasks with samples of size 1, usually the difference between the two kinds of tasks is clear. For instance, it makes a difference whether, for a given student population, one looks at the distribution of the heights of the students in a given classroom (frequency distribution) or at the distribution of the average heights of students in many classrooms (sampling distribution). To argue about sampling distributions means to argue about something that cannot be directly seen or experienced. Whereas the height of one student can be seen directly, the mean height of all students in a class is something that takes additional effort to calculate and to imagine. Of course, nowadays, sampling distributions are part of everyday life as, for instance, in the media (e.g., average incomes) or in diagnosis (e.g., summary scores of IQ tests), but it is hard to think of examples in the contemporary natural environment; and it is still harder to imagine something of that sort in the Stone Age, a time period closer to the focus of theories in evolutionary psychology. Therefore, if our mind was shaped after the environment in the course of evolution, it seems plausible not to expect intuitions about such tasks.[6]

Sampling distributions, however, play a most important role in statistical inference, such as in the understanding of confidence intervals and significance testing. Consistent with the results of Kahneman and Tversky (1972) and those of later researchers, even experienced psychologists seem to have great problems understanding the impact of sample size on sampling distributions (e.g., Oakes, 1986; Sedlmeier & Kılınç, 2004; Tversky & Kahneman, 1971). A case in point is the sensitivity to statistical power. In 1962, Cohen found that researchers did not seem to pay any attention to the chances they had of obtaining a significant test result for a plausible population effect (Cohen, 1962). Twenty-five years later, this state of affairs had rather worsened (Sedlmeier & Gigerenzer, 1989). One possible explanation for the common result that sample size is not taken into account in judgments about sampling distributions may be that because people lack adequate intuitions about sampling distributions, they treat them as if they were frequency distributions, whose variance does not depend on sample size.[7]

Problems with the understanding of sampling distributions can be alleviated somewhat by using good visualizations (Sedlmeier, 1998). An even more successful route to a better understanding of how sample size affects sampling distributions seems to be the use of the size-confidence intuition in training (Sedlmeier, 1999). A computer program that demonstrates how

---

[6] From an abstract point of view, one might argue that each given event or object consists of a sample of features and so should be seen as an aggregate event or object itself. In this view, the distributions of events or objects might justifiably be called sampling distributions. Usually, however, distributions are not about objects in their entirety but about values of variables (e.g., height, weight, or age, but not persons), in which case the distinction between the distribution of heights over persons and the distribution of mean heights over groups (samples) of persons seems quite uncontroversial.

[7] This statement is, of course, only approximately true because the best estimate of a population variance is the sum of squares of the sample values divided by $n - 1$ instead of n, but this difference is negligible for practical purposes for all but very small samples.

sampling distributions arise from repeated sampling of frequency distributions (for which the size-confidence intuition works) is currently being tested in German high schools (Sedlmeier & Köhlers, 2001; Wassner et al., 2002).

SUMMARY AND CONCLUSION

This chapter offers a positive perspective: In principle, people intuitively know about the impact of sample size on judgments about aggregate values. I have postulated that this intuitive knowledge can be explained as a byproduct of associative learning. Intuitive judgments of relative frequency become more exact as sample size increases; and the confidence in these judgments also increases with sample size, although not linearly. The increase of confidence with increasing sample size has been termed size-confidence intuition and both this intuition and the growing accuracy of estimates correspond to the empirical law of large numbers. These intuitive judgments are, however, only valid if the sample they rely on is representative of the task at hand. Biased samples lead to biased judgments and (if the associative memory model described here is correct) the reason for biased samples may lie in selective sampling or in factors that influence encoding processes and the building of memory representations. Intuitions do not cover judgments about sampling distributions but can be used effectively in training programs about such sampling distribution tasks.

References

Cleveland, W. S. (1979). Robust locally weighted regression and smoothing scatterplots. *Journal of the American Statistical Association, 74,* 829–836.

Cohen, J. (1962). The statistical power of abnormal-social psychological research. *Journal of Abnormal and Social Psychology, 65,* 145–153.

Engel, J., & Sedlmeier, P. (2004). Zum Verständnis von Zufall und Variabilität in empirischen Daten bei Schülern. [School students' understanding of chance and variability in empirical data] *Unterrichtswissenschaft, 32,* 169–191.

Erlick, D. E. (1964). Absolute judgments of discrete quantities randomly distributed over time. *Journal of Experimental Psychology, 57,* 475–482.

Evans, J. St. B. T., & Dusoir, A. E. (1977). Proportionality and sample size as factors in intuitive statistical judgement. *Acta Psychologica, 41,* 129–137.

Evans, J. St. B. T., & Pollard, P. (1985). Intuitive statistical inferences about normally distributed data. *Acta Psychologica, 60,* 57–71.

Fiedler, K. (1991). The tricky nature of skewed frequency tables: An information loss account of distinctiveness-based illusory correlations. *Journal of Personality and Social Psychology, 60,* 24–36.

Fiedler, K. (1996). Explaining and simulating judgment biases as an aggregation phenomenon in probabilistic, multiple-cue environments. *Psychological Review, 103,* 193–214.

Fiedler, K. (2000). Beware of samples! A cognitive–ecological sampling approach to judgment biases. *Psychological Review, 107,* 659–676.

Fiedler, K., Brinkmann, B., Betsch, R., & Wild, B. (2000). A sampling approach to biases in conditional probability judgments: Beyond baserate neglect and statistical format. *Journal of Experimental Psychology: General, 129,* 1–20.

Fiedler, K., Walther, E., Freytag, P., & Plessner, H. (2002). Judgment biases in a simulated classroom – A cognitive-environmental approach. *Organizational Behavior and Human Decision Processes, 88,* 527–561.

Fischbein, E. (1975). *The intuitive sources of probabilistic thinking in children.* Reidel: Dordrecht-Holland.

Haberstroh, S., & Betsch, T. (2002). Online strategies versus memory-based strategies in frequency estimation. In P. Sedlmeier & T. Betsch (Eds.), *Etc. Frequency processing and cognition* (pp. 205–220). Oxford: Oxford University Press.

Hamilton, D. L., & Gifford, R. K. (1976). Illusory correlation in interpersonal perception: A cognitive basis of stereotypic judgments. *Journal of Experimental Social Psychology, 12,* 392–407.

Hogarth, R. (2001). *Educating intuition.* Chicago: University of Chicago Press.

Kahneman, D., & Tversky, A. (1972). Subjective probability: A judgment of representativeness. *Cognitive Psychology, 3,* 430–454.

Kahneman, D., & Tversky, A. (1982). The psychology of preferences. *Scientific American, 246,* 160–173.

Klayman, J., & Ha, Y.-W. (1989). Hypothesis testing in rule discovery: Strategy, structure and content. *Journal of Experimental Psychology: Learning, Memory and Cognition, 15,* 596–604.

Mazzoni, G., & Memon, A. (2003). Imagination can create false autobiographical memories. *Psychological Science, 14,* 186–188.

Nisbett, R. E., Krantz, D. H., Jepson, C., & Kunda, Z. (1983). The use of statistical heuristics in everyday inductive reasoning. *Psychological Review, 90,* 339–363.

Oakes, M. (1986). *Statistical inference: A commentary for the social and behavioral sciences.* New York: Wiley.

Reagan, R. T. (1989). Variations on a seminal demonstration of people's insensitivity to sample size. *Organizational Behavior and Human Decision Processes, 43,* 52–57.

Ruscio, J. (2003). Comparing Bayes's theorem to frequency-based approaches to teaching Bayesian reasoning. *Teaching of Psychology, 30,* 325–328.

Sedlmeier, P. (1998). The distribution matters: Two types of sample-size tasks. *Journal of Behavioral Decision Making, 11,* 281–301.

Sedlmeier, P. (1999). *Improving statistical reasoning: Theoretical models and practical implications.* Mahwah, NJ: Lawrence Erlbaum.

Sedlmeier, P. (2002). Associative learning and frequency judgments: The PASS model. In P. Sedlmeier & T. Betsch (Eds.), *Etc. Frequency processing and cognition* (pp. 137–152). Oxford: Oxford University Press.

Sedlmeier, P. (2003). Simulation of estimates of relative frequency and probability for serially presented patterns. Unpublished raw data.

Sedlmeier, P. (2005). From associations to intuitive judgment and decision making: Implicitly learning from experience. In T. Betsch & S. Haberstroh (Eds.), *Experience-based decision making* (pp. 83–99). Mahwah, NJ: Lawrence Erlbaum.

Sedlmeier, P., & Gigerenzer, G. (1989). Do studies of statistical power have an effect on the power of studies? *Psychological Bulletin, 107,* 309–316.

Sedlmeier, P., & Gigerenzer, G. (1997). Intuitions about sample size: The empirical law of large numbers. *Journal of Behavioral Decision Making, 10,* 33–51.

Sedlmeier, P., & Gigerenzer, G. (2000). Was Bernoulli wrong? On intuitions about sample size. *Journal of Behavioral Decision Making, 13,* 133–139.

Sedlmeier, P., & Gigerenzer, G. (2001). Teaching Bayesian reasoning in less than two hours. *Journal of Experimental Psychology: General, 130,* 380–400.

Sedlmeier, P., Hertwig, R., & Gigerenzer, G. (1998). Are judgments of the positional frequencies of letters systematically biased due to availability? *Journal of Experimental Psychology: Learning, Memory, and Cognition, 24,* 754–770.

Sedlmeier, P., & Kılınç, B. (2004). The hazards of underspecified models: the case of symmetry in everyday predictions. *Psychological Review, 111,* 770–780.

Sedlmeier, P., & Köhlers, D. (2001). *Wahrscheinlichkeiten im Alltag: Statistik ohne Formeln.* [Probabilities in everyday life: statistics without formula]. Braunschweig: Westermann (textbook with CD).

Sedlmeier, P., & Renkewitz, F. (2004). *Attentional processes and judgments about frequency: Does more attention lead to higher estimates?* Manuscript submitted for publication.

Shanks, D. R. (1995). *The psychology of associative learning.* Cambridge, UK: Cambridge University Press.

Tversky, A., & Kahneman, D. (1971). Belief in the law of small numbers. *Psychological Bulletin, 73,* 105–110.

Wassner, C., Martignon, L., & Sedlmeier, P. (2002). Entscheidungsfindung unter Unsicherheit als fächerübergreifende Kompetenz. *Zeitschrift für Pädagogik, 45* (Suppl.), 35–50.

Wender, K. F., & Rothkegel, R. (2000). Subitizing and its subprocesses. *Psychological Research, 64,* 81–92.

4

The Role of Information Sampling in Risky Choice

Ralph Hertwig, Greg Barron, Elke U. Weber,
and Ido Erev

Life is a gamble. True to the cliché, we can rarely be certain of the consequences of our everyday actions. Presumably because of life's lack of certainty, much of the psychological research on decision making under risk
is devoted to the study of choices between monetary gambles or prospects.
With few exceptions, the outcomes and outcome probabilities of such gambles are explicitly described to the decision maker before he or she makes
a choice. How often are life's gambles described in this way? When we
decide whether to back up our computer's hard drive, cross a busy street,
or go out on a date, we often do not know the complete range of the possible outcomes, let alone their probabilities. Yet we routinely make such
decisions, usually without difficulty. We do so by drawing upon our experience with the relevant prospects, for instance, by recalling the outcomes
of previous choices. Though effective, sampling from experience brings
with it the potential to generate a skewed picture of the risky prospects
we face. As we will argue here, this potential is greatest when the outcome
probabilities are small and relatively few outcomes have been experienced
by the decision maker. Under these circumstances, decisions derived from
experience are likely to be systematically different from those made in full
knowledge of the outcomes and outcome probabilities.

In this chapter, we investigate risky decisions made on the basis of sampled experience. Three findings deserve emphasis. First, when outcome information is sampled, the likelihood of rare events is often underestimated.
Second, when the sampling process is sequential, more recent outcomes
tend to receive more weight than do earlier outcomes. Third, in decisions
from experience, the option that appears best in light of the experienced
sample tends to be the one selected. To account for these effects, we propose a learning mechanism that models how people update their estimates
of a risky prospect's value in light of newly sampled outcomes.

INTRODUCTION

CALL ME ISHMAEL. Some years ago – never mind how long precisely – having little or no money in my purse, and nothing particular to interest me on shore, I thought I would sail about a little and see the watery part of the world. (Melville, 1851/1993, p. 1)

Thus begins *Moby Dick*, and with it the voyage of the Pequod and her captain, Ahab – a voyage of which Ishmael will be the sole survivor. With its myriad metaphors, symbols, and characters, Herman Melville's great novel has been interpreted as everything from a saga about Promethean determination and undying hatred to allegories about the Gold Rush, man's quest for knowledge, or man's will to master nature. Yet *Moby Dick* is also a fascinating factual chronicle of nineteenth-century whaling as revealed in the hardships, routines, and adventures of the men aboard the Pequod. As Melville (1851/1993, pp. 321–322) portrayed it, whale hunting was a high-stakes gamble:

But here be it premised, that owing to the unwearied activity with which of late they have been hunted over all four oceans, the Sperm Whales, instead of almost invariably sailing in small detached companies, as in former times, are now frequently met with in extensive herds, sometimes embracing so great a multitude, that it would almost seem as if numerous nations of them had sworn solemn league and covenant for mutual assistance and protection. To this aggregation of the Sperm Whale into such immense caravans, may be imputed the circumstance that even in the best cruising grounds, you may now sometimes sail for weeks and months together, without being greeted by a single spout; and then be suddenly saluted by what sometimes seems thousands on thousands.

The whalers aboard the Pequod were forced to make decisions about, for instance, the course of their ship under conditions that can be described in decision-theoretic terms as ignorance of the options' outcomes and probabilities. Although they hoped that sailing in known cruising grounds would increase their chance of being at the right place at the right time, they knew neither how large the herds they came across would be nor the probability of such encounters. Instead of guessing, however, the whalers most likely relied on their experience from previous hunting expeditions. Moreover, they were able to update their expectations in light of new samples of information. In this view, each day on which they cruised waters that the sperm whale was believed to frequent could be seen as a draw from a payoff distribution. After sequentially accumulating experience about a specific cruising ground, they would decide to stay or to abandon it.

Of course, decisions in which knowledge about the outcomes and outcome probabilities is scant are not unique to nineteenth-century whaling. In fact, ignorance or at least partial ignorance of outcomes and their probabilities may be the rule rather than the exception in many everyday decisions.

Like Melville's whalers, decision makers then can rely on past experience and continue to sample information about such prospects before making a decision. We have referred to this category of decisions as *decisions from experience* (Hertwig et al., 2004).

As the metaphor of life as a gamble implies, decisions from experience represent a bet on the future that is based on one's necessarily limited experience of the past. Indeed, the gambling metaphor has inspired a good deal of the psychological research on decision making. On the basis of a review of the early literature such as Savage's (1954) *The Foundations of Statistics,* Goldstein and Weber (1997, p. 569) concluded the following:

Gambling decisions were believed to be prototypical of virtually *all* decisions. Almost any contemplated action, for example, where to go to school, what job to take, or whom to marry, will have consequences that cannot be predicted with certainty but which vary in their likelihood and desirability. In this sense, then, real-life decisions have the same structure as gambles. Moreover, because this applies to virtually all decisions, *life is a gamble.*

Given this background, it is not surprising that monetary gambles became the fruit fly of decision research (Lopes, 1983). Just as biologists use the *Drosophila* as a model organism to study, for instance, genetic inheritance, so psychologists use the monetary gamble as a model task to study, for instance, decision making under risk. A monetary gamble is a well-defined alternative consisting of n outcomes ($n = 1, 2, 3, \ldots, k$) and those outcomes' probabilities. In the typical notation, the gamble "$32, .1; 0 otherwise" yields an outcome of $32 with a 10% chance, and an outcome of 0 with a 90% chance. A dilemma is introduced by presenting people with two such gambles and asking them to say which one they prefer. For instance, they may be asked to choose between the following two gambles:

$32, .1; 0 otherwise $3 for sure.

If people maximize expected value, they will decide against the sure gamble because the two-outcome gamble offers the higher expected value, $3.20.[1]

Now compare this choice with the choices faced by the whalers aboard the Pequod or, to turn to a present-day example, with those faced by overweight people considering whether to go on a diet. Although the diet has the potential to help them reduce their weight, it may have undesirable side effects, the nature and probability of which are relatively unknown. Consider the Atkins diet. Its recommendation to shun carbohydrates and eat any amount of fat contradicts the predominant view in the field of nutrition

---

[1] The principle of choosing the option with the highest expected value (EV) is expressed as $EV = \sum p_i x_i$, where p_i and x_i are the probability and the amount of money, respectively, of each possible outcome ($i = 1, \ldots, n$) of a gamble.

that a weight-loss diet should be low in fat and high in starch. Two recent studies published in the *Annals of Internal Medicine*, however, appear to validate the Atkins approach (Stern et al., 2004; Yancy et al., 2004). These studies found that the diet results not only in weight loss but also in cardiovascular health benefits and improved blood sugar, prompting some nutritionists to warn of the diet's unknown health effects in the long term (Big News, 2004). Thus, even when rigorous research results are available, life is indeed a gamble. Because risky prospects do not necessarily come with convenient descriptions of all possible outcomes and outcome probabilities, people must often rely on their experience to choose among them.

What do we know about the psychology underlying such experience-based decisions? Almost nothing. This is because researchers have focused almost exclusively on what we have called *decisions from description* (Hertwig et al., 2004). The goal of this chapter is to explore the important class of decisions from experience. As we demonstrate shortly, investigating decisions from experience does not require researchers to abandon their Drosophila. Choices among monetary gambles can be constructed such that people sample information from the payoff distributions – that is, garner experience of the outcomes and outcome probabilities – before they select among them.

DECISIONS FROM EXPERIENCE IN MONETARY GAMBLES

How much information do decision makers sample from payoff distributions before they feel ready to choose among them? Will their choices differ from those based on complete descriptions of the gambles? To address these questions, Hertwig et al. (2004) presented students at the Technion (Haifa, Israel) with the six risky choice problems displayed in Table 4.1. The two gambles in each problem differed from one another with regard to both expected value and variability. Of the six problems, four required choosing among positive prospects (gains), and two required choosing among negative prospects (losses).

Unlike in most studies of risky choice, respondents were not given the gambles' outcomes and probabilities. Instead, they were afforded the opportunity to sample information about the properties of the two gambles in each problem. Specifically, each respondent was shown two boxes on a computer screen and was told that each box represented a payoff distribution. Clicking on a given box triggered random sampling (with replacement) of an outcome from its distribution. Respondents could sample outcomes from the boxes as often as and in whatever order they desired. They were encouraged, however, to sample until they felt confident enough to decide from which of the two boxes they would prefer to draw given that their next draw would have real monetary consequences. After they had stopped sampling and indicated their choice, they proceeded to the

TABLE 4.1. *Summary of the Choice Problems and Results*

Choice Problem	Gambles[a]		Expected Value		Percentage Choosing H		Prediction for H Choices[b]		Difference Between Groups[c]
	H	L	H	L	Description Group	Experience Group	Rare Event	H Choices	
1	4, <u>8</u>	3	3.2	3	36	88	0, .2	Higher	−52(z = 3.79; p = .000)
2	4, <u>2</u>	3, .25	.8	.75	64	44	4, .2	Lower	+20(z = 1.42, p = .176)
3	−3	<u>−32, 1</u>	−3	−3.2	64	28	−32, .1	Lower	−36(z = 2.55, p = .005)
4	−3	<u>−4, 8</u>	−3	−3.2	28	56	0, .2	Higher	+28(z = 2.01, p = .022)
5	<u>32, 1</u>	3	3.2	3	48	20	32, .1	Lower	+28(z = 2.09, p = .018)
6	<u>32, .025</u>	3, .25	.8	.75	64	12	32, .025	Lower	−52(z = 3.79, p = .000)

[a] Underlining indicates the gamble including the rare event. *H*, gamble with the higher expected value; *L*, gamble with the lower expected value.

[b] The entries in this column identify the rare event and indicate whether the percentage of respondents choosing the *H* gamble was expected to be higher or lower in the experience group than in the description group, assuming underweighting of rare events in the experience group.

[c] This column shows the percentage of *H* choices in the experience group minus the percentage of *H* choices in the description group, along with the z statistic testing whether the difference between the two sample proportions is significantly different from zero.

Source: "Decisions from experience and the effect of rare events in risky choice," by R. Hertwig, G. Barron, E.U. Weber and I. Erev, 2004, *Psychological Science,* 15, p. 536. Copyright 2004 by the American Psychological Society. Reprinted with permission.

next problem. After making all six choices, they played out their preferred gambles and received the associated payoffs. In both groups, participants received a $4.50 show-up fee and 2¢ for each point won (e.g., the outcome 32 was worth 64¢).

Let us turn to Problem 1 in Table 4.1 to illustrate the procedure. This problem offers a choice between two payoff distributions: a sure-thing distribution in which every draw yields 3, and a risky-prospect distribution in which a draw results in 4 or 0, with the former outcome being four times as likely as the latter outcome (.8 versus .2). One respondent in Hertwig et al.'s (2004) study sampled seven times from the sure-thing distribution, each time encountering 3, and sampled eight times from the risky-prospect distribution, finding 0 once and 4 seven times. Which one did she choose? She selected the risky prospect, as did 88% of respondents.

How does this context compare with one in which people do not need to develop their own representations of the gambles? Hertwig et al.'s (2004) study addressed this question by including a second group of respondents. Whereas those in the aforementioned *experience* group ($n = 50$) learned about each gamble's outcomes and outcome probabilities by *sampling* information from the payoff distributions, respondents in the description group ($n = 50$) received complete information (using the notation introduced in the foregoing) about each gamble's outcomes and outcome probabilities. Table 4.1, which shows the choices by group for each of the six problems, illustrates the striking differences between decisions from descriptions and decisions from experience. In Problems 2, 3, and 6 as well as in Problem 5, respondents in the description group were more likely to select the gamble with the higher expected value, H, than were respondents in the experience group; in Problems 1 and 4, the pattern was reversed. Except in Problem 2, all differences in the proportion of H choices were statistically significant.

Hertwig et al. (2004) proposed that the crucial difference between decisions from description and decisions from experience resides in the psychological impact of rare events on each type of decision. According to the dominant psychological theory of decision making under risk, *prospect theory* (Kahneman & Tversky, 1979; Tversky & Kahneman, 1992), people choose between risky prospects as if small-probability events receive more weight than they deserve given their objective probabilities and as if large-probability events receive less weight than they deserve given their objective probabilities.[2] Prospect theory's weighting function, however, is

---

[2] Prospect theory's decision-weight function plots weights that range from 0 to 1 against objective probabilities: Points above the 45° diagonal signal overweighting; that is, they represent weights that exceed the outcomes' objective probability of occurrence. Points below the 45° diagonal signal the opposite pattern. Note that such decision weights are not assumed to reflect any explicit judgment of the subjective probability of outcomes. Instead, weights are inferred from choices and provide a measure of an outcome's impact on a decision.

based exclusively on studies in which respondents received explicit descriptions of the gambles' outcomes and probabilities. Consequently, Hertwig et al. proposed that in decisions from description people choose between risky gambles as if they *overweight* low-probability events relative to their objective probability of occurrence, consistent with prospect theory, whereas in decisions from experience they choose as if they *underweight* low-probability events.[3]

To see how this pattern of weighting of rare events is consistent with the observed choices, look again at Problem 1, in which the rare event (i.e., 0) occurs with a probability of .2. If in decisions from description the rare outcome has more than its due impact on the attractiveness of the option, then respondents will tend to prefer the sure outcome 3 over the risky prospect 4 with probability .8. Indeed, 64% of respondents in the description group chose the sure thing in Problem 1 (Table 4.1). If in decisions from experience the rare outcome has less than its due impact on the attractiveness of the option, then respondents will tend to prefer the risky prospect to the sure thing. Indeed, the large majority of respondents in the experience group selected the risky prospect in Problem 1 (Table 4.1).

More generally, Table 4.1 shows that, in each of the six problems, the pattern of *H* choices in the description and experience groups was consistent with that expected assuming that people choose as if they underweight rare events in decisions from experience and overweight rare events in decisions from description (see the Prediction for H choices column in Table 4.1). Comparison of Problems 1 and 2 highlights this difference in the weighting of rare events. Linear weighting of probabilities implies that a given person will have the same preference in Problem 1 as in Problem 2 (i.e., the person will prefer *H* or *L* in both problems, depending on his or her utility function). In contrast, overweighting of rare events ($p < .25$) implies more *H* choices in Problem 2 than in Problem 1, and underweighting of rare events implies the opposite preference. Kahneman and Tversky (1979) observed 65% *H* choices in Problem 2 and 20% *H* choices in Problem 1. They interpreted this result, a violation of expected utility theory known as the Allais paradox, as an instance of overweighting of rare events (Kahneman & Tversky, 1979). [For other violations that prospect theory explains in terms of overweighting of rare events, see Camerer (2000) and Tversky & Kahneman (1992).] As Table 4.1 shows, the majority choices in the description group replicate those observed by Kahneman and Tversky, whereas those in the experience group show the reverse pattern.

Although people in the experience and description groups responded to structurally identical choice problems, their choices were dramatically

---

[3] Here we define a low-probability, rare event in a preliminary and somewhat arbitrary fashion as an event with a probability less than or equal to .2.

different. Respondents in the experience group sequentially sampled outcomes from the respective payoff distributions and chose between options as if they underweighted rare events. Respondents in the description group received complete outcome and probability information and chose between options as if they overweighted rare events. Hertwig et al.'s (2004) findings corroborate those observed in a small set of other studies that investigated decisions from experience (e.g., Barkan, Zohar, & Erev, 1998; Barron & Erev, 2003; Weber, Shafir, & Blais, 2004). The question is, why do people choose as if they underweight rare events in decisions from experience?

INFORMATION SEARCH IN DECISIONS FROM EXPERIENCE

Decisions from experience depend on what outcomes are sampled before the decision is made. Therefore, any account of the experience–description distinction ought to consider how much information people sample from the gambles' payoff distributions and how they combine the sampled information into a judgment of the gambles' attractiveness. We first address the issue of sample size and then investigate the impact of recency on the updating of sampled information.

Sample Size Matters

Figure 4.1 shows the median number of draws per problem in the experience group. Two observations are noteworthy. First, people sampled almost equally often from each payoff distribution in each problem. Second, the total number of draws per problem was relatively small, with a median of 15 draws.

Whatever the reasons for people's limited search effort,[4] it has an obvious consequence: The smaller the number of draws, the larger the probability that a decision maker will never come across rare events, remaining ignorant of their existence. Table 4.2 presents the number of respondents who never encountered the rare event in the experience group. For illustration, consider Problem 5. Here the median respondent sampled seven cards from the payoff distribution that offered 32 with probability .1 (and 0 otherwise). As a consequence, 68% of respondents (17 out of 25) never encountered the good but rare outcome 32. Across the six problems, the rare event was not encountered in 44% of all sampling sequences.

[4] One explanation involves short-term memory limits that provide a natural stopping rule for information acquisition (Kareev, 2000). In fact, more than half of respondents in Hertwig et al.'s (2004) experience group sampled exclusively from one payoff distribution before switching to the other one. The median number of draws from each option was around seven, a number often associated with the capacity of short-term memory.

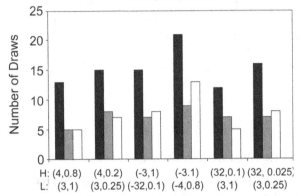

- ■ Total Number
- ▨ Draws (Alternative H)
- ◻ Draws (Alternative L)

FIGURE 4.1. Median number of draws in the experience group in each of the six choice problems (see Table 4.1). *H* and *L* represent the gambles with the higher and lower expected value, respectively. From "Decisions from experience and the effect of rare events in risky choice," by R. Hertwig, G. Barron, E. U. Weber and I. Erev, 2004, *Psychological Science*, 15, p. 537. Copyright 2004 by the American Psychological Society. Reprinted with permission.

Ignorance of the rare event appears to have a clear impact on the subsequent choice (see the Choosing Rare-Event Gamble columns in Table 4.2). When the rare event was "good" relative to other outcomes, as in Problems 2, 4, 5, and 6, not encountering it virtually assured that respondents did not select the gamble involving the rare event, whereas encountering it at least once raised the chance of choosing the rare-event gamble to about 50%. When the rare event was "bad" relative to other outcomes, as in Problems 1 and 3, the opposite was true. When respondents never encountered the rare event, they always selected the gamble involving the rare event. In contrast, encountering it at least once reduced the chance of choosing the rare-event gamble to about 66%.

As well as increasing the risk of never encountering the rare event, drawing a small sample makes it more probable that one will encounter the rare event less frequently than expected given its objective probability. This is because the binomial distribution for the number of times a particular outcome will be observed in n independent trials is markedly skewed when p is small (i.e, the event is rare) and n is small (i.e., few outcomes are sampled). For such distributions, one is more likely to encounter the rare event less frequently than expected (np) than more frequently than expected. For illustration, let us assume that 1,000 people sample from a distribution in which the critical event has a probability of .1 and estimate the event's probability to be the proportion in whatever sample they

TABLE 4.2. *Sample History and Choice*

Choice Problem	Gambles[a]		Rare Event Never Seen	Choosing Rare-Event Gamble	Rare Event Encountered (at Least Once)	Choosing Rare-Event Gamble
	H	*L*				
1	4, .8 0, .2 *bad*	3	11	11	14	11
2	4, .2 *good* 0, .8	3, .25 0, .75	7	0	18	11
3	−3	−32, .1 *bad* 0, .9	10	10	15	8
4	−3	−4, .8 0, .2 *good*	2	0	23	11
5	32, .1 *good* 0, .9	3	17	1	8	4
6	32, .025 *good* 0, .985	3, .25 0, .75	19	1	6	2

[a] Underlining indicates the rare event. Each rare event is indicated as either good or bad relative to other outcomes. *H*, gamble with the higher expected value; *L*, gamble with the lower expected value.

observe. Each person samples 20 times. Of the 1,000 people, 285 will observe the critical event twice and thus are likely to estimate its probability accurately. Another 392 will never observe the critical event or will observe it only once and thus will probably underestimate p. The remaining people (323) will encounter the crucial event 3, 4, 5, ..., or 20 times and thus are likely to overestimate its probability. Note that, averaged across the 1,000 people, the estimated probability of the rare event will equal its probability in the population (i.e., .1) because the sample proportion is an unbiased estimator of the proportion in the population. However, for small samples, more people will encounter the rare event less frequently than expected rather than more frequently than expected.

Hertwig et al. (2004) averaged the number of times that respondents encountered the rare event across all six problems in the experience group and found that, consistent with the analysis of the binomial distribution, 78% of respondents encountered the rare event less frequently than expected (i.e., fewer than np times) whereas 22% of respondents encountered the rare event as frequently or more frequently than expected. In addition, the experienced frequency of the critical event had a clear impact on choices. When the rare event was "good" (e.g., 32 in Problem 5), the option involving it was selected in only 23% of cases in which it was encountered less frequently than expected. The same option was selected in 58% of cases in which it was encountered as frequently or more frequently than expected. Similarly, when the rare event was "bad" (e.g., 0 in Problem 1), the option involving it was selected in 92% of cases in which it was encountered less frequently than expected but in only 50% of cases in which it was encountered as frequently or more frequently than expected.

These results show that, in the small samples drawn by respondents in Hertwig et al.'s (2004) study, the experienced relative frequencies of rare events were on average smaller than the rare events' objective probabilities. It is therefore plausible that, even if people correctly recalled all the experienced outcomes and computed the probabilities based on the outcomes' frequencies in the sample, they tended to *underestimate* the probability of rare events. In this context, underestimation does not necessarily imply psychological distortion but rather a kind of information sampling (i.e., drawing small samples) that gives rise to a distorted picture of the world.

Given that many of the respondents received input that would give them a skewed picture of the objective probabilities, how good were their choices? To address this question, we computed the expected value of each gamble for each respondent *on the basis of the outcomes that he or she observed*. Viewed in light of the samples they drew, respondents generally acted like expected value maximizers: Across the six problems, respondents chose the gamble that according to the sample promised the higher expected value in 74% of cases. Only 44% of choices, however, maximized the expected value as calculated on the basis of the objective probabilities.

Sampling Order Matters

Decisions from experience require people to update their impression of the desirability of a gamble by combining newly sampled outcomes with previously sampled outcomes. If all observations received an equal weight – that is, a weight of $1/n$, where n is the number of observations sampled from a payoff distribution – then the order in which observations are sampled would not matter. Both the expected value calculus and the expected utility calculus are impervious to order effects. Research on memory (e.g., Atkinson & Shiffrin, 1968) and belief-updating (e.g., Hogarth & Einhorn, 1992), however, show that the order in which evidence is presented to people matters even in tasks where it should be irrelevant (e.g., in free recall of the content of lists or in overall impressions of the value of choice options when outcomes are randomly sampled).

One order phenomenon is the *recency* effect, according to which observations made late in a sequence of observations receive more weight than they deserve (i.e., more than $1/n$). Such an effect would result in choices consistent with underweighting of rare events. To see why, let us assume the most extreme form of recency, in which the attractiveness of a gamble is determined solely by the most recently sampled outcome. Let us further assume that each of 100 people sample from two distributions: distribution A, in which the rare event, 32, has a probability of .1 and the common event, 0, has a probability of .9, and distribution B, in which 3 occurs for certain. After any number of draws, only 10 of the 100 people, on average, will have encountered 32 in their most recent draw from distribution A; the rest will have encountered 0 in that draw. Moreover, all of them will have encountered 3 in the most recent draw from distribution B. If only the most recent draw per distribution matters to a gamble's attractiveness, it is likely that the majority of participants will prefer distribution B over A, resulting in a modal choice that is consistent with underweighting of rare events. In other words, in decisions resulting from continuous updating of outcome probabilities, the recency effect can produce choices that make people look as if they underweight rare events.

To examine whether recency affected decisions in the experience group, Hertwig et al. (2004) split, for each choice problem and each respondent, the outcomes sampled from each option according to whether they fell in the first half or the second half of the sample sequence. After computing the options' average payoffs according to the information observed in the first half and the second half of the sequence, respectively, the authors then predicted each person's choice in each problem on the basis of these average payoffs. The predictive power of the payoffs computed from the second half of the sequence clearly outperformed that of those computed from the first half. Whereas the first half predicted, on average, 59% of the choices, the second half predicted 75% of the choices; the difference between the two proportions was statistically significant ($t(49) = -3.1$, $p = .003$; two-tailed).

Hertwig et al.'s findings suggest the existence of two phenomena. First, people may underestimate the objective probabilities of rare events in small samples, and, second, they may overweight the impact of recent events. As demonstrated, both phenomena can contribute independently to choices consistent with underweighting of rare events. Let us clarify, however, that we do not assume that people explicitly estimate outcome probabilities or explicitly combine outcomes and probability estimates in a fashion suggested by expected value theory, expected utility theory or prospect theory. Instead, we propose that people go through a cognitively far less taxing process of evaluating and updating the value of a gamble, resulting in choices that appear to reflect underweighting of rare events. In what follows, we propose a model of such a process.

INFORMATION INTEGRATION IN DECISIONS FROM EXPERIENCE: THE VALUE-UPDATING MODEL

How do people integrate sequentially sampled outcomes? Many learning studies point to a weighted adjustment process in which a previous impression of the value of an option is combined with a newly sampled outcome (see, e.g., Bush & Mosteller, 1955; March, 1996). We propose a specific variant of a weighted adjustment mechanism and apply it to value updating in decisions from experience. In the *value-updating model*, a learner is assumed to update his or her estimate of the value of a gamble after each new draw from the payoff distribution by computing a weighted average of the previously estimated value and the value of the most recently experienced outcome. The model has three building blocks:

> *Sampling rule:* During sampling, each trial consists of a random draw from one of the two payoff distributions. The person stops searching after a total of s draws (with each payoff distribution being sampled $s/2$ times).
>
> *Choice rule:* When making the choice, the person selects the gamble with the highest value as determined on the basis of the sampled outcomes.
>
> *Updating rule:* In algebraic terms, the updating mechanism can be written $A_j(t) = (1 - \omega_t)A_j(t - 1) + (\omega_t)v(x_t)$, where $A_j(t)$ represents the value of gamble j after t samples are drawn. This value is the weighted average of the value of the gamble based on the $t - 1$ previously

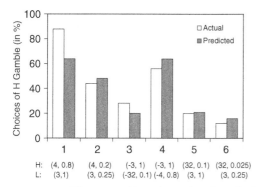

FIGURE 4.2. The proportion of people who selected *H*, the gamble with the higher expected value, in each choice problem and the corresponding proportion predicted by the value-adjustment model.

drawn outcomes, $A_j(t-1)$, and the value of the outcome obtained in the most recently drawn outcome, x_t. The weight accorded to the new outcome is $\omega_t = (1/t)^{\varphi}$, where φ is a recency parameter. When $\varphi = 1$, all outcomes are equally weighted. When $\varphi < 1$, more recent outcomes receive more weight, implying a recency effect. When $\varphi > 1$, earlier outcomes receive more weight, implying a *primacy* effect. Finally, $v(.)$ is prospect theory's value function, which is assumed to be concave for gains and convex for losses (Kahneman & Tverksy, 1979):

$$v(x_i) = \begin{cases} x_i^{\alpha} & \text{if } x_i \geq 0, \\ \lambda|x_i|^{\alpha} & \text{if } x_i < 0. \end{cases}$$

Although the value-updating model has four parameters, only the recency parameter, φ, needs to be estimated from the data. The parameters α from the gain value function and λ from the loss value function are assumed to take the values estimated by Tversky and Kahneman (1992; $\alpha = .88; \lambda = 2.25$), and the number of draws, s, is taken to be the empirically observed mean sample size for each choice problem (see Figure 4.1).

We fit the model to the choices observed in Hertwig et al.'s study (2004). Figure 4.2 shows the actual and predicted (best mean-squared fit) choice proportions. The mean-squared deviation score between the proportions was .012, and the correlation between the actual and the predicted proportions was .91. With the exception of Problem 1, the predictions closely tracked the observations. Moreover, the φ parameter was estimated to be 0.29. As expected, this value implies a recency effect.

In conclusion, the value-updating model suggests one way in which sampled outcomes are integrated into an impression of the value of a gamble in the context of decisions from experience. In future studies of decisions from experience, this mechanism is flexible enough to take different degrees of sampling (i.e., sample sizes) into account (through s), and it allows for quantification of the recency effect.

CONCLUSIONS

In a review of all published studies examining risky choices made by human respondents between a sure option and a two-outcome gamble of equal expected value, Weber et al. (2004) assessed the prevalence of decisions from description in the laboratory. Every single choice involved a decision from description. There is no doubt that important situations outside the laboratory call for decisions from description. Consider, for example, parents who must decide whether their child should receive the diphtheria, tetanus, and pertussis (DTaP) vaccine. By researching the DTaP vaccine at the National Immunization Program Web site, for instance, parents will learn that the possible side effects are high fever (a temperature of more than 105°F) and seizures and that up to 1 child in 1,000 suffers from the former and about 1 child in 14,000 suffers from the latter as a result of immunization.

Yet circumstances often deny us access to thorough descriptions of options' outcomes and their probabilities. One way to fill the void is to sequentially sample information and update one's mental representation of the situation. According to an article in *The New Yorker* (Dubner, 2004), this is how a thief named Blane Nordahl managed to steal millions of dollars worth of silver from scores of stately homes on the east coast of the United States. Burglary is fraught with risks, most of which cannot be estimated by consulting the Internet. Arguably the most important risk for a burglar to assess is that of getting caught. Nordahl learned from experience that it was dangerous to have partners in crime because there was a high probability that, if they got caught, they would turn him in. So he worked alone. Experience also taught him to steal silver – which, unlike jewelry, for example, tends to be stored on the ground floor of a home – and to work during the night rather than the day. Of course, the samples of experience on which risky choices are based can be vicarious as well as first-hand. Quoted in the same article in *The New Yorker*, Malcolm X's recollections of his early career as a burglar provide an example of decisions from vicarious experience:

I learned from some of the pros, and from my experience, how important it was to be careful and plan.... Burglary, properly executed, though it had its dangers, offered the maximum chances of success with the minimum risk.... And if through some slip-up you were caught, later, by the police, there was never a positive eyewitness (Dubner, 2004, p. 76).

In other contexts (e.g., financial investment), decision makers have access both to comprehensive statistics and to personal experience. Expert and novice decision makers often differ in the degree to which they rely on one type of information versus the other. For instance, in the domain of insurance, experts (i.e., actuaries) rely almost exclusively on summary statistics when pricing policies, whereas novices (i.e., insurance buyers) often use personal experience with the hazard to decide whether a price is fair. This can make people reluctant to buy actuarially fair or advantageous insurance policies for rare hazards like floods (Kunreuther et al., 1978). The difference between decisions from experience and decisions from description is one of a number of possible contributors to expert–novice differences in judgment and choice (for others, see Slovic & Weber, 2002).

The experience–description distinction can also shed light on some striking similarities between the behavior of humans and other animals (see also Weber et al., 2004). Because animals do not share the human ability to process symbolic representations of risky prospects, all their decisions are decisions from experience. In a study of the foraging decisions made by bees, Real (1991) observed that "bumblebees underperceive rare events and overperceive common events" (p. 985). To explain why bees' "probability bias" diverges from that postulated in humans (in prospect theory's weighting function), Real cited, among other factors, the fact that bees' samples from payoff distributions (i.e., foraging patches) are truncated by memory constraints. When humans are placed in situations where they must make decisions based on experience and their experience is limited (by memory constraints or by a small number of encounters), they, like bumblebees, behave as if they underestimate the probability of rare events.

A Mere Mention Lends Weight

Why people choose as if they overweight rare events in decisions from description remains an open question in the literature (for a discussion, see Gonzalez & Wu, 1999), and we have no counterpart to the value-updating model to offer as an explanation. We suspect, however, that the overweighting of rare events is one of several phenomena in which merely presenting a proposition appears to increase its subjective truth and psychological weight. For instance, merely asking people whether a particular person is a crook makes them more likely to believe that such a person is one (e.g., Wegner et al., 1981); merely considering a proposition enhances its subjective truth (Fiedler, 2000; Fiedler et al., 1996), and merely imagining the occurrence of positive events (e.g., winning a free trip) or negative events (e.g., being arrested) increases the subjective likelihood of those events (Gregory, Cialdini, & Carpenter, 1982). We hypothesize that merely presenting a rare event may increase its psychological weight in a similar

way. The propositional representations of gambles in decisions from description – for instance, "32 with probability .1; 0 otherwise" – put more equal
emphasis on the two possible outcomes than the discrepancy between their
actual probabilities of occurrence warrants. If attention translates into decision weight as some research suggests (Weber & Kirsner, 1996), then, other
things being equal, the psychological weights of rare and common events
will indeed be closer to one another than they should be. Thus, the effect
of mere presentation might explain underweighting of common events as
well as overweighting of rare events in decisions from description.

The mere presentation account raises interesting questions. For example,
would overweighting of rare events in decisions from description decrease
or even vanish if the gambles were described in an analogical way? The
structure of analogical representations reflects more the structure of what is
represented than do propositional representations. For instance, the option
"32 with probability .1; 0 otherwise" can be expressed as follows: "a random
draw from the population {0, 0, 0, 0, 0, 32, 0, 0, 0, 0}." In this representation,
information regarding the relative frequency of the option's outcomes can
be read off directly. Moreover, to the extent that more attention needs to
be allocated to the processing of the frequent event than the rare event,
the resulting decision weights may more accurately reflect the objective
probabilities.

Small Samples Show Less Variability

Reliance on small samples not only plays a role in decisions from experience but also contributes to the fact that people perceive the world as
less variable than it actually is (Kareev, Arnon & Horwitz-Zeliger, 2002).
In fact, underestimating the variance of populations is equivalent to underweighting rare events. For instance, in Problem 6, about two-thirds of
respondents encountered only the outcome 0 when they drew from the
payoff distribution involving the rare event (Table 4.2). If they had used
sample variability or lack thereof as an estimate of population variability
without correcting for sample size, then they would have underestimated
the true variability in that option's payoff distribution. Consistent with
this conclusion, Kareev et al. showed that people tend to perceive sample
variability as smaller than it is and that reliance on small samples (the size
of which is often related to working-memory capacity) causes the underestimation.

Kareev et al. (2002) suggested that time constraints, limits in memory
capacity, and lack of available data are key conditions under which people perceive variability as smaller than it is (see also Kareev et al. in this
volume). These may also be the conditions under which decisions from
experience and decisions from description diverge. We think that even if

decision makers derive their decisions from large samples of outcomes, however, the psychological impact of a rare event may still deviate from its objective probability because recency could still amplify the impact of recently sampled outcomes.

Primacy and Recency Join Forces

In our analysis based on the value-updating model, we found evidence only for recency. However, whether it is later or earlier observations in a sequential updating process that get more weight is a subject of dispute. Some researchers have found evidence of primacy (Nisbett & Ross, 1980, p. 172), others of recency (Davis, 1984), and still others of both effects (Anderson, 1981). In an effort to make sense out of this heterogeneous set of findings, Hogarth and Einhorn (1992) proposed a belief-adjustment model that predicts different types of order effects as a function of the nature of the task (e.g., judging the value of an option after each outcome is drawn or only after sampling is complete), the complexity of the outcome information [e.g., a single value, as in Hertwig et al.'s (2004) study, or a page of text describing a person], and the length of sampling. The key point here is that both primacy and recency effects will cause rare events to receive less weight than they deserve given their objective probability of occurrence. Because of the very rarity of rare events, most people are less likely to encounter rare events than frequent events both at the beginning and at the end of a sequence.

EPILOGUE

Let us return to the decks of the Pequod. On the second day of their three-day chase of Moby Dick, the chief mate, Starbuck, implored Ahab:

Never, never wilt thou capture him, old man – In Jesus' name no more of this, that's worse than devil's madness.... Shall we keep chasing this murderous fish till he swamps the last man? Shall we be dragged by him to the bottom of the sea? Shall we be towed by him to the infernal world? Oh, oh, – Impiety and blasphemy to hunt more! (Melville, 1851/1993, p. 467)

Ahab ignored the plea. In Melville's portrayal, to hunt down and kill Moby Dick was nothing less than Ahab's destiny. He had no choice but to continue the quest, destroying himself, his ship, and his crew in the process. Seen from the perspective of decisions from experience, one may speculate that Ahab sampled the hunt and the kill of "this murderous fish" in his imagination innumerable times. And, in our imaginations at least, every one of us is master of the odds.

References

Anderson, N. H. (1981). *Foundations of information integration theory.* New York: Academic Press.

Atkinson, R. C., & Shiffrin, R. M. (1968). Human memory: A proposed system and its control processes. In K. W. Spence & J. T. Spence (Eds.), *The psychology of learning and motivation: Advances in research and theory* (Vol 2., 89–195). New York: Academic Press.

Barkan, R., Zohar, D., & Erev, I. (1998). Accidents and decision making under uncertainty: A comparison of four models. *Organizational Behavior and Human Decision Processes, 74,* 118–144.

Barron, G., & Erev, I. (2003). Small feedback-based decisions and their limited correspondence to description-based decisions. *Journal of Behavioral Decision Making, 16,* 215–233.

Big News. (2004, May 22). *The Economist, 371,* 87.

Bush, R., & Mosteller, F. (1955). *Stochastic models for learning.* New York: Wiley.

Camerer, C. F. (2000). Prospect theory in the wild: Evidence from the field. In D. Kahneman & A. Tversky (Eds.), *Choices, values, and frames* (pp. 288–300). Cambridge, UK: Cambridge University Press.

Davis, J. H. (1984). Order in the courtroom. In D. J. Miller, D. G. Blackman, & A. J. Chapman (Eds.), *Perspectives in psychology and law.* New York: Wiley.

Dubner, S. J. (2004, May 17). The silver thief. *The New Yorker, 74*–85.

Fiedler, K. (2000). On mere considering: The subjective experience of truth. In H. Bless & J. P. Forgas (Eds.), *The message within: The role of subjective experience in social cognition and behavior* (pp. 13–36).

Fiedler, K., Armbruster, T., Nickel, S., Walther, E., & Asbeck, J. (1996). Constructive biases in social judgment: Experiments on the self-verification of question contents. *Journal of Personality and Social Psychology, 71,* 861–873.

Goldstein, W. M., & Weber, E. U. (1997). Content and discontent: Indications and implications of domain specificity in preferential decision making. In W. M. Goldstein & R. M. Hogarth (Eds.), *Research on judgment and decision making: Currents, connections, and controversies* (pp. 566–617). New York: Cambridge University Press.

Gonzalez, R., & Wu, G. (1999). On the shape of the probability weighting function. *Cognitive Psychology, 38,* 129–166.

Gregory, W. L., Cialdini, R. B., & Carpenter, K. M. (1982). Self-relevant scenarios as mediators of likelihood estimates and compliance: Does imagining make it so? *Journal of Personality and Social Psychology, 43,* 89–99.

Hertwig, R., Barron, G., Weber, E. U., & Erev, I. (2004). Decisions from experience and the effect of rare events in risky choice. *Psychological Science, 15,* 534–539.

Hogarth, R. M., & Einhorn, H. J. (1992). Order effects in belief updating: The belief-adjustment model. *Cognitive Psychology, 24,* 1–55.

Kahneman, D., & Tversky, A. (1979). Prospect theory: An analysis of decision under risk. *Econometrica, 47,* 263–291.

Kareev, Y. (2000). Seven (indeed, plus or minus two) and the detection of correlations. *Psychological Review, 107,* 397–402.

Kareev, Y., Arnon, S., & Horwitz-Zeliger, R. (2002). On the misperception of variability. *Journal of Experimental Psychology: General, 131*, 287–297.

Kunreuther, H., Ginsberg, R., Miller, L., Sagi, P., Slovic, P., Borkin, B., & Katz, N. (1978). *Disaster insurance protection: Public policy lessons.* New York: Wiley.

Lopes, L. L. (1983). Some thoughts on the psychological concept of risk. *Journal of Experimental Psychology: Human Perception and Performance, 9*, 137–144.

March, J. G. (1996). Learning to be risk averse. *Psychological Review, 103*, 309–319.

Melville, H. (1993). *Moby Dick.* New York: Barnes & Noble Books. (Original work published 1851.)

Nisbett, R., & Ross, L. (1980). *Human inference: Strategies and shortcomings of human judgment.* Englewood Cliffs, NJ: Prentice-Hall.

Real, L. A. (1991). Animal choice behavior and the evolution of cognitive architecture. *Science, 253*, 980–986.

Savage, L. J. (1954). *The foundations of statistics.* New York: Wiley.

Slovic, P., & Weber, E. U. (2002). *Perception of risk posed by extreme events.* White paper for conference "Risk Management Strategies in an Uncertain World," Palisades, NY, April 12, 2002.

Stern, L., Iqbal, N., Seshadri, P., Chicano, K. L., Daily, D. A., McGrory, J., Williams, M., Gracely, E. J., Samaha, F. F. (2004). The effects of low-carbohydrate versus conventional weight loss diets in severely obese adults: One-year follow-up of a randomized trial. *Annals of Internal Medicine, 140*, 778–785.

Tversky, A., & Kahneman, D. (1992). Advances in prospect theory: Cumulative representation of uncertainty. *Journal of Risk and Uncertainty, 5*, 297–323.

Weber, E. U., & Kirsner, B. (1996). Reasons for rank-dependent utility evaluation. *Journal of Risk and Uncertainty, 14*, 41–61.

Weber, E. U., Shafir, S., & Blais, A.-R. (2004). Predicting risk-sensitivity in humans and lower animals: Risk as variance or coefficient of variation. *Psychological Review, 111*, 430–445.

Wegner, D. M., Wenzlaff, R., Kerker, R. M., & Beattie, A. E. (1981). Incrimination through innuendo: Can media questions become public answers? *Journal of Personality and Social Psychology, 40*, 822–832.

Yancy, W. S., Jr., Olsen, M. K., Guyton, J. R., Bakst, R. P., Westman, E. C. (2004). A low-carbohydrate, ketogenic diet versus a low-fat diet to treat obesity and hyperlipidemia. *Annals of Internal Medicine, 140*, 769–777.

5

Less Is More in Covariation Detection – Or Is It?

Peter Juslin, Klaus Fiedler, and Nick Chater

INTRODUCTION

One of the more celebrated conclusions in cognitive psychology refers to the limited computational capacity of controlled thought, as typically epitomized by Miller's (1956) estimate of a short-term-memory holding capacity of "seven-plus-or-minus-two" chunks. That people can only keep a limited amount of information active for controlled processing at any moment in time has inspired humbling conclusions in regard to problem solving (Newell & Simon, 1972), reasoning (Evans, Newstead, & Byrne, 1993), and, perhaps especially, judgment and decision making (Gilovich, Griffin, & Kahneman, 2002). This limitation is often raised as a main obstacle to people's attainment of classical rationality, suggesting that at best people can aspire to bounded rationality (Simon, 1990). In the context of this volume the implication is that at any moment in time controlled processes of thought can only access a small sample of observations.

The default interpretation seems to be to emphasize the liabilities of these limitations and to regard the current state in the evolutionary development as representing at best a local maximum. Organisms are thus restricted to limited samples of information, although there is agreement that on normative grounds as much information as possible is needed to optimize judgments and decisions. More rarely is the question raised of whether there can be a functional significance attached to apparent cognitive limitations. It has, however, been proposed that capacity limitations may improve the efficiency of foraging behavior (Thuijsman et al., 1995) or boost the development of perceptual development (Turkewitz & Kenny, 1982) or language learning (Elman, 1993). It has been proposed that the limited number of chunks (here conceived of as three to four chunks per Broadbent, 1975, rather than seven) is the optimal number for efficient retrieval from memory (MacGregor, 1987) and that limited attention binds perceptual features into objects to solve the binding problem in perceptual organization (Humphreys & Heinke, 1998).

A particularly intriguing argument has recently been presented by Kareev (2000): The limitation to a small sample may have evolved because it provides an adaptive advantage for the early detection of useful correlations in the environment. Kareev proposes that, because of the skew of the sampling distribution of certain correlation measures (e.g., the ϕ coefficient), the hit rate may actually decrease with larger sample size. That is, the correlations observed in small samples more often exceed the population correlation than correlations estimated from larger samples drawn from the same population. At least superficially – and in contrast to the other arguments for an adaptive benefit of limited capacity – this argument seems to fly in the face of logic (since a larger sample already contains the smaller subsample) and statistics (since a larger sample allows more accurate estimation of a parameter) (Juslin & Olsson, 2005).

The purpose of this chapter is to bring together the different viewpoints in a scientific controversy elicited by the argument presented in Kareev (2000). We will distinguish between two varieties of the claim for an adaptive benefit of small samples. *Global adaptive advantage* implies that – all things considered – the organism is better off with a small sample than a large one. This is the advantage required if capacity limitations have *evolved* for a functional purpose. An advantage of a small sample in a specific circumstance can only drive natural selection if it is not paid for by a larger disadvantage in another equally relevant circumstance. *Local adaptive advantage* implies that there are specific conditions that can be identified where a smaller sample is advantageous. If these conditions obtain in real environments and the organism can detect when they hold, they open the possibility that the organism adapts by limited sampling in these circumstances.

In the first section of this chapter, the original less-is-more argument from Kareev (2000) is presented. In the second section we evaluate the claim for a global adaptive advantage of smaller samples, although – naturally given the abstractness and generality of the notion – this is no trivial feat. We provide arguments framed in logic and information theory (Shannon, 1948) as well as results from Monte Carlo simulations. In the third section, we analyze conditions where it can be argued that smaller samples provide an adaptive advantage. Moreover, theoretical assumptions will be delineated under which it can even be argued that a global small-sample advantage might obtain. Whether this seemingly impossible case is realistic, to be sure, depends on the viability of those theoretical assumptions. Finally, we offer some conclusions about the small-sample advantage separately by the contributors to the chapter. Be warned: Do not expect us to agree!

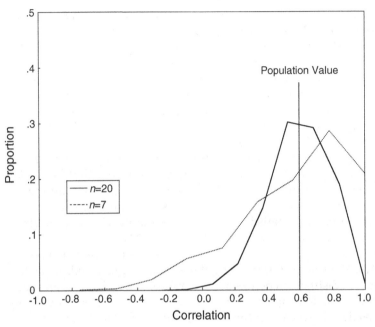

FIGURE 5.1. Sampling distributions for the sample correlation ϕ when the population correlation ρ is .6 for sample sizes (n) 7 and 20. Adapted from Juslin & Olsson (2005) with permission from the publisher.

SMALL SAMPLES AND EARLY DETECTION OF CORRELATION

The point of departure in Kareev (2000) is that the sampling distribution for the correlation coefficient is skewed (i.e., for $\rho \neq 0$), and more so for smaller samples. This is illustrated in Figure 5.1, which presents the sampling distributions for the phi-coefficient (ϕ) at sample sizes (n) of 7 or 20 generated by Monte Carlo simulation, where the population correlation (ρ) is .6. [Here we concentrate on the correlation between binary variables, as in Kareev (2000). The arguments, however, are equally relevant to continuous variables as discussed in Kareev (1995).]

Figure 5.1 corresponds to a person making n bivariate observations of two binary variables. For example, assume that the observations are patients. The first variable may be *smoking* versus *not smoking,* and the second variable *lung cancer* versus *no lung cancer.* In this population, the correlation (ρ) is .6 between smoking and lung cancer (i.e., smoking implies an increased risk of lung cancer). Naïve people observe patients from this population, benefiting either from a sample size of 7 or 20. Each person considers the sample and the threshold for detecting a correlation in the sample is a ϕ-coefficient equal to or larger than .6. Which people, those with sample size 7 or 20, are more likely to detect an above-threshold correlation between smoking and lung cancer?

Because of the skewed sampling distribution at small n, most ϕ are higher than the threshold set equal to the population correlation ρ (Figure 5.1). This means that more people with sample size 7 (57%) than sample size 20 (53%) will observe a correlation (ϕ) between smoking and lung cancer that is above the threshold (here the threshold coincides with the population value, $\rho = .6$, but this is not crucial to the effect). Small samples thereby appear to amplify the probability of observing a ϕ that is more extreme than the population correlation (ρ). If large ϕ are more easily detected, small samples may be beneficial in the sense of increasing the chances of detecting useful correlations in the environment (i.e., the Hit rate).

There is little doubt that this effect can confer smaller samples a local advantage, in the sense that in conditions with (say) a nonzero positive population correlation, the ϕ-coefficient is most often more extreme than the population correlation, increasing the likelihood that it passes a detection threshold (as also verified in experiments by Kareev, Lieberman, & Lev, 1997). In the notation introduced in the following, because of the sampling distribution for the correlation coefficient, the probability $p(c|C)$ of concluding that there is a correlation (c) given the existence of a population correlation (C) – the Hit rate (MacMillan & Creelman, 1991) – can be higher for smaller samples.

The problem with interpreting this as implying a global advantage of small samples is apparent by considering that in a situation where there is always a useful positive correlation, it is a local advantage to ignore the sample data altogether and to always claim a positive correlation. Yet – unless the organism has means to detect when this condition obtains (which makes the detection redundant) – no one would claim that ignoring data and always claiming a positive correlation is adaptive in a more general sense. This complication was acknowledged by Kareev (2000), who noted that to some extent this advantage may be offset by an increased rate of false alarms, but he defended the claim that small samples may afford an evolutionary advantage over larger samples:

[A] biased estimate may better serve the functioning of the organism than an unbiased one. By providing such a biased picture, capacity limitations may have evolved so as to protect organisms from missing strong correlations and to help them handle the daunting tasks of induction. Because correlations involving binary variables often require a high level of analysis and play a special role in cognitive functioning, the magic of the number seven may lie in the special advantage it provides for the early detection of such correlations. (Kareev, 2000, p. 401)

In the rest of this chapter we evaluate the scope and conditions for this proposed benefit of smaller samples. In the next section we evaluate the argument that a smaller sample size is an adaptive advantage in the global sense and that capacity limitations may have evolved because it improves our ability to detect correlations. After reviewing these arguments mainly provided by the chapter authors who are skeptical about the idea of global adaptive advantage of small samples (PJ & NC), we review new arguments by Fiedler and Kareev (2004) that there nonetheless are conditions that support claims for a local, or even a global, adaptive advantage of smaller samples.

CAN LESS KNOWLEDGE BE AN ADVANTAGE?

On logical grounds alone, the claim that a smaller sample is more informative than a larger sample is startling considering that a large sample size dominates a small sample size n in the sense that you can always get n observations from $n + 1$ observations by throwing one observation away.[1] One could enter some cost involved in the storage or computation of a larger sample, but this argument is orthogonal to the argument in Kareev (2000) in the sense that this cost would be an argument for small samples regardless of the statistical point raised by Kareev. Moreover, whether the argument is relevant is questionable because to *explain* cognitive limitations as having a rational basis, adverting to cognitive limitations as part of the explanation is inappropriate. So, if we disregard the potential costs involved in storage and computation of a larger sample, we need to show that attending to yet another observation effectively degrades your performance.

In this section we consider arguments involving mathematical results and computer simulations suggesting that the argument for a global advantage cannot be sustained and that the magical number 7 ± 2 cannot have evolved because it benefits the early detection of correlations; in this respect, less is actually *less*. We start by considering formal arguments derived in the context of information theory (Shannon, 1948), then turn to results of computer simulations that directly address the case of correlation detection.

There are two broad contexts in which we can ask the question of whether sampling more data is, or is not, likely to be beneficial to an agent (Chater, Crocker, & Pickering, 1998). In disinterested inquiry the agent is concerned with acquiring information about some domain, potentially by collecting data, but does not have any particular interests or goals that guide the process. Disinterested inquiry is, therefore, an idealization

---

[1] Alternatively, from a sample of size $n + 1$ you can get $\binom{n+1}{n}$ samples of size n. Apparently, to claim that a sample size of 5 is more useful than a sample size of 6 is to claim that one sample of 5 observations is better than six samples of 5 observations (although, of course, the value of these six samples is limited by the lack of independence).

of exploratory or curiosity-driven behavior in humans and animals, corresponding to "pure" science. Here, then, lies the challenge that Kareev raises: Are there situations in which a search for knowledge is likely to be impeded by sampling more data? The second situation is where an interested agent has to act; and the goal of learning from data is to choose actions that yield the highest payoff – that is, the actions with the highest utility or that optimize a cost–benefit trade-off. Here the question is the following: Are there situations in which an agent can rationally decide not to sample fresh data, on the grounds that this extra sampling is likely to lead to poorer decisions? Drawing on information theory we consider these two cases in turn.

Disinterested Inquiry[2]

In disinterested inquiry, the goal of the agent is to learn as much as possible about some aspect of the world, with no concern for how this learning might be applied in practice. In this context, the standard measure for the amount of knowledge about the world is *information*, in the technical sense used in communication theory (Shannon, 1948). This measure of the value of data has been explored in statistics (Berger, 1985; Good, 1950; Lindley, 1956), machine learning (Mackay, 1992), and cognitive psychology (Corter & Gluck, 1992; Chater, Pickering & Crocker, 1998; Oaksford & Chater, 1994). The question of whether sampling further data can be irrational translates into the question of whether an agent can expect that, on sampling further data, it will be less informed than it is at present. Can the *expected* informational value of new samples be negative?

Suppose we are interested in the value of some variable $\theta \in \boldsymbol{\theta}$, where θ represents the state of the system about which we wish to learn and can range over almost any kind of value; it might correspond to a scalar parameter, a vector, a symbolically represented grammar, or a scientific hypothesis concerning some domain. Consider an agent a, who may, or may not, sample from some data source $\boldsymbol{D}$. If the data source is sampled, then a has prior expectations concerning possible outcomes $D \in \boldsymbol{D}$. We assume only that a has some prior probability distribution over the possible values of θ and $\boldsymbol{D}$; call this $P_a(\theta, \boldsymbol{D})$. Notably, these assumptions are quite weak. We do not assume anything about the *quality* of the agent's probability assignments – the agent may display penetrating insight or staggering ignorance. We have also said nothing to indicate whether there is any connection between θ and $\boldsymbol{D}$. Of course, if they are completely independent, then $\boldsymbol{D}$ will give no information about θ (as might apply if we were to attempt to use astrological data to predict political events). However, in this

---

[2] These proofs were derived by Nick Chater.

case, sampling from D will not provide negative information, on average; samples of any size, small or large, will have no value.

Given this extremely general setup, we can show that the expected value of the information obtained by sampling from D is nonnegative: A rational agent should never decline to sample information. The standard measure of information (Shannon, 1948), the amount of uncertainty that the agent's probability distribution $P_a(\theta)$ embodies about the variable, θ, is the entropy, given by

$$H(\theta) = \sum_{\theta \in \theta} P_a(\theta) \log_2 \frac{1}{P_a(\theta)}. \tag{5.1}$$

The amount of uncertainty after some $D \in D$ have been sampled is simply the entropy of the agent's probability distribution over θ, in the light of data D:

$$H(\theta|D) = \sum_{\theta \in \theta} P_a(\theta|D) \log_2 \frac{1}{P_a(\theta|D)}. \tag{5.2}$$

The informational value of D is measured by the difference between those two entropies – that is, the extent to which the sampling of D has led to a *decrease* in the agent's uncertainty about θ:

$$H(\theta) - H(\theta|D). \tag{5.3}$$

If Eq. 5.3 is positive, then D has *reduced* uncertainty and hence led to a *gain* of information about θ. Of course, for specific data D Eq. 5.3 can be negative (i.e., specific data D can serve to increase the uncertainty about a hypothesis H). This happens, for example, when the jury is almost sure that Smith, rather than Jones, is the murderer, until Smith's DNA analysis is revealed to be suspect. The new evidence now leaves the jury more uncertain than it was before. The question is whether the *expected* value of Eq. 5.3 is negative. Can, on average, less be more? Can the jury ever rationally decide not to hear any more evidence, because more evidence will, on average, leave it more uncertain than it is at present? To show this is not possible, we need to show that

$$E_a(H(\theta) - H(\theta|D)) \geq 0, \tag{5.4}$$

where E_a is the expectation over θ and D, given the agent's prior distribution over $P_a(\theta, D)$. This is proved in Appendix A. Thus, from the standpoint of disinterested inquiry, a rational agent should never prefer a small sample over a larger sample – that is, less is never more. On average, fresh evidence will reduce, or at least not increase, the jury's uncertainty – and hence the jury should not prefer less evidence to more.

We have so far looked at the problem from an "inside" perspective, from the agent's viewpoint. We might equally well consider the position from an objective "outside" standpoint. Can an outsider, viewing the agent, conclude that the agent is better informed if it does not sample further data (whether the agent realizes this or not)? It turns out that, also from an outsider's perspective, the expected value of the information gained from the agent sampling further data cannot be negative (see Appendix B). Therefore, from the perspective of disinterested inquiry it seems that acquiring more information has a positive expected value, whether considered from the "inside" perspective of the agent itself or from the "outside" perspective from which the truth is known. We conclude that in the context of information gain in disinterested inquiry there are no grounds for claiming that a smaller sample is a global adaptive advantage.

Deciding on a Course of Action

In a second context, inquiry is valuable only insofar as it leads to actions that have, on the whole, better outcomes. Here, the standard measure of the value of data is the degree to which it increases the expected utility of actions. This approach, that concerns the costs and benefits of actions based on different sample sizes, is grounded in utility theory (von Neumann & Morgenstern, 1944) and decision theory more generally. As in disinterested inquiry, we consider the problem both from the "inside" perspective of a rational agent who attempts to maximize expected utility and from the "outside" perspective on an organism that attempts to detect covariation.

In a decision problem, the value of data can be treated as the additional utility that can be obtained by choosing actions in the light of that new information, rather than without it. As in disinterested inquiry, specific data can have a negative effect. Suppose, for example, that from a position of ignorance, an agent decides on an action that, by happy chance, is optimal. Sampling further may suggest some other action – the resulting change in action, although perfectly rational, will lead to the agent obtaining less utility. But can this happen *in expectation*? Can a rational agent ever be justified in declining to sample data on the grounds that taking account of data is likely to harm the agent's performance? For example, can a "rational" surgeon ever say, "I've decided to operate – now do not tell me any more about the patient, or I may change my mind, with terrible results." We show that this cannot happen. Intuitively, the idea is to compare the case where the agent adapts its behavior after sampling from D with an agent who decides on a course of action before sampling from D and sticks with that choice come what may. (This is, of course, equivalent to never sampling from D at all.) It seems intuitively plausible that tuning one's choice of action in the light of the data that are sampled cannot systematically harm performance – the surgeon will, on the whole, do better to learn more about the patient and take this into account in making his or her decision.

Consider, as before, that we are interested in some variable θ, which represents the state of some aspect of the world, and that we can sample, if we choose, some data source D, which may be informative about this state. Specifically, suppose that the agent a has a prior probability distribution over θ and D, $P_a(\theta, D)$, which captures the agent's view of the dependence between the the data and the state of the world. Now the agent must choose one of a set of actions $A \in A$ (for concreteness, we may visualize this choice as between a finite set of actions; but this is not necessary); the goodness of the outcome of the action may depend on the state of the world. For example, suppose the action is whether to operate, the relevant aspect of the world is whether the patient has cancer; and the data concern various symptoms and lab tests. Can a rational surgeon really be justified in declining to hear the result of a new test, fearing that he or she will, on average, make a worse decision having heard it?

To formalize this, suppose that the agent's beliefs about how the value of the variable $\theta \in \theta$ relates to the utility of action $A \in A$ is expressed by $U_a(A, \theta)$. Then, prior to sampling from D, an expected utility maximizing agent will choose $A_i \in A$ such that, for all $A_j \in A$,

$$EU_a(A_i) \geq EU_a(A_j). \tag{5.5}$$

That is, it will choose the action with the highest expected utility (or choose one such action arbitrarily, in case of ties). The expected utility of an action is, by definition, just the utility of the action, for each state of the world, weighted by the probabilities of each state of the world. In symbols, we write this as

$$EU_a(A_i) = \sum_{\theta \in \theta} P_a(\theta) U_a(A_i, \theta). \tag{5.6}$$

By elementary probability theory, this can be written successively as

$$\sum_{\theta \in \theta, D \in D} P_a(\theta, D) U_a(A_i, \theta) \tag{5.7}$$

and as

$$\sum_{D \in D} P_a(D) \sum_{\theta \in \theta} P_a(\theta | D) U_a(A_i, \theta). \tag{5.8}$$

Now suppose that some $D \in D$ have been sampled. Let us write $EU_{a|D}(A_{kD})$ for the expected value of the action A_{kD}, now that agent a knows data D (and these data may potentially convey information about θ). Analogously with Eq. 5.6, we can write

$$EU_{a|D}(A_{kD}) = \sum_{\theta \in \theta} P_a(\theta | D) U_a(A_{kD}, \theta). \tag{5.9}$$

As before, we assume that action A_{kD} is chosen to maximize expected utility given data D, so analogous to Eq. 5.5 we choose A_{kD} such that, $\forall A_i \in \boldsymbol{A}$,

$$EU_{a|D}(A_k) \geq EU_{a|D}(A_i), \tag{5.10}$$

and, in particular, this implies that A_k cannot be *expected* to do worse than the A_i that was chosen before D was sampled, once we know that D has been sampled – otherwise, there would be no justification for switching from A_i to A_k (although, as we have noted, in any particular case, A_k may *actually* do worse).

Now, the *expected* value of an action, chosen after sampling $\boldsymbol{D}$, is

$$\sum_{D \in \boldsymbol{D}} P_a(D)EU_{a|D}(A_{KD}), \tag{5.11}$$

where the specific action A_{kD} may, of course, depend on which data D are observed.

Appendix C shows that this expected value is no lower than the expected value before $\boldsymbol{D}$ has been sampled and hence that a rational agent concerned with decision making can have no justification in declining to sample from $\boldsymbol{D}$. Intuitively, the point is that the expected value of an action before the data are sampled is necessarily the same as the expectation of the value of that action in the light of the data, once it has been sampled (but we do not yet know which data these will be). However, then this expectation can be equalled or bettered by allowing ourselves the freedom to choose to switch actions in the light of some particular data, if they arrive, if this gives us a higher expected utility.

Thus, when we consider the problem from the inside perspective of a rational agent there can be no utility-based rational justification for preferring a small data sample over a larger one. On average, large samples suggest better decisions.

Detection of Correlation in Simulated Environments

For the "outside" perspective, general theoretical results analogous to those we have considered already are not possible because there are obviously situations in which sampling further data is expected (from the outside) to have a negative impact on the expected utility. These are situations in which the agent is, by happy accident, performing the right action. If further data are sampled, there is a possibility that the agent's updated beliefs might persuade it to alter its actions – and this can only be a bad thing, given that the current action is optimal. Of course, such situations would appear to be rare: There appear to be many more situations in which further data will help an agent to correct its incorrect choice of actions. Rather than attempting to address the "outside" perspective in generality, we focus on the specific case of correlation detection.

One major reason why Juslin and Olsson (2005) arrived at conclusions different from that of Kareev (2000) is that, unlike Kareev, Juslin and Olsson base their argument on signal-detection theory, which states that the expected value of a response strategy – a decision criterion – is a function both of the hit rate *and* the base rate of signal (MacMillan & Creelman, 1991). Within this framework, it can be assumed that adaptive costs and benefits of correlation detection are not determined by the probability of successful detection as conditioned on the presence of a useful correlation, but by the *posterior probability* of a hit after drawing a data-based decision. The hit rate may in certain circumstances be a negative function of sample size, but this need not imply that the posterior probability of a hit is a negative function of sample size. According to Juslin and Olsson, the adaptive costs and benefits are determined by the posterior probabilities.

In Juslin and Olsson (2005) the approach was to consider plausible models of the environment to which the organism needs to adapt and to ascertain the posterior probabilities of the various outcomes that may ensue from covariation detection with different sample sizes. Denote the state of nature where the population correlation satisfies some criterion C and the state where it does not satisfy the criterion $\overline{C}$. For example, the organism may only consider correlations of an absolute magnitude larger than .5 useful enough to merit attention (C) and correlations between $-.5$ and .5 to be too weak to support useful predictions ($\overline{C}$). In a detection problem, the organism encounters sample data with the intent to determine whether C or $\overline{C}$ holds. Denote the conclusion that C holds by c and the conclusion that $\overline{C}$ holds by $\overline{c}$.

Following standard terminology in detection theory (Macmillan & Creelman, 1991) Juslin and Olsson distinguished between four classes of outcomes: a Hit is the correct detection of a correlation (including a correct belief about its sign) (c & C); a Miss is the incorrect conclusion that there is no correlation ($\overline{c}$. & C); a False Alarm is the incorrect conclusion that there is a correlation when no correlation exists (c & $\overline{C}$); and a Correct Rejection is the correct conclusion that there is no correlation when none exists ($\overline{c}$ & $\overline{C}$).

Kareev notes that the sampling distribution for correlations implies that the Hit rate – the probability $p(c|C)$ of concluding that there is a correlation given a correlation – sometimes decreases with sample size. Recall that, as illustrated in Figure 5.1, if a population correlation ρ of, say, absolute magnitude larger than .6 is considered an important event to discover, the hit rate $p(c|C)$ decreases from .57 at sample size 7 to .53 at sample size 20. However, the Hit rate per se does not determine the adaptive value; it represents a propensity to respond given that one is in a specific state, but it ignores the probability of this state. The insight captured in Bayes's theorem is that the posterior probability is a function of both likelihood (e.g., $p(c|C)$) and prior probability.

As noted in connection with Eq. 5.6, the expected utility of an action is, by definition, the utility of the action, for each state of the world, weighted by the probabilities of each state of the world. The expected value EV_n of detection with sample size n therefore amounts to

$$EV_n = p_n(c \,\&\, C) \cdot v(c \,\&\, C) + p_n(\overline{c} \,\&\, C) \cdot v(\overline{c} \,\&\, C)$$
$$+ p_n(c \,\&\, \overline{C}) \cdot v(c \,\&\, \overline{C}) + p_n(\overline{c} \,\&\, \overline{C}) \cdot v(\overline{c} \,\&\, \overline{C}), \qquad (5.12)$$

where $p_n(*)$ is the probability of outcome "*" at sample size n, and $v(*)$ is the adaptive value of the outcome. Equation 5.12 simply states that the overall adaptive value of a sample size (i.e., for covariation detection) is the value of all possible outcomes weighted by their probability or relative frequency of occurrence in the environment. The formal basis for the argument in Kareev (2000) is the possibility that the value of a Hit is large, whereas either the value of other outcomes are negligible and/or the effect of smaller sample size does not affect the probabilities of the other outcomes in such a way that the overall expected value is decreased by the smaller sample size. For example, if the probability of a Hit ($p_n(C \,\&\, c)$) decreases with sample size, and the value of Hits ($v(C \,\&\, c)$) is large while the value of the other outcomes is zero, the overall expected value in Eq. 5.12 could decrease with increasing sample size (per the effect in Figure 5.1).

Equation 5.12 theory is equivalent to

$$EV_n = p(C) \cdot p_n(c|C) \cdot v(c \,\&\, C) + p(C) \cdot p_n(\overline{c}|C) \cdot v(\overline{c} \,\&\, C)$$
$$+ p(\overline{C}) \cdot p_n(c|\overline{C}) \cdot v(c \,\&\, \overline{C}) + p(\overline{C}) \cdot p_n(\overline{c}|\overline{C}) \cdot v(\overline{c} \,\&\, \overline{C}), \qquad (5.13)$$

where it is obvious that the expected value is a function of both the likelihood of a conclusion given a state and the prior probability of the state. For example, the posterior probability of a Hit is a function both of the Hit rate discussed by Kareev (2000) and illustrated in Figure 5.1 (i.e., $p(c|C)$) and the prior probability or base rate of useful correlations in the environment (i.e., $p(C)$). Equation 5.13 also makes evident that the increased Hit rate at smaller sample size must not be purchased at the cost of a decreased rate of Correct Rejection or increased rate of False Alarms and Misses. The challenge raised in Juslin and Olsson (2005) was to demonstrate that (1) there exist conditions where a smaller sample yields a larger expected value also when we consider the prior distribution of correlations and the effect on all relevant outcomes and (2) these conditions are prevalent enough to be an important adaptive and evolutionary consideration.

As first steps toward such explorations, Juslin and Olsson (2005) used Monte Carlo simulations to investigate the posterior probabilities of the four possible outcomes in a covariation detection problem in the case of an organism that adapts to the distributions of correlations in three different environments. One environment was of "high predictability" with most correlations close to either 1 or −1; one environment was of "low predictability" with most correlations close to 0; and, finally, one environment had a uniform distribution of correlations between −1 and 1. The idea was that an organism that adapts to an unknown environment: (a) is confronted with an unknown true distribution of correlations between a large set of variables in the environment, (b) observes samples of the relationship between pairs of variables and this provides the basis for inferring the population correlation, and (c) has to act (somehow) contingently on the estimated correlations. These simulations addressed the posterior probabilities of all four outcomes in Eq. 5.13.

The result of one Monte Carlo simulation is illustrated in Figure 5.2. (Here we illustrate the results for the environment with a uniform distribution of correlations but the results for the other two environments were similar.) In this simulation it is assumed that the organism inhabits an environment where any correlation in the interval [−1, 1] is equally likely to obtain prior to observation (a uniform "prior" probability for ρ). The

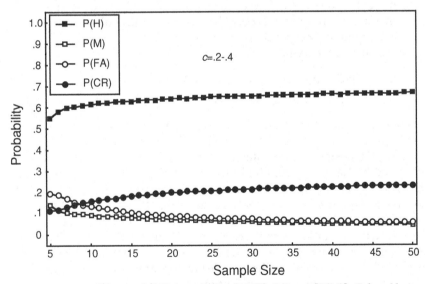

FIGURE 5.2. Posterior probabilities of Hits [$P(H)$], Misses [$P(M)$], False Alarms [$P(FA)$], and Correct Rejections [$P(CR)$] as a function of sample size for a uniform distribution of ρ and criteria c for a useful correlation uniformly distributed between .2 and .4. Adapted from Juslin & Olsson (2005) with permission from the publisher.

criterion for considering an observed correlation ϕ to be potentially useful is assumed to be uniformly distributed in the interval .2.–4 (i.e., the criterion is assumed to vary from iteration to iteration as a function of several factors, including specific payoff contingencies). A ϕ that exceeds the criterion in absolute value corresponds to a sample correlation large enough to be considered sufficient either for taking some instrumental decision (e.g., to stop smoking) or for taking some epistemic decision, such as deciding to observe additional samples. The marginal frequency of both events is .5/.5.

Each iteration with the simulation thus involves (a) a random sampling of a population correlation ρ from the uniform distribution in the environment, (b) a random sampling of n bivariate observations from the cross table defined by the ρ sampled in the first step to generate a sample correlation ϕ, and (c) a random sampling of a criterion between .2 and .4 to evaluate whether the computed ϕ is considered a "signal." The result of each iteration is a sampled ρ, a sampled ϕ, and a sampled c. Each such configuration is coded in terms of one of four categories: Hit (correct detection of a useful correlation), Miss (missing a useful correlation), False Alarm (false belief in a useful correlation that does not exist), and Correct Reject (correctly concluding that there is no useful correlation). [See Juslin & Olsson (2005) for further details on the simulations.]

The result of this simulation is presented in Figure 5.2. It is evident that the posterior probabilities of Hit and Correct Rejection are monotonically increasing functions of sample size. If this condition holds the expected adaptive value of a smaller sample size cannot exceed the expected adaptive value of a larger sample size. This result (as well as the other results reported in Juslin & Olsson, 2005) would suggest that evolution cannot have selected for a limited sample size on the grounds that it improves the overall ability to detect correlations. Yet, if the result of the same simulation is plotted as a sampling distribution for ϕ, we get the sampling distribution in Figure 5.1, where the Hit rate is a decreasing function of sample size at small sample size.

Kareev (2000) also noted that small samples sometimes do not allow estimation of any correlation. For example, in a random sample of five people all five may happen to be nonsmokers, making it impossible to detect a correlation between smoking and lung cancer. As demonstrated in Juslin and Olsson (2005), at equal marginal frequencies for the events considered (e.g., smoking and nonsmoking is equally common) the proportion of undefined correlations is only 3% at sample size 7. With sample size 7 and marginal frequencies of 1/.9 (e.g., if 90% of the population are nonsmokers) the proportion of undefined correlations is in excess of 68%. With larger sample sizes the probability that an observed sample can form the effective basis for inferring a correlation increases. Juslin and Olsson propose that this frequent inability to detect and estimate covariation when the sample size is small is a nontrivial disadvantage of small samples.

Interim Conclusion 1

We have considered the claim that small samples are a global adaptive advantage for the detection of useful correlations from a number of perspectives. Small samples appear to be logically dominated by large samples and it can thus be proven that a rational agent should never decline to sample additional data, neither in the context of disinterested inquiry nor in a decision context. Once the problem of adaptive benefit is considered in terms of the posterior probabilities there seems to be no evidence for a general benefit of smaller samples.

Needless to say, strong arguments about the global adaptive advantages that can drive evolutionary selection are inherently (and notoriously) difficult to make. The arguments drawing on information theory, for example, are confined to epistemic and decision-theoretic considerations and the results of the Monte Carlo simulations are concerned specifically with correlation detection. Capacity limitations have consequences in many other behavioral domains too. Nonetheless, the arguments in this section undermine Kareev's strong claim that small samples are a general adaptive advantage for covariation detection and that evolution has selected for limited capacity on these grounds.

CAN LESS NEVER BE MORE? A NEW LOOK AT ADVANTAGES OF SMALL SAMPLES

Should this interim conclusion be already the final conclusion? Should we irreversibly bury Kareev's (2000) intriguing notion that small samples can outperform large samples? The normative arguments presented thus far (see also Juslin & Olsson, 2005) as well as the computer simulations conducted in a signal-detection framework seem to suggest the unqualified conclusion that neither the expected informativeness nor the expected utility of decisions can be higher for small samples than for large ones. Indeed, the simulation results showing increasing hit rates and correct-rejection rates with increasing sample size corroborated what a sound normative analysis, based on information theory and Bayes theorem, suggests on analytical grounds alone. According to the aforementioned analysis, then, there seems to be little doubt that the amount of potentially useful information that is inherent in a sample cannot decrease with increasing sample size. It appears that the consensus underlying this disillusioning but healthy message can be hardly ignored.

Nevertheless, consensus need not imply ultimate truth, and so it remains tempting to ask whether decisions informed by small information samples can never be superior in expected value to decisions informed by large samples. The crucial difference in this revised question is between *expected value of all information in a sample* and *expected value of a decision algorithm that utilizes samples of varying size*. Although it may be true that the expected value of all information potentially inherent in a sample increases with sample size, it may nevertheless be the case that decision makers (or, more generally, organisms) use decision algorithms that utilize only part of the available information and that small samples unfold their advantage within this actually utilized part of information. This could even be a global advantage – to keep within the previous terminology – if it turns out that such a decision algorithm can be applied across diverse problem situations, rather than being restricted to local problem contexts.

In the present section, then, a new attempt will be undertaken to save the appeal and the intriguing biological and psychological implications of the claim that "less may be more," based on a new model advanced by Fiedler and Kareev (2004). Of course, this new attempt is not made to refute the logical or mathematical consistency of the foregoing normative analysis, which shall not be cast into doubt. The typical problem with normative proofs is not that they are analytically wrong but that they may be based on assumptions, which can be replaced by alternative assumptions, leading to different conclusions. So let us see what possible alternative assumptions can be proposed to make the point that certain decision algorithms may render small samples more adaptively useful than large samples – even though the message conveyed so far remains valid. A related question we will have to face is the following: How realistic are the assumptions constructed to save the idea of a small-sample advantage?

Looking for Alternative Assumptions Leading to Small-Sample Advantage

To delineate the conditions under which an advantage of small samples over large samples is possible, we need to revise the assumptions of how to measure adaptive value and how to define "hits" and "false alarms" appropriately, within an embedding decision model or action theory. In the normative analysis presented so far, it has been tacitly presupposed that – expressed in signal-detection terms – the criterion event, or "signal", that is crucial to distinguish between correct and incorrect trials, causing adaptive benefits and costs, respectively, is the presence of a "true" contingency C in the environment. Thus, within the underlying decision model, a "hit" occurs when an actually existing C confirms a decision based on a sample contingency c, and a "false alarm" means that c informs a decision that is at variance with C. The reference set for computing the hit and false-alarm rates is the set of all contingencies C encountered in the environment that are considered useful. That is, whenever an environmental C

is strong enough to be of sufficient utility, that contingency must be manifested in a sample c and lead to an appropriate decision; otherwise, that instance is counted as an error (i.e., a miss). In calculating the benefits and costs resulting from all correct and incorrect decisions, it has been assumed that each and every existing contingency C (i.e., each difference between two decision options, persons, products, actions, etc.) has to be counted. The reference set for computing adaptive value is the totality of all existing contingencies C as well as all noncontingencies $\overline{C}$ (considered useful), conceived to be complementary, as if all those cases were rare opportunities and as if it were important for adaptation not to miss any existing C.

An alternative way of interpreting the interaction between environmental contingencies C and decisions informed by sample decisions c is to assume that the information environment is replete with potentially useful contingencies C, which are abundantly available all the time, rather than being a rare good. There are countless ways in which potential decision objects differ, only a small subset of which are used for sampling and lead to decisions. Thus, the rare good, or crucial ingredient, for adaptive behavior is not whether C exists, but whether an organism takes up the affordance, draws a sample c, and makes a decision that is in line with C. In other words, the graduator for adaptive behavior – that is, for a hit defined as C & c – in such a revised cognitive–environmental model is whether c confirms C, not whether C confirms c. It does not matter how many C opportunities go unnoticed, or fail to inform correct decisions, assuming that the number of such unutilized chances is virtually infinite. Moreover, counting all cases of not existing contingencies $\overline{C}$, (i.e., all cases where decision options do not differ) would appear to be even less relevant from the standpoint of the revised decision model, for the number of $\overline{C}$ cases can be considered infinite, which implies that the rate of false alarms and the resulting costs must also be infinite. It would thus appear unrealistic to expect that organisms respond to each and every C and $\overline{C}$ in the world and that all these cases enter the calculation of adaptive value. Rather, to calculate benefits and costs, one only has to consider the subset of cases in which an organism happens to be exposed to a relevant c and actually happens to make a decision. As we shall see soon, that subset not only may be greatly restricted but may be selectively biased in favor of a small-sample advantage.

For an illustrative example, consider an elementary problem of personnel selection, corresponding to a binary contingency problem, namely the choice between two applicants, A_1 and A_2, based in the 2×2 distribution of positive and negative characteristics in the two persons. The sample correlation c is the observed 2×2 distribution in the data; the environmental correlation C can be conceived as the limiting distribution that would result if an infinite number of observations could be gathered about the assets and deficits of A_1 and A_2. Let A_1 be defined as superior to A_2; that is, in the universe of all observations, the proportion of positive characteristics is higher for A_1 than for A_2,

$$p(+/A_1) > p(+/A_2), \tag{5.14}$$

and let this case define a contingency $0 < C = p(+/A_1) - p(+/A_2)$ that has a positive sign. Assume also that many real contingencies are quite modest in size, so it may be useful to correctly assess all contingencies that deviate regularly from zero.

As the job market is replete with pairs of applicants that differ in qualification (by some nonzero amount), the criterion for successful adaptation cannot be to capture all existing contingencies (i.e., inequalities between all pairs of applicants in the world). In other words, it may not be relevant to compute the posterior distribution of hits, integrating across all possible states of nature, weighted by their prior probabilities. Because the environment offers a virtually infinite number of contingencies C, and also renders a virtually infinite number of observable contingencies c available, it may be worthless to assume an *optimization* criterion, which calls for the selection of the absolutely best applicants that exist. More realistically, a more modest, *satisficing* criterion may be to assume that there are many reasonable ways of picking up pairs of applicants and that choosing the better one from each pair will result in reasonable – though not optimal – selection decisions. Within such a satisficing framework, adaptive value can be conceived as the sum of benefits and costs across the subset of binary choice pairs for which samples are actually gathered and that actually lead to manifest decisions.

In which respect should the subset of cases in which decisions are actually made differ from the (infinite) universe of all contingencies C and $\overline{C}$ that exist in the environment? The most reasonable answer in a satisficing theory framework is that decision makers (or, more generally, organisms) will make decisions whenever the observed contingency c *in the sample* is strong and clear-cut enough to warrant correct choices but refrain from decisions when an observed c is below that threshold. Given that there are plenty of environmental contingencies C, offering many different routes to good decisions, organisms can afford to discard many opportunities in which the evidence c is below threshold. To the extent that the subset of cases that enter the calculation of adaptive value is constrained in this way, including only clear-cut sample contingencies c, the normative advantage of large samples may already shrink. If an observed difference is strong enough, even small samples can be expected to reveal which of two choice options is better. However, beyond this argument about the reduced danger of incorrect decisions, it can even be shown that small samples are more likely than large samples to inform correct above-threshold decisions. This is simply because c is more likely to exceed high thresholds when sample size is small rather than large and because this advantage more than out-weighs the residual disadvantage of small samples in terms of inaccuracy.

Consider again the example of two decision options, of which A_1 dominates A_2; that is, choosing A_1 is correct and choosing A_2 is wrong; the sign of C is thus by definition positive. Now assume that a decision is only made when the observed sample contingency exceeds some threshold T. That is, only when a positive c (defined as observed superiority of A_1) is higher than $+T$ will the decision maker choose A_1. Clearly, if T is high enough, say $+.5$, saying that the observed proportion of positive characteristics is .5 higher in A_1 than in A_2, then even rather small samples will almost always result in correct choices. Moreover, to arrive at a false alarm, the observed c would have to be more negative than $-T$ (i.e., $-.5$), which is quite unlikely. In all other, intermediate cases of sample correlations c, lying between $+T$ and $-T$, no decision will be made, causing neither benefits nor costs. Rather, the organism will look for a new sample, or try to increase the sample, until an above-threshold decision sample contingency is found and a decision can be made. Thus, depending on the value of T, quite a large proportion of all existing contingencies C and even many observed sample contingencies c will not lead to decisions and thus will not enter the calculation of benefits and costs. Instead, the organism may draw a new sample, or increase the sample, or make no decision at all.

Note that within such a satisficing framework, the definitions of hits and false alarms differ fundamentally from Juslin and Olsson's (2005) definitions, as specified in the preceding section. In their normative analysis, a decision was counted as incorrect when c is above some threshold while $\underline{C}$ is below the threshold (false positive) or when $\underline{c}$ is below and $\underline{C}$ above the threshold (false negative). Thus, even when $\underline{c}$ and $\underline{C}$ are of the same sign, suggesting the same choice, the resulting decision would be considered wrong if $\underline{c}$ and $\underline{C}$ fall on different sides of the threshold. In the revised, satisficing framework, in contrast, it is not counted as a false negative when the "true" contingency $\underline{C}$ is above the threshold while the observed sample contingency $\underline{c}$ is below the threshold. Similarly, when the observed $\underline{c}$ is above the threshold although the actual contingency $\underline{C}$ is below the threshold, but of the same sign, this is not counted as a false positive, or false alarm, but rather as a hit, because the decision is still "correct" (i.e., A_1 is chosen). Thus, the definitions of hits and false alarms are crucially different from those of the normative analysis presented in the previous section. A hit (i.e., a correct decision in favor of) requires that c is above threshold, $\underline{c} > +T$ while $C > 0$ is of the same sign. A false alarm occurs only when $c > +\underline{T}$, although $C < 0$ is of opposite sign. Additionally, the present, revised model also diverges from the work of Juslin and Olsson (2005) in not assuming that the decision threshold is identical with the size of a true contingency $\underline{C}$ that is considered useful. Rather, the decision threshold only refers to the contingency observed in the sample; it must

Probability

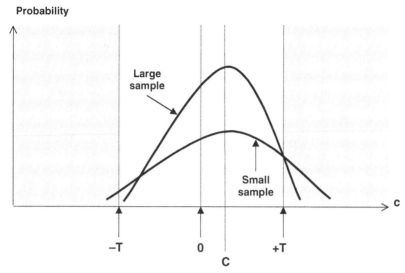

FIGURE 5.3. Distribution of sample contingencies drawn from a population with a true contingency C. *T*, decision threshold.

be independent of the population contingency, which is unknown to the organism and therefore cannot enter the decision algorithm.

Given these revised assumptions, there exists a small-sample advantage – conceived in terms of benefits and costs resulting from hits and false alarms, respectively – especially for values of $\underline{T}$ that are substantially higher than $\underline{C}$. To the extent that organisms (decision makers) use rather strict, clear-cut decision thresholds $\underline{T}$ and naturally existing contingencies $\underline{C}$ that are often weaker, small samples can be expected to result in more correct above-threshold decisions than larger samples. Figure 5.3 illustrates how this strong claim is possible. The two graphs in the figure represent the sampling distributions of contingencies observed in large and small samples drawn from the same universe. We refrain from indicating the precise sample size but confine ourselves to the ordinal assumption that $N_{large} > N_{small}$ to highlight the generality of the argument. Crucial is only the assumption that the dispersion of the sampling distribution is higher for N_{small} than for N_{large}. Unlike Kareev (2000, cf. Figure 5.1) we are not concerned here with the *skew* of the sampling distribution that only holds for high correlations, but with the *dispersion* of sampling distributions that is always stronger for small samples than for large ones. The present argument is thus more general than the original argument of Kareev (2000). Note that the decision threshold *T* in Figure 5.3 is assumed to be higher than C and that hit and false-alarm rates are represented by the shaded areas under the right- and left-tail distributions, respectively, which exceed $\pm T$.

It is easy to see, graphically, and to understand, statistically, that both hit rates and false-alarm rates must be higher for small samples than for large ones owing to the higher dispersion. However, most importantly, this inequality is more pronounced for the hit rates on the right than for the false-alarm rates on the left. That is, the difference in p(hit) for N_{small} and for N_{large} is larger than the corresponding difference for false alarms. Apparently, then, at least under the assumptions unfolded in this section, when decision making is contingent on the rare good of informative, above-threshold samples, it can be shown that small samples can outperform large samples in the very sense we have explicated here.

What can be shown graphically (with reference to Figure 5.3), of course, can also be shown in computer simulations to hold quantitatively. The demonstration is simple and straightforward. From a large 2×2 population distribution of positive ($+$) and negative ($-$) characteristics of two options, A_1 and A_2, we let the computer draw repeated samples of different size, N_{large} and N_{small}. By definition, A_1 is better than A_2, so that $p(+/A_1) \geq p(+/A_2)$, which implies that the population contingency C is positive, operationalized by the Δ rule as $C = p(+/A_1) - p(+/A_2) \geq 0$. For each sample drawn, the sample contingency is computed and a correct choice (for A_1) leading to a hit is made when $c_{sample} > +T$, whereas a wrong decision leading to a false alarm is made (for A_2) when $c_{sample} < -T$.

Table 5.1 gives the frequencies of hits out of 10,000 simulations resulting for various levels of the actual environmental correlation C and for various levels of the decision criterion $\underline{T}$, exemplified for $N_{large} = 24$ and $N_{small} = 12$. Similar results can be shown to hold for other sample sizes. Obviously, for many combinations of $\underline{C}$ and $\underline{c}$, small samples gain more in hits than they lose in false alarms, relative to large samples. In particular, the key assumption is that small samples unfold their advantages to the extent that $\underline{c}$ exceeds $\underline{C}$. In other words, when existing contingencies are moderate or small and when organisms make decisions and choices contingent on clear-cut contingencies observed in samples, they will fare better when they rely on small rather than large samples.

Whether this is a local advantage, or even a global one, depends on empirical assumptions about the world, that is, on whether it is realistic to assume that environmental contingencies are not too strong and that organisms – at least in certain states or decision situations – base their decisions on rather strong thresholds, in a higher range than the population contingencies themselves. However, if this assumption holds, then we have found a decision algorithm that produces higher differences of hits minus false alarms for small samples than for large ones – just because the rare good of above-threshold sample contingencies is more likely to be obtained with small samples.

Is it appropriate to evaluate decision success and adaptive value in terms of the difference between hit and false-alarm rates, rather than their ratio?

TABLE 5.1. *Number of Hits and Number of False Alarms (Shaded Area), Relative to Varying Decision Criteria T, Obtained in Computer Simulations with Small ($n_{small} = 12$) and Large ($n_{large} = 24$) Samples Drawn from Populations in Which the Contingency Is $\Delta = .1, .2, .3, .4,$ or $.5$*

T	Small Sample ($N_{small} = 12$)					Large Sample ($N_{large} = 24$)					Difference				
	$\Delta = .1$	.2	.3	.4	.5	.1	.2	.3	.4	.5	.1	.2	.3	.4	.5
−.9		8	23	63	123	271	0	0	0	0	11	8	23	63	123
−.8	46	99	229	436	868	0	2	13	48	175	46	97	216	388	693
−.7	153	288	628	1053	1865	6	27	115	343	967	147	261	513	710	898
−.6	382	751	1367	2151	3353	49	124	390	1042	2291	333	627	977	1109	1062
−.5	637	1120	1921	2921	4217	147	390	1105	2322	4259	490	730	816	599	−42
−.4	1391	2163	3425	4653	6157	395	974	2190	3888	6134	996	1189	1235	765	23
−.3	2037	3083	4384	5695	7080	1062	2240	3985	5937	7886	975	843	399	−242	−806
−.2	3362	4630	6024	7303	8416	2553	4266	6282	7942	9198	809	364	−258	−639	−782
−.1	4639	5881	7211	8269	9047	4235	6185	7839	9053	9706	404	−304	−628	−784	−659
0.0	5532	6743	7961	8844	9433	6067	7785	8955	9634	9902	−535	−1042	−994	−790	−469
+.1	2843	1838	1036	474	222	2341	1102	464	110	28	502	736	572	364	194
+.2	2118	1303	700	289	129	1214	470	141	27	5	904	833	559	262	124
+.3	1070	587	282	110	25	406	130	33	4	1	664	457	249	106	24
+.4	681	351	155	48	8	134	36	9	3	0	547	315	146	45	8
+.5	364	158	67	21	1	32	4	1	1	0	332	154	66	20	1
+.6	139	59	22	8	1	9	0	0	0	0	130	59	22	8	1
+.7	67	16	9	3	1	2	0	0	0	0	65	16	9	3	1
+.8	9	1	0	1	0	1	0	0	0	0	8	1	0	1	0
+.9	1	0	0	0	0	0	0	0	0	0	1	0	0	0	0
1.0	0	0	0	0	0	0	0	0	0	0	0	0	0	0	0

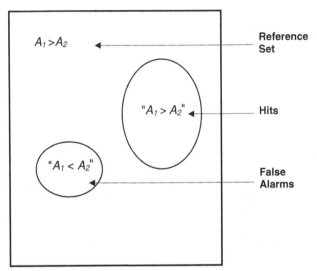

FIGURE 5.4. Graphical illustration of relative set sizes representing hit rate and false-alarm rate, relative to the total reference set in which one decisions option, A_1, dominates the other, A_2.

The answer is yes, as illustrated in Figure 5.4. The entire set is defined by the condition that $A_1 > A_2$, with the ">" sign denoting that A_1 dominates A_2. The subset of cases when the decision maker decides, on the basis of an above-threshold sample, that "$A_1 > A_2$" represents the proportion of hits, whereas the case when the decision says "$A_1 < A_2$" represents false alarms. The remaining part of the entire set does not lead to any decision or to any payoffs. To calculate the expected value (EV) of decisions, one can use any version of the rule $EV = p(\text{correct decision}) - \text{benefit} + p(\text{incorrect decision}) - \text{cost}$. For simplicity, we may insert the same value $\pm v$ for benefits and costs, or we may ignore payoffs fully and assume that $|v| = 1$, counting only the relative frequencies of correct and incorrect decisions. The expected (adaptive) value of decisions turns out to be an additive (rather than a multiplicative) function of hits and false alarms:

$$EV = p(\text{"}A_1 > A_2\text{"}) \cdot p(A_1 \geq A_2) \cdot v$$
$$+ p(\text{"}A_1 < A_2\text{"}) \cdot p(A_1 \geq A_2) \cdot (-v)$$
$$+ p(\text{"no choice"}) \cdot p(A_1 \geq A_2) \cdot 0. \tag{5.15}$$

Because $p(A_1 \geq A_2) = 1$, this reduces to

$$EV = p(\text{"}A_1 > A_2\text{"}) \cdot v + p(\text{"}A_1 > A_2\text{"})(-v); \tag{5.16}$$

that is, the expected value is only a function of the differential number of hits minus false alarms, which is often higher for small samples than for large ones.

Interim Conclusion 2

We have demonstrated that it is possible to construct boundary conditions under which small samples will inform better decision than large samples. The crucial question, then, remaining to be answered is whether these boundary conditions are realistic, rather than far-fetched exceptions.

CONCLUSIONS (SEPARATELY BY THE AUTHORS)

Peter Juslin

The argument in Kareev (2000) – no doubt exciting and thought-provoking – has inspired a new look at an old and received wisdom. Ultimately, however, I do not think that the small-sample argument is sustainable: As far as I can tell a larger sample (if properly processed) is more useful for the detection of correlations and the capacity limitation of working memory cannot have evolved because it provides an adaptive benefit for the detection of correlations.

Based on Fiedler and Kareev (2004) it was proposed that the organism may rely on "decision algorithms that utilize only part of the available information and that small samples unfold their advantage within this actually utilized part of the information." I remain unconvinced by the specifics of this argument, basically because I do not find it well justified to ignore the nondecision trials (the area between $-T$ and $+T$ in Figure 5.3). In a more traditional detection theory analysis it would be appropriate to consider that a high criterion T comes at a cost of an increased rate of missed correlations. This means that in Figure 5.3, for example, the appropriate way to benefit from the fact that a large sample is more informative about covariation is to use the larger sample size, but to decrease the level of the criterion T. That the larger sample is more informative is evident from Table 5.1 – the probability that a decision made is correct increases monotonically with sample size (i.e., as defined by the ratio between the number of decisions in the correct direction and the total number of decisions made).

Importantly, the conflict between these results and the results arguing against an adaptive benefit of small samples is more apparent than real. The results of the simulations in Fiedler and Kareev (and Anderson et al., 2005) can be interpreted in two different ways: (a) There are conditions when a smaller sample is genuinely more informative for making appropriate decisions. (b) A larger sample is always more informative with regard to the true state of the world, but the threshold required for optimal use of a sample differs depending on its sample size.

I believe that the latter interpretation is appropriate. Consistently with this conclusion, in Anderson et al. (2005) the criterion-specific small-sample advantage (similar to the one obtained in Fiedler & Kareev, 2004) disappeared when the criterion was allowed to vary so as to keep the false-alarm (or type I error) rate constant. In this case a larger sample was always superior. This is not to imply that the results in Fiedler and Kareev (or Anderson et al.) are uninteresting. They do suggest that under specific circumstances an organism *constrained to use a fixed decision threshold* may attain a higher expected value with a smaller sample, although my hypothesis is that an even larger expected value is obtained with a larger sample and an appropriately adapted decision threshold. The idea that people are unable to use optimal thresholds is plausible. If we take this and other plausible *computational limitations* into account it is conceivable that a smaller sample is an optimal (or good enough) satisfaction of the computational and the environmental constraints that delimit the process, but this is quite a different story. Unless we already assume some initial computational constraint on the organism's ability to benefit from the larger information contained in a larger sample, there appears to be no way in which a smaller sample can be more informative about the covariation that obtains in the real world.

Nick Chater

Like Juslin, I believe that the proposal that less information can provide better evidence is intriguing but not ultimately sustainable. A rational agent with more information can always expect to do at least as well, according to any reasonable criterion, as an agent with less information; because the agent with more information can, in the worst-case scenario, throw that information away. Indeed, we have seen a range of mathematical results that indicate that, in general, more information is beneficial, whether we are concerned with distinterestedly learning more about the environment or attempting to maximize the expected utility of our actions.

Thus, from this purely rational point of view, less information cannot be more. However, once cognitive constraints are introduced, too much information can certainly be harmful. Given too much information, we may be unable to sift through it successfully and extract the most relevant pieces of information that we require. It is easier to recognize a familiar face on a blank field, than in a crowd of people; it is easier to remember a word if it is presented alone, rather than in a list of other words. Indeed, throughout the fields of attention and memory research, the effects of information overloading are widely studied. In the face of cognitive limitations, there is no doubt that less can be more, but the key claim of the less-is-more viewpoint as described in this chapter has the opposite logic: that some of the limits of the cognitive system are present because they exploit normative, rational advantages of having less, rather than more, information. My conclusion is that this direction of argument is not sustainable, because, despite an interesting case to the contrary, less information really is less; and more information really is more.

Klaus Fiedler

There can be no doubt that the total amount of information that is inherent and potentially available in a sample increases monotonically with sample size. That is, any additional piece of information – provided it is randomly drawn from an underlying population – cannot reduce the information that is available about that population. This was the uncontested message of Juslin and Olsson's (2005) recently published note and of the proof delineated in the first part of this chapter. Granting this as a matter of course, however, in the last part of the chapter I have tried to explain that, psychologically, this does not preclude the fascinating less-is-more effect first envisaged by Kareev (2000) and later elaborated by Fiedler and Kareev (2004). What seems to be impossibe on logical grounds becomes possible, nevertheless, just because decision makers or organisms in general cannot be expected to utilize all information that is potentially available. Many existing contingencies (or noncontingencies) will never be sampled and many samples may not lead to decisions, if decision processes are contingent on a threshold, that is, if decisions arise only from those samples in which a clearly visible (above-threshold) contingency is present. Under such conditions, it can be shown – both in computer simulations and in decision experiments – that small samples enable more decisions than large samples, when decision thresholds are quite high relative to the size of the contingencies that exist in the world. Under such conditions, small samples facilitate obtaining above-threshold sample contingencies that exceed the true contingency in the universe. Moreover, it can be shown that the resulting advantage of small samples in terms of hits is higher than the disadvantage in terms of false alarms (cf. Table 5.1). Thus, regardless of the general proof that information value increases with the number of observations sampled – if the totality of the entire sampling distribution is considered – a decision process that only (or predominantly) uses extreme samples gives an advantage to small samples.

One may ask, to be sure, whether these conditions hold in reality. Is it realistic to assume, for instance, that there are two decision thresholds, an upper and a lower threshold, with no decisions made in between, and can that threshold be assumed to be constant across samples sizes? This, of course, is an empirical question. An empirical answer is not available at the moment, but we do have experimental data to claim that at least under certain task conditions, the small-sample condition is possible. For instance, consider a recent experiment (Fiedler, Kareev, & Renn, 2004) in which participants were exposed to samples of positive and negative characteristics (represented by smilies and frownies) observed in pairs of applicants – quite analogous to the contingency example used in this chapter. Those samples were drawn from populations where the actual dominance of the better applicant amounted to contingencies of variable size ($C = .1$ versus $.2$ versus $.4$). Within each participant, the correlations were computed across 60 decision trials between sample size and correctness of the decision drawn. As it turned out, these correlations were negative within the vast majority of participants and highly significant for the entire experiment. Although we do not know to what extent these findings can be generalized, the experiment at least yields an existence proof, demonstrating that the conditions that produce the less-is-more effect we have depicted do exist in principle. After all, the job-application task used in this experiment has content validity for a decision task that can take place in many real organizations.

There is no contradiction at all between the two positions juxtaposed in this chapter. Both positions can be reconciled if one realizes that the information utlized in a psychological decision algorithm is not the same as the total amount of information that is logically inherent in a sample. For another, more compelling illustration of this point, consider the two curves in Figure 5.5, which may represent the performance of two co-workers. After sufficient time, both people will reach the same performance assymptote but A_1 reaches the assymptote earlier than A_2. After a long observation period, no difference will be visible, but after a short period – a smaller sample, as it were – the difference will be evident. Apparently, it is but a matter of imagination to think of decision algorithms that render the seemingly impossible less-is-more phenomenon possible. It is a matter of future research to figure out the precise boundary conditions under which this actually occurs in the real world.

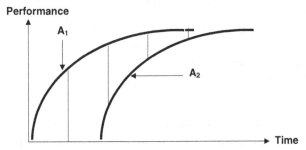

FIGURE 5.5. Performance curves for two co-workers, A_1 and A_2, whose performance differs only at the beginning, but not at the end, of a work period.

APPENDIX A

Expanding the left-hand side of Eq. 5.4, we get

$$E_a(H(\boldsymbol{\theta}) - H(\boldsymbol{\theta}|\boldsymbol{D})) = \sum_{D \in \boldsymbol{D}} P_a(D)[H(\boldsymbol{\theta}) - H(\boldsymbol{\theta}|D)] \tag{A.1}$$

$$= \sum_{D \in \boldsymbol{D}} P_a(D) \left\{ \sum_{\theta \in \boldsymbol{\theta}} P_a(\theta) \log_2 \frac{1}{P_a(\theta)} \right.$$

$$\left. - \sum_{\theta \in \boldsymbol{\theta}} P_a(\theta|D) \log_2 \frac{1}{P_a(\theta|D)} \right\}. \tag{A.2}$$

Here, as throughout, we will assume that θ is countable and hence use summation signs to range over the variable; if θ is continuous, then the summations can be replaced by integrals, without any other change to the argument.

Simple probability theory and algebra gives the following derivation, which begins by modifying Eq. A.2 to give

$$\sum_{\theta \in \boldsymbol{\theta}} P_a(\theta) \log_2 \frac{1}{P_a(\theta)} \sum_{D \in \boldsymbol{D}} P_a(D) - \sum_{D \in \boldsymbol{D}} (D) \sum_{\theta \in \boldsymbol{\theta}} P_a(\theta|D) \log_2 \frac{1}{P_a(\theta|D)}.$$

$$\tag{A.3}$$

Note that $\sum_{D \in \boldsymbol{D}} P_a(D) = 1$ and $P_a(\theta) = \sum_{D \in \boldsymbol{D}} P_a(\theta, D)$. This latter equality implies that $P_a(\theta) = \sum_{D \in \boldsymbol{D}} P_a(D) P_a(\theta|D)$, which can be substituted into the left-hand side of Eq. A3 and rearranged to give

$$\sum_{D \in \boldsymbol{D}} P_a(D) \left\{ \sum_{\theta \in \boldsymbol{\theta}} P_a(\theta|D) \left[\log_2 \frac{1}{P_a(\theta)} - \log_2 \frac{1}{P_a(\theta|D)} \right] \right\} \tag{A.4}$$

$$= \sum_{D \in \boldsymbol{D}} P_a(D) \sum_{\theta \in \boldsymbol{\theta}} P_a(\theta|D) \log_2 \frac{P_a(\theta|D)}{P_a(\theta)}. \tag{A.5}$$

We recall that the Kullback–Liebler distance $KL(Q||R)$ between two probability distributions $Q(x)$ and $R(x)$, is defined as

$$KL(Q||R) = \sum_{x} Q(x) \log_2 \frac{Q(x)}{R(x)} \geq 0. \tag{A.6}$$

The KL distance cannot be negative, for any distributions Q and R (Cover & Thomas, 1991). We can rewrite Eq. A.5 as

$$\sum_{D \in \boldsymbol{D}} P_a(D) KL(P_a(\theta|D)||P_a(\theta)) \geq 0, \tag{A.7}$$

where the final inequality follows from the nonnegativity of the KL distance. This derivation (A.1–A.7) shows that $E_a(H(\boldsymbol{\theta}) - H(\boldsymbol{\theta})|\boldsymbol{D})) \geq 0$.

APPENDIX B

The goal of this derivation is to show that, from an objective outsider's perspective, the expected information gain from an agent sampling further data cannot be negative. Let us suppose that the task of the agent is to learn about the true value of a variable, from some data D. This variable can be as complex as we like – for example, corresponding to a scalar parameter, a vector, grammar, scientific hypothesis, or whatever. The "objective outsider" knows the true value θ of $\boldsymbol{\theta}$ and hence the true probabilities $P(\boldsymbol{D})$ of the possible data sets that the agent might sample (i.e., speaking somewhat metaphorically, from an objective perspective, we know how the world really is as well as the true probability of each set of data that might be obtained from the world).

The agent, however, does not know the true value of $\boldsymbol{\theta}$. Its subjective beliefs about $\boldsymbol{\theta}$ correspond to a probability distribution $P(\boldsymbol{\theta})$ (if $\boldsymbol{\theta}$ ranges over real values, $P(\boldsymbol{\theta})$ is, of course, a probability density function, but we ignore this here). In sampling data D, the agent's probability assignment to the true value θ will shift from $P(\theta)$ to $P(\theta|D)$. Thus, the uncertainty concerning the true hypothesis, θ, will shift from an initial value of $\log_2 P(\theta)$ to $\log_2 P(\theta|D)$ (where uncertainties associated with an state of affairs are, from standard information theory, the logs of the reciprocals of the probabilities associated with that state of affairs). Thus, the information gain is

$$\log_2 P(\theta) - \log_2 P(\theta|D). \tag{B.1}$$

This information gain can be positive or negative. It will be negative if the data are misleading – that is, if the data turn out to disfavor θ, even though θ is true. Intuitively, it seems that this should only occur by "bad luck" – that is, it seems that the expected value of information about θ should be positive, if we sample from data where the true paramater value is θ. This intuition proves to be correct, as we now see. The expected value of the information gain, where we sample data from θ, is

$$\sum_{D \in \boldsymbol{D}} P(D) \left[\log_2 P(\theta) - \log_2 P(\theta|D) \right], \tag{B.2}$$

which can be rearranged as follows:

$$\sum_{D \in \boldsymbol{D}} P(D) \left[\log_2 \left(\frac{P(\theta)}{P(\theta|D)} \right) \right], \tag{B.3}$$

and using one form of Bayes's theorem,

$$\frac{P(\theta)}{P(\theta|D)} = \frac{P(D)}{P(D|\theta)},$$

we can reexpress Eq. B.3 as

$$\sum_{D \in D} P(D) \log_2 \left(\frac{P(D)}{P(D|\theta)} \right), \tag{B.4}$$

which is just the Kullback–Liebler distance between the probability distributions $P(D)$ and $P(D|\theta)$:

$$KL(P(D|\theta)||P(D)) \tag{B.5}$$

However, as we noted in Appendix A, the Kullback–Liebler distance is known to be nonnegative between an arbitrary pair of probability distributions. This means that the information obtained from an agent a, sampling data in a world in which θ is the true parameter value, will be expected to be nonnegative, whether from the agent a's own perspective (as shown in Appendix A), or from the perspective of an outsider who knows that θ is the true parameter value, as shown here.

APPENDIX C

We aim to show that the expected utility of an action, which is to be taken in the light of as yet unsampled data D, is greater than the expected utility of an action taken in ignorance of D. Expressing this in symbols, we aim to show that,

$$\sum_{D \in D} P_a(D) EU_{a|D}(A_{KD}) \geq EU_a(A_i), \tag{C.1}$$

where A_{KD} is an "optimal" action in the light of data D, and A_i is an optimal action in the absence of sampling any data. We know that, because A_{KD} is the optimal action given D,

$$\sum_{D \in D} P_a(D) EU_{a|D}(A_{KD}) \geq \sum_{D \in D} P_a(D) EU_{a|D}(A_i) \tag{C.2}$$

where A_i is any alternative action. The term $EU_{a|D}(A_i)$ on the right hand side of Eq. C.2 is the expected utility of action A_i, for agent a, given data D. We can expand this by summing over all possible values of θ:

$$EU_{a|D}(A_i) = \sum_{\theta \in \theta} P_a(\theta|D) U_a(A_i, \theta). \tag{C.3}$$

Hence, the right hand side of Eq. C.2 can be expanded as

$$\sum_{D \in D, \theta \in \theta} P_a(D) P_a(\theta|D) U_a(A_i, \theta) = \sum_{D \in D, \theta \in \theta} P_a(\theta, D) U_a(A_i, \theta). \tag{C.4}$$

Because the summation over D has no impact on $EU_a(A_i, \theta)$, once θ is fixed, this sum can be collapsed to give

$$\sum_{\theta \in \theta} P_a EU_a(A_i, \theta) = EU_a(A_i), \tag{C.5}$$

where the final equality follows from Eq. 5.6. This completes the proof.

References

Anderson, R. B., Doherty, M. E., Berg, N. D., & Friedrich, J. C. (2005). Sample size and the detection of correlation – A signal-detection account: A comment on Kareev (2000) and Juslin and Olsson (2005). *Psychological Review, 112*, 268–279.

Berger, J. O. (1985). *Statistical decision theory and Bayesian analysis*. New York: Springer-Verlag.

Broadbent, D. E. (1975). The magic number seven after fifteen years. In A. Kennedy & A. Wilkes (Eds.), *Studies in long term memory*, (pp. 3–18). London: Wiley.

Chater, N., Crocker, M., & Pickering, M. (1998). The rational analysis of inquiry: The case of parsing. In: N. Chater & M. Oaksford (Eds), *Rational models of cognition* (pp. 441–468). Oxford: Oxford University Press.

Corter, J., & Gluck, M. (1992). Explaining basic categories: Feature predictability and information. *Psychological Bulletin, 111*, 291–303.

Elman, J. L. (1993). Learning and development in neural networks: the importance of starting small. *Cognition, 48*, 71–99.

Evans, J., Newstead, S., & Byrne, R. (1993). *Human reasoning: The psychology of deduction*. Potomac, Maryland: Lawrence Erlbaum Associates.

Fiedler, K., & Kareev, Y. (2004). *Does decision quality (always) increase with the size of information samples? Some vicissitudes in applying the Law of Large Numbers.* Manuscript submitted for publication. Discussion Paper #347, Center for the Study of Rationality, The Hebrew University, Jerusalem.

Gilovich, T., Griffin, D., & Kahneman, D. (2002). *Heuristics and biases: The psychology of intuitive judgment*. Cambridge: Cambridge University Press.

Good, I. J. (1950). *Probability and the weighing of evidence*. London: Charles Griffin.

Humphreys, G. W., & Heinke, D. (1998). Spatial representation and selection in the brain: Neuropsychological and computational constraints. *Visual Cognition, 5*, 9–47.

Juslin, P., & Olsson, H. (2005). *Capacity limitations and the detection of correlations: Comment on Kareev (2000). Psychological Review, 112*, 256–267.

Kareev, Y. (1995). Through a narrow window: working memory capacity and the detection of correlation. *Cognition, 56*, 263–269.

Kareev, Y. (2000). Seven (indeed, plus or minus two) and the detection of correlation. *Psychological Review, 107*, 397–402.

Kareev, Y., Lieberman, I., & Lev, M. (1997). Through a narrow window: Sample size and the perception of correlation. *Journal of Experimental Psychology: General, 126*, 278–287.

Lindley. D. V. (1956). On a measure of the information provided by an experiment. *Annals of Mathematical Statistics, 27*, 986–1005.

MacGregor, J. N. (1987). Short term memory capacity: Limitation or optimization? *Psychological Review, 94,* 107–108.

MacKay, D. J. C. (1992). Information-based objective functions for active data selection. *Neural Computation, 4,* 590–604.

MacMillan, N. A., & Creelman, C. D. (1991). *Detection theory: A user's guide.* New York: Cambridge University Press.

Miller, G. A. (1956). The magical number seven, plus or minus two: Some limits on our capacity for processing information. *Psychological Review, 63,* 81–97.

Newell, A., & Simon, H. A. (1972). *Human problem solving.* Englewood Cliffs, NJ: Prentice-Hall.

Oaksford, M., & Chater, N. (1994). A rational analysis of the selection task as optimal data selection. *Psychological Review, 101,* 608–631.

Simon, H. A. (1990). Invariants of human behavior. *Annual Review of Psychology, 41,* 1–19.

Shannon, C. E. (1948). A mathematical theory of communication. *The Bell System Technical Journal, 27,* 379–423, 623–656.

Thuijsman, F., Peleg, B. Amitai, M., & Shmida, A. (1995). Automata, matching and foraging behavior of bees. *Journal of Theoretical Biology, 175,* 305–316.

Turkewitz, G., & Kenny, P. A. (1982). Limitations on input as a basis for neural organization and perceptual development: A preliminary theoretical statement. *Developmental Psychobiology, 15,* 357–368.

von Neumann, J., & Morgenstern, O. (1944). *Theory of games and economic behavior.* Princeton, NJ: Princeton University Press.

PART III

BIASED AND UNBIASED JUDGMENTS FROM BIASED SAMPLES

6

Subjective Validity Judgments as an Index of Sensitivity to Sampling Bias

Peter Freytag and Klaus Fiedler

The 1990s are associated with keywords like *new economy* and *information society*. The increased importance of information technologies underlying these phenomena was due primarily to the transformation of the Internet into a medium of mass communication, but the rising number of 24/7 TV stations and the popularity of cellular phones contributed as well. As one side effect of these developments, people in the early twenty-first century are confronted with an unprecedented flood of statistical information. Statistics on almost any topic are available anywhere at any time, whether on the latest unemployment rates, recent trends in climate change, or the odds ratio that a football match will end in a draw when the away team leads by two goals at halftime. At least to anyone trained in statistics, however, the question presents itself whether people have the skills necessary to deal with the available information critically. After all, citizens, politicians, consumers, and managers still have to decide on their own whether to trust a statistic and the conclusions derived from it.

A glance at the major findings accumulated by the heuristics and biases program during the past decades suffices to substantiate a skeptic's position. People have been shown, among other things, to prefer individuating information to base-rate information when estimating conditional probabilities (Bar-Hillel, 1980; Kahneman & Tversky, 1972), to collapse data over levels of significant third variables (Schaller, 1992a,b), and to confuse the conditional probability of a criterion event (e.g., becoming a drug addict) given a predictor event (e.g., having consumed marijuana) with its inverse (Evans, 1989; Sedlmeier, 1999). Despite some evidence for ameliorating effects of formal training (e.g., Schaller et al., 1996), supportive information presentation formats (e.g., Gigerenzer & Hoffrage, 1995), or the explicit abolition of misleading conversational norms (e.g., Schwarz, 1994), the bottom line of research on statistical reasoning reads that people are ill-prepared for the information society for they utilize the available information in inappropriate ways.

THE SAMPLING APPROACH TO JUDGMENT BIASES

A different perspective on the utilization of statistical information has been advanced recently by the sampling approach to judgment biases (Fiedler, 2000a). Whereas previous accounts of judgment biases have stressed the importance of biased information *processing*, the sampling approach posits that many normatively biased judgments originate in biased information *generation*. Drawing on the statistical metaphor of a sample drawn from a population, the key assumption of the sampling approach states that judgments are mediated by samples of observations drawn from the population of observations to which they refer. When the sampling process is biased in some way, judgments will be biased even when the information sampled is processed without bias. In contrast to a biased processing perspective, the sample metaphor implies that accurate, unbiased processing will lead to biased judgments even though, or exactly because, people are sensitive to the information conveyed by a (biased) sample.

The sampling approach has been successful in providing a parsimonious explanation for a variety of well-established judgment biases such as illusory correlations (Fiedler, 2000b; Fiedler et al., 2002a), positive testing (Fiedler et al., 1999), Simpson's paradox (Fiedler et al., 2003; Fiedler et al., 2002b), and base-rate neglect (Fiedler et al., 2000). Its innovative potential, however, derives from a shift in attention to the role of information generation processes. As illustrated in Figure 6.1, the shift promotes a distinction between biases that refer to discrepancies (a) between population parameters and sample statistics on one hand (i.e., sampling biases) and (b) between sample statistics and subjective parameter estimates on the other (i.e., judgment biases). More importantly, however, the shift has stimulated research on the ability of lay people to deal with biases that originate in the different stages of information processing. In this chapter, we

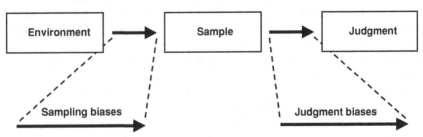

FIGURE 6.1. Schematic representation of Fiedler's (2000a) sampling metaphor. Judgments about properties of the environment are mediated by the properties of samples drawn from the universe of the potentially available information. Sampling biases are defined relative to population statistics, whereas judgment biases are defined relative to sample statistics.

concentrate on sensitivity to sampling biases and explore the potential of a subjective validity paradigm for its assessment.

THE INFORMATION SEARCH PARADIGM

Most research cited in the original formulation of the sampling approach (Fiedler, 2000a) used an information search paradigm (ISP) that renders obvious both the mediator function of samples and the difficulties people have in understanding the implications of sampling biases. Studies using the ISP ask participants to assess a contingency by searching a database that provides information about the joint occurrence of the levels of two dichotomous variables. Depending on the way people search the database, the resulting sample statistics either reflect population parameters accurately or provide a biased impression of them. Studies based on the ISP share as a common denominator that subjective parameter estimates resemble sample statistics closely, whereas evidence of judgment biases proper is often missing completely. However, as judgments reflect sample statistics irrespective of their fit with population statistics, sampling biases carry over to judgments. Apparently, people are insensitive to biases in samples they generated themselves.

For instance, in experiments on positive testing, Fiedler et al. (1999) had participants determine the aggression style of men and women by sampling information from a database that provided confirming or disconfirming feedback on acts of overt or covert aggression. As expected, participants engaged in positive testing in their search for information. When testing the counterstereotypical hypothesis that women aggress overtly and that men aggress covertly, for example, they directed more requests at the database on overt as opposed to covert aggression by women – and *vice versa* for men. Although the database responded with a fixed confirmation rate for each combination of gender and aggression style, participants considered the hypothesis tests true. Obviously, hypothesis confirmation in this setting has been driven by the larger sample size for hypothesis-consistent as opposed to hypothesis-inconsistent information. Participants gave the same proportion of confirming to disconfirming information more weight when it was based on a larger sample, presumably because the same confirmation rate appeared more reliable as sample size increased.

The intriguing point, however, is that participants were unaware that they had generated the crucial differences in sample size themselves (Fiedler et al., 2000). Another demonstration of the ignorance regarding the implications of information generation processes can be taken from research on base-rate neglect (Fiedler et al., 2000). The relevant data point here is that the well-known tendency to provide inflated estimates for the conditional probability of a rare criterion event disappears when people engage in appropriate sampling strategies in the ISP. In an experimental task based on the prominent breast cancer problem (Eddy, 1982), for instance, Fiedler et al. (2000) had participants assess the conditional probability of suffering from breast cancer given a positive mammography by scanning index-card boxes that coordinated breast cancer and mammography data of individual in-patients. Participants chose between a box sorted by mammography data (i.e., positive *versus* negative) and a box sorted by breast cancer data (i.e., present *versus* absent), and they studied as many index cards as they wanted in a self-determined way. Subsequently, participants were asked to estimate the conditional probability of breast cancer given a positive mammography – that is, $p(BC+ \mid M+)$.

The only (quasi-) experimental manipulation consisted in the index-card box choices. In either box, the base rate of breast cancer (i.e., $p(BC+)$ was 4%, the hit rate for the detection of breast cancer by mammography (i.e., $p(M+ \mid BC+)$) was 80%, and the false-alarm rate for a positive mammography in the absence of breast cancer (i.e., $p(M+ \mid BC-)$) was about 16.7%. Thus, all participants were confronted with the same population. However, as participants tended to sample approximately equal numbers of index cards from both levels of the variable by which their box was sorted, the resulting sample statistics differed dramatically. Obviously, drawing equal numbers of M+ and M− cases does not affect the proportion of breast cancer cases in the sample systematically, whereas drawing equal numbers of BC+ and BC− cases does: In fact, it increases the proportion of breast cancer cases in the sample to 50% as compared to only 4% in the population. Consequently, judgments of $p(BC+ \mid M+)$ were inflated when the information search was contingent on the criterion (i.e., breast cancer), but they were quite close to the population parameter when the information search was contingent on the predictor (i.e., mammography).

Analyses revealed that judgments resembled sample statistics irrespective of their representativity of population parameters (Fiedler et al., 2000). To the degree that the sample represented the population correctly, estimates approximated the population parameter of 16.7% for $p(BC+ \mid M+)$. However, to the degree that the proportion of breast cancer cases was inflated in the sample, the corresponding estimates were inflated, too. Again, obvious sampling biases carried over as the sample statistics were taken at face value. Participants whose index-card box was sorted by the criterion did not realize that it was their own decision to sample about 50% BC+ cases that raised the probability of $p(BC+ \mid M+)$ in the data.

The list of factors contributing to biased information generation in human contingency assessment could be prolonged easily (see Crocker, 1981, for a stimulating overview). In view of the limited insight people have into the effects of selective information search on their impressions of the environment, one may ask whether the sampling approach only shifted the reasons for pessimism about statistical reasoning from data integration to data generation. No matter on which stage of information processing research is focused, people are unaware of biased information processing, let alone their contribution to it. Such a pessimistic view, however, misses the fact that the ISP is a complex task. Participants have to understand the goal of information generation, to select an appropriate sampling strategy, and to monitor information as they scan the database. Moreover, people might hold different lay theories about information generation that affect performance in the ISP and that cannot be controlled experimentally, such as the preference for equal sample sizes previously discussed. Thus, it would be premature to conclude that people are insensitive to sampling biases altogether. Rather, we would like to suggest that a direct test of sensitivity to sampling biases requires experimental manipulations of the information generation process that are independent of cognitively demanding information search processes.

THE SUBJECTIVE VALIDITY PARADIGM

In the remainder of this chapter, we explore the potential of a subjective validity paradigm (SVP) for such a direct test. The SVP replaces active information search by a detailed description of the sampling procedure applied by a fictitious scientist who examines the contingency of two dichotomous variables. For instance, in studies on the assessment of conditional probabilities, participants are provided with information about the population parameters before the sampling procedure is described, so that discrepancies between population parameters and sample statistics are fully revealed to them. The dependent variables comprise a global classification of the fictitious study as valid or invalid (coded as +1 *versus* −1, respectively) and a rating of participants' confidence in their assessment on a graphical rating scale whose endpoints are labeled "not confident" (coded as 0) and "very confident" (coded as 100). Multiplying validity judgments by confidence ratings, a *subjective validity index* is created that ranges from −100 to +100. If participants were sensitive to sampling bias, the index values should increase as the degree of sampling bias decreases.

A pilot study based on the SVP has been reported by Fiedler et al. (2000). Participants were asked to evaluate the validity of a fictitious study concerned with the estimation of the conditional probability of a criterion event, breast cancer, given a predictor event, positive mammography, that is, $p(BC+ \mid M+)$. In the paragraph preceding a detailed description of the sampling procedure, participants learned that the base rate of the criterion event, breast cancer, was 4% in the population. The sample drawn by the scientist comprised an equal number of 300 cases with and without positive mammography in a *1:1 predictor sampling* condition, an equal number of 300 cases with and without breast cancer in a *1:1 criterion sampling* condition, and 900 BC− and 300 BC+ cases in a *1:3 criterion sampling* condition. Right below the description of the sampling procedure, a 2 × 2 table gave the resulting joint frequencies for breast cancer and mammography data. In the 1:1 criterion sampling condition, for instance, 240 M+/BC+ cases, 60 M−/BC+ cases, 59 M+/BC− cases, and 241 M−/BC− cases emerged. Before filling in the dependent measures, participants finally learned that the crucial estimate of $p(BC+ \mid M+)$ was calculated (correctly) by dividing the number of BC+/M+ cases by the number of all M+ cases.

Despite the obviousness of the discrepancy between population and sample parameters in the criterion sampling conditions, no evidence for enhanced sensitivity to sampling bias could be found. All of the mean subjective validity index scores were substantially positive, and the predictor sampling condition yielded a slightly higher score only. From a normative perspective, however, one would have expected very high scores for predictor sampling, moderate scores for 1:3 criterion sampling, and very low scores for 1:1 criterion sampling. The deficient sensitivity to sampling bias became fully transparent when comparing the different criterion sampling conditions: Participants judged the conclusions derived from criterion sampling *more* valid when they were based on equal samples rather than unequal samples. This difference was actually more pronounced than the aforementioned difference in favor of predictor sampling. Apparently, participants did not understand that estimates of $p(X \mid Y)$ must be misleading when X is overrepresented in a sample. Notably, even formally trained participants who were professionally concerned with medical statistics were insensitive to these biases in the data generation process. When asked how the study could be improved methodologically, they only called for larger samples and true randomness.

The following sections will serve to illustrate the way in which diverse aspects of people's sensitivity to sampling biases can be studied using the SVP. To this end, we will provide a synopsis of recent findings obtained using the SVP in diverse areas such as the estimation of conditional probabilities, contingency assessment, and causal attribution. For the sake of clarity, we will be mainly concerned with the SVP as a tool for studying sampling bias sensitivity in the domain of base-rate neglect. As an experimental tool, however, the SVP can be easily adapted to capture the effects of sampling procedures on other meaningful conclusions drawn from empirical findings. Using a modified version of the SVP in experiments on causal attribution, for instance, Fiedler et al. (2004a) found lay people to confuse effect size and causal power in judgments of causal influence. In these experiments, the descriptions of fictitious studies varied in the strength of the experimental manipulation applied and in the strength of the effects obtained − and the dependent measures asked for an assessment of the strength of the causal influence of the independent variable. In contrast to the logical rule that the same effect size is indicative of a stronger causal influence when based on a weak rather than a strong manipulation, causal influence judgments were primarily determined by effect size but were insensitive to data generation. Under some conditions, participants even reversed the logic of causal power by crediting more causal influence to strong manipulations rather than to weak ones that brought about effects identical in effect size. In general, then, the implementation of the SVP in different domains requires us to make the crucial features of sampling processes salient to participants (e.g., base-rate conservation and sampling of extreme groups) and to administer dependent measures that capture the methodological insight in question. [For an example in the domain of stimulus sampling, see Wells & Windschitl (1999).]

The intriguing findings of the SVP pilot study reviewed here prompted a series of experiments. On one hand, we were interested in identifying the necessary conditions for normatively correct appraisals of data generation processes. Would performance improve if sample statistics matched population statistics perfectly? Would participants eventually realize the discrepancy between population and sample statistics when confronted with different samples drawn from the same population? And to what degree might training sensitize participants to such discrepancies? On the other hand, we were also interested in exploring the degree to which different features of the SVP determine responses to the validity measure. Would performance in the SVP vary with manipulations located at the data generation versus data integration stages? Would performance vary with manipulations pertaining to the appropriate sampling strategy? Would performance in the SVP turn out to depend on prior beliefs? To address these questions, we extended the SVP in various ways without changing its basic structure.

Sensitivity to Information Generation versus Information Integration

In view of the pervasive tendency to regard the conclusions derived from prefabricated data sets as valid irrespective of their appropriateness, we felt it was necessary to demonstrate that participants in studies using the SVP do not simply accept any conclusion uncritically. To this end, a first study (Fiedler et al., 2004b, Experiment 1) included a manipulation located at the information integration stage, in addition to the basic manipulation addressing information generation. Building on Evans's (1989) observation that people tend to confuse the conditional probability of $p(X \mid Y)$ with its inverse, $p(Y \mid X)$, we manipulated whether our fictitious researcher engaged in a similar confusion of different conditional probabilities. For instance, when computing an estimate of $p(BC+ \mid M+)$, the researcher computed either the proportion of $BC+ \cap M+$ cases among all $M+$ cases (correct calculation condition) or the proportion of $BC+ \cap M+$ cases among all $BC+$ cases (incorrect calculation condition). If performance in the SVP constituted nothing but uncritical compliance with the information given, no effect of the within-participants variable calculation should be observed.

Two additional modifications of the basic SVP procedure introduced in Fiedler et al. (2000) served to elucidate the origin and scope of the ignorance regarding information generation. First, we replaced the comparison between 1:1 criterion and 1:3 criterion sampling by a more natural comparison between 1:1 criterion and proportional criterion sampling, a sampling procedure that exactly reproduced the population base rate of the criterion event in the sample. For instance, in an investigation of $p(BC+ \mid M+)$, participants learned that the base rate of breast cancer was 4% in the population. As in the pilot study, the fictitious researcher in a 1:1 condition drew an equal number of 500 BC+ and BC− cases, inflating the proportion of BC+ cases in the sample. In a proportional condition, however, the researcher drew only 40 BC+ cases and 960 BC− cases, conserving the proportion of BC+ cases in the sample and providing an accurate estimate of $p(BC+ \mid M+)$.

Second, in an attempt to highlight the importance of sampling bias sensitivity to the emergence of unbiased impressions of the environment, we included a measure of participants' real-world beliefs regarding the focal probability estimates (e.g., $p(BC+ \mid M+)$). In the event that participants should use the estimates derived in fictitious studies as an anchor, sensitivity to sampling bias would prove to be an important prerequisite to the critical use of statistical information in everyday life. To make sure that participants did not confuse the real-world belief measure with a manipulation check, instructions stated explicitly that personal beliefs might differ from the result of the studies they had just evaluated and that we were interested in their subjective assessment of the critical probabilities. If anything, these instructions ought to work in favor of contrast effects, and any correspondence of validity scores and real-world beliefs should not be discarded as a simple anchoring effect.

The calculation and sampling variables were manipulated within participants using a Latin square design. Thus, each participant worked on four SVP tasks, with each task operationalizing a different combination of the two within-participant variables. In addition to the relation between mammography and breast cancer, participants worked on fictitious studies pertaining to the relationship between depression and suicide, measles and prenatal damage, and sex role conflicts and anorexia, respectively. The base rate of the criterion event, its conditional probability, and the size of the samples drawn varied across problems, but all were based on skewed distributions for the rare criterion events. Thus, 1:1 criterion sampling always inflated the proportion of the criterion event and its conditional probability.

Consistent with the pattern obtained regularly in studies using the ISP, participants were sensitive to biases in information integration but were insensitive to biases in information generation. For information integration, a main effect for calculation ruled out the possibility that participants in the SVP comply with the conclusions drawn by fictitious scientists uncritically:

If the scientists confused $p(X|Y)$ with $p(Y|X)$ in calculating conditional probabilities, subjective validity index scores were rather low. If calculation was correct, subjective validity index scores increased markedly. For information generation, however, no improvement in sensitivity to sampling biases could be found. Neither the sampling main effect nor the interaction of sampling and calculation was significant. Again, participants did not realize the severe bias inherent to 1:1 criterion sampling. On the contrary, validity scores tended to be higher for 1:1 samples than for proportional samples, although only the latter provide normatively correct probability estimates. Interestingly, the highest validity ratings were obtained for correctly computed probabilities based on 1:1 sampling. Thus, incorrect calculation or proportional samples alone were sufficient to attenuate confidence in the research findings.

In line with this interpretation, the effects of the experimental manipulations on the real-world belief measure resembled the effects obtained for the subjective validity index. Real-world beliefs increased when biased sampling procedures resulted in strong conditional probabilities for a rare criterion event. For instance, when a researcher had included an inflated proportion of BC+ cases in the sample, yielding a high conditional probability of $p(BC+|M+)$, participants reported that they expected similarly high conditional probabilities to obtain in reality. However, the effects of biased sampling carried over to real-world beliefs only when calculation was correct. As is evident from the significant sampling–calculation interaction, real-world beliefs increased only when an inflated conditional probability had been computed correctly, but not when computation was flawed. Thus, in contrast to a simple anchoring explanation, research findings were accepted as a proxy for personal beliefs only when participants did not doubt the data analysis strategy.

Sensitivity by Design

The fact that the very exposure to proportional sampling per se did not sensitize judges to sampling bias was puzzling. The perfect match between population and sample parameters was not sufficient to turn participants' attention to the issue of base-rate conservation. In an attempt to increase the salience of the discrepancy between population and sample statistics, we ran another study (Fiedler et al., 2004b, Experiment 2) using a within-participants design suited to highlight the experimentally controlled variables (Fischhoff, Slovic, & Lichtenstein, 1979). Using the breast cancer problem only, participants in the crucial sensitivity condition of this study were confronted with the diverging findings obtained by two different scientists: One scientist used the proportional sampling procedure, concluding that $p(BC+|M+)$ was low, and the other scientist used the 1:1 criterion sampling procedure of the preceding study, concluding that $p(BC+|M+)$ was

Scientist A		
	Breast cancer	No Breast cancer
Positive Mammography	32	188
Negative Mammography	8	772

Scientist B		
	Breast cancer	No Breast cancer
Positive Mammography	400	98
Negative Mammography	100	402

FIGURE 6.2. Stimulus display used in the sensitivity condition of Fiedler et al. (2004b), Experiment 2. The left-hand sample conserves the low base rate of the criterion event, breast cancer, in the population, whereas the right-hand sample inflates the base rate of the criterion event from 4% in the population to 50% in the sample.

high (see Figure 6.2). Participants had to indicate which of the two studies appeared more valid to them and filled in the subjective validity measures for this study only. To quantify the effects of within-participants designs, we included two additional conditions in which participants evaluated either the 1:1 criterion sampling procedure or the proportional sampling procedure only.

Two equally likely scenarios regarding the within-participants design were considered. On one hand, performance in the SVP might improve in the sensitivity condition, because the confrontation with the diverging effects of different sampling procedures might turn attention to their differential appropriateness in terms of base-rate conservation. At least in the studies using the ISP, improved performance following sensitization to discrepancies between population parameters and sample statistics has already been demonstrated. For instance, Plessner et al. (2001) asked football coaches to assess the contingency between the performance of a fictitious team and the participation of a particular player, Peter F., by scanning a database sorted either by team performance (criterion sampling) or by participation of Peter F. (predictor sampling). Team performance was predominantly poor, so the criterion event had a low base rate. As expected, coaches in either sampling condition sampled approximately even numbers of cases from both levels of the variable by which the database was sorted. Consequently, coaches in the criterion sampling condition overestimated the probability of good performance given the participation of Peter F.; when coaches were asked to reconsider the proportion of good-performance cases in the database versus in the sample, however, coaches in the criterion sampling condition revised their estimates p(team performed well | Peter F. in team) appropriately.

In view of the pervasive preference for the 1:1 criterion sampling procedure in the preceding studies, on the other hand, one could as well expect the insensitivity to sampling bias to be exacerbated in the sensitivity condition. In the event that a preference for equal sample sizes is the driving force behind the effects obtained in the preceding experiments, for instance, actively choosing between one sampling procedure complying to this preference and another one violating it may lead to even higher subjective validity index scores favoring 1:1 criterion sampling.

Apart from the exploratory assessment of within-subject manipulations in the SVP, we also examined an explanation of the previous findings in terms of prior beliefs. If fictitious studies yielding high estimates for the a posteriori probability of rare criterion events (e.g., studies based on 1:1 criterion sampling) resembled prior beliefs more closely than fictitious studies yielding low estimates (e.g., studies based on proportional sampling), the preference for methodologically flawed studies would simply be due to their fit with prior beliefs. Likewise, from a representativeness perspective (Kahneman & Tversky, 1972) one may argue that high a posteriori probabilities are more representative of what lay people expect in predictions of significant events, such as risk of suicide. From this point of view, a 1:1 criterion sampling procedure would even have a built-in advantage over more appropriate sampling procedures, independent of any considerations regarding data generation proper. Prior beliefs were assessed before exposing participants to the SVP task.

Again, participants favored 1:1 criterion sampling over proportional sampling. In the sensitivity condition, this tendency was evident from an overwhelming preference for the study using 1:1 criterion sampling: More than 90% opted for this study as the more valid one. By coding choices favoring 1:1 criterion sampling versus proportional sampling as +1 versus −1, respectively, the corresponding subjective validity index score was moderately high. Thus, sensitizing participants to differences between sampling procedures did not improve their understanding of sampling bias. Comparisons revealed that the preference for equal sample sizes was even more pronounced when the relevance of sampling differences was made salient. That is, in the 1:1 criterion sampling and the proportional sampling conditions, the general tendency to accept the studies as valid reached levels similar to those in preceding studies; however, subjective validity scores were not as high as in the within-participants condition, which apparently pushed the subjective validity index in the wrong direction.

Further analyses revealed that alternative explanations in terms of prior beliefs deserve to be investigated in more detail. Irrespective of experimental condition, participants expected about two out of three women with a positive mammography to suffer from breast cancer, an estimate approximated far better by 1:1 criterion sampling than by proportional sampling. Indeed, prior beliefs were substantially correlated with validity ratings in both the 1:1 criterion ($r = .56$) and the proportional sampling conditions

($r = .27$). In the sensitivity condition, this correlation was eliminated as the result of the almost perfect preference for 1:1 sampling. These results support the idea that participants may have based their evaluation on the fit with prior beliefs alone. Based on the present data, however, we cannot determine whether expectancy congruence or a heuristic preference for equal samples determined performance primarily.

The Matter-of-Factness of Sample Statistics

Apparently, the preference for sampling procedures involving equal sample sizes – irrespective of their appropriateness – is a pervasive phenomenon. In each of the preceding studies, participants took sample statistics at face value, although the SVP provides almost optimal conditions for the detection of sampling bias in typical base-rate-neglect scenarios. The small proportion of a rare event in a population is stated explicitly and the large proportion of that same event in a sample drawn from that population is mentioned next. In fact, had we found bias reduction under these conditions, it would not have been surprising if reviewers would have argued that these findings may not reflect insight into the dynamics of sampling processes, but blatant demand instead. However, participants paid no attention to the issue of base-rate conservation. Moreover, not only did they fail to check for the congruence of population versus sample statistics, but they also incorporated the conclusions suggested by a single data point in their personal assessment of meaningful real-world phenomena.

One reason for the uncritical stance on the information given may be that people are used to getting along very well in many settings by relying on their experience, without ruminating about possible alternative views or consulting others for their opinion. In fact, the idea that the same real-world relation may appear different when viewed from a different perspective does not become apparent to us usually unless we are confronted with a diverging position on a topic. However, just as language developed late in evolution, perspective taking and reflection about alternative hypotheses may be recent developments. The typical situation that may have determined cognitive evolution may rather have been one in which the only information available was that provided by the organism's learning history. In line with this idea, people have been shown to have a hard time when confronted with tasks involving sampling distributions (Kahneman & Tversky, 1972; Sedlmeier, this volume), that is, tasks that require considering the hypothetical outcome of sampling the same population over and over again.

From this perspective, two routes toward improved performance in the SVP suggest themselves: (1) Evaluations of sampling procedures may become more accurate when people are made sensitive to the crucial issue of base-rate conservation by teaching and (2) sensitivity to sampling bias may increase as manipulations of the appropriate sampling procedure provide an opportunity to actively experience that different sampling procedures are appropriate under different conditions. We explored these ideas by first establishing, as a premise, that carry-over effects to real-world beliefs can be reduced by post hoc debriefing on the appropriateness of sampling procedures (Fiedler et al., 2004b, Experiment 3). In addition, we developed a formal training unit intended to sensitize participants to the importance of base-rate conservation (Fiedler et al., 2004b, Experiment 4). In the same study, we also examined whether variation in the appropriateness of equal versus unequal samples would suffice to promote sensitivity to sampling bias.

Feedback on the Appropriateness of Sampling Procedures

What will happen when participants are exposed to biased samples and – after they have formed an erroneous judgment – have it explained to them in which way the original sample was biased? This question amounts to asking whether people are able to understand the pathways by which sampling processes shape perception and to change their assessment accordingly. Relying again on the breast cancer problem alone, three experimental conditions were created. All participants worked on the 1:1 criterion sampling version of the task. The only between-participants variable referred to the type of feedback participants received after filling in the subjective validity index, but before furnishing the real-world belief measure. In a *correct feedback* condition, participants learned that an expert panel had considered the methodology of the fictitious study to be in line with the state of the art. In an *incorrect feedback* condition, the experts were said to have called the validity of the study into question without referring to specific flaws in its methodology. In a *sampling-bias feedback* condition, finally, participants received the feedback of the incorrect feedback condition, supplemented, however, by an appropriate explanation referring to the overrepresentation of BC+ cases in the sample caused by the inclusion of an equal number of BC+ and BC− cases.

As in preceding studies, there was a moderately high tendency to endorse the methodology used by the scientist, with acceptance rates ranging from about 60% in the incorrect feedback condition to 70% in the sampling-bias feedback condition. As confidence ratings were high on average, the mean subjective validity index scores were significantly positive and did not differ between experimental conditions. Thus, the stage was set for a test of the effect of the feedback manipulation on possible carry-over effects to real-world beliefs. As expected, positive feedback led to stronger carry-over effects than negative feedback. No differences could be found between the incorrect feedback and sampling-bias feedback conditions, respectively. This pattern corroborates the assumption that participants base real-world beliefs on the results of fictitious studies only when they have no reason to doubt the underlying procedure – just as they did in response to the correct versus incorrect calculation of conditional probabilities at the information integration stage of our initial study. However, because there was no surplus benefit from a genuine sampling-bias feedback, these findings also demonstrate that true insight in the mechanics of sampling biases may be difficult to obtain.

Sensitivity by Experience

If it is possible to correct biases in information generation externally by debriefing, it might also be possible to find a treatment that helps prevent erroneous validity judgments in the first place. The aim of a final study was to investigate whether a sensitivity training can promote an effective learning transfer to similar problems. To set true sensitivity to sampling bias apart from superficial rule learning, we also widened the range of SVP tasks. Note that in all of the preceding studies, the criterion events were rare in the population, so that 1:1 criterion sampling procedures necessarily inflated the proportion of the criterion in the sample. Invariant boundary conditions, however, may suggest that equal samples are always misleading. To avoid any effects of what may be termed a learning environment specificity, we included problems with rare as well as common criterion events. Under these conditions, sampling bias is manifest in equal sample sizes in some SVP tasks but in unequal sample sizes in others. Thus, sensitivity to sampling bias would be evident if participants understood which sampling procedure was appropriate under which conditions.

To this end, different constellations of normatively required and actually employed sampling procedures had to be manipulated within participants. In the SVP part of the study, participants worked on a series of four SVP tasks. The fictitious scientists drew a sample of equal size in two of the SVP tasks; samples of unequal size were drawn in the remaining tasks. However, each sampling procedure was appropriate in one case but not in the other. Moreover, to avoid any impact of prior beliefs, all tasks were assigned to the artificial setting of the evaluation of a novel pain reliever. The tasks requiring equal samples referred to *common* effects of the pain reliever (e.g., mild side effects), whereas the tasks requiring unequal samples referred to *rare* effects (e.g., severe side effects).

For instance, in one study a status labeled "completely pain free" served as a criterion event, a rather rare and unlikely event calling for unequal samples. The crucial paragraph in the outline of the study read ". . . independent from a specific treatment, not more than 5% (i.e., 5 out of 100 people) are completely pain free." Depending on the condition, the fictitious scientist either drew a sample of 500 people who were completely pain free and 500 who were not – or the scientist correctly conserved the low base rate and drew only 50 people who were pain free and 950 who were not. Consequently, the scientist either overestimated the effectiveness of the pain reliever or arrived at a realistic estimate.

The *sensitivity training* aimed at sensitizing participants to the importance of the base-rate conservation. To this end, participants first estimated the conditional probability of blue eyes (i.e., B+ *versus* B−) in men and women (i.e., M *versus* W). They were then presented with a 2×2 table showing the joint frequencies for each combination of gender and blue eyes versus other eye colors. The table yielded an estimate for the conditional probability of blue eyes in women of $p(B + | W) = .70$. Clearly, this estimate collided with people's real-world beliefs. On the next page, the obvious flaw was revealed: The fact that in reality not ten out of twenty people have blue eyes. Next, participants were presented with another 2×2 table in which the number of people with other eye colors was raised to 100, resulting in a downward shift of $p(B + | W) = .19$, a value that is more congruent with people's world knowledge. To further strengthen this insight, participants were asked to estimate what would happen to $p(B + | W)$ if the number of people with other eye color was raised to 1,000.

Apart from this sensitivity training, we included another manipulation intended to increase the attention to sampling procedures that was based on *suspicion*. The cover story simply informed participants that the following studies were conducted to test the effectiveness of a new pain reliever and that the scientists conducting the evaluation studies were all employed by the company that had developed the pain reliever. Both the sensitivity training and the suspicion manipulations were administered between participants.

Prior to analyses, each SVP task was recoded according to the specific sampling procedure applied by the scientist and according to the *appropriateness* of the sampling procedure. Using the individual SVP tasks as the unit of analysis, the subjective validity judgment scores were then regressed on the effect coded variables "sensitivity training" (training *versus* no training), "suspicion" (suspicion *versus* no suspicion), "sampling procedure" (unequal samples *versus* equal samples), and "appropriateness" (appropriate *versus* inappropriate). We also included three interaction terms to test the predicted interaction hypotheses involving the different debiasing treatments. A "training × correctness" term served to test whether participants in the training condition were more sensitive to sampling bias, a "suspicion × correctness" term served to test the corresponding hypothesis for the suspicion manipulation, and an "appropriateness × reason" term served to test whether participants who gave reasons for their validity judgments in a free format were sensitive to the appropriateness of the sampling procedures.

These analyses yielded some evidence that participants were sensitive to sampling bias. Apart from the regression weight for sampling procedure – reflecting the well-established advantage of equal as opposed to unequal samples – the only significant regression weight in the equation pertained to the appropriateness variable: Subjective validity judgments increased when the scientists selected the sampling procedure appropriate for the task at hand. This effect was not qualified by interactions involving the sensitivity training and suspicion manipulations, respectively. Apparently, the variation in the underlying problem type – that is, whether equal versus unequal samples were appropriate – helped participants realize that different kinds of problems require different solutions and that equal samples are appropriate only under certain conditions. In sharp contrast to the pessimistic impression conveyed by those SVP studies involving a single problem type only, these results suggest that people are capable of detecting sampling biases when they actively experience variation in the fit of population parameters and sampling procedures.

STATUS OF THE SVP AS AN EXPERIMENTAL TOOL

The present report served to explore the potential of the subjective SVP as an experimental tool for the assessment of sampling biases that is independent of self-determined information search processes. The findings obtained so far are encouraging. By implementing a wide range of experimental manipulations addressing data generation and data integration processes, variation in the sensitivity to sampling bias ranging from complete ignorance to at least partial insight could be demonstrated. Although some questions could not be answered exhaustively in the present series of experiments such as, for example, the interplay of sampling procedures and prior beliefs, the flexibility of the SVP makes it a promising arena for studies concerned with competitive tests among rival hypotheses regarding the origins of insensitivity to sampling bias.

Among the multitude of findings, two effects will probably attract attention in future research. First, we found a pervasive tendency to regard findings obtained using equal samples as more valid than findings based on unequal samples. Although this advantage of equal samples may derive from rather heuristic fairness concerns, we need to consider the possibility that the relatively small sample sizes in most of the proportional sampling conditions appeared impoverished to our participants. Thus, one may speculate whether results would have looked different if the proportional samples had not comprised 40 BC+ versus 960 BC− cases, but 4,000 BC+ versus 96,000 BC− cases instead. Future research may examine such reasons for the acceptance versus rejection of data-based conclusions that are rooted in aspects unrelated to sampling procedures.

Second, the fact that within-participants manipulations of the appropriateness of equal versus unequal samples brought about sensitivity to sampling bias suggests a promising way to increase sampling bias sensitivity. Of course, most statistical information in everyday life is presented in isolation, thereby confronting lay people with the task of overcoming the matter-of-factness of the information selected by advertisers, politicians, and acquaintances. However, to the degree that the within-participants manipulation in the last experiment reported here could be implemented in what may be termed a critical-observer training, people may understand that the conclusions drawn from any sample are not informative unless the correspondence of population and sample statistics in the crucial variables is clearly demonstrated.

At least the idea that the flood of information impinging on people in the early twenty-first century calls for refined and effective training seems to be shared outside psychology, too. Noam Chomsky introduced his recent critique of the mass media by confessing that "My personal feeling is that citizens of the democratic societies should undertake a course of intellectual self-defense to protect themselves from manipulation and control" (Chomsky, 2002, p. 8). Studies based on the SVP might contribute significantly to the curriculum. For instance, magazines and newspapers may consider demonstrating the congruence (or divergence) of the population versus sample statistics for statistically founded claims not only by global allusions to representativeness but also by the inclusion of graphics that picture the composition of the population and of the sample on which a claim is based.

For example, it would be very instructive if the popular bar diagrams used to visualize the proportion of people fulfilling some criterion were split routinely by the most relevant third variables. For instance, it is not rare for news reports on criminal acts committed by immigrants to pit the proportion of immigrants in the total population against the proportion of immigrants in the deviant population. Comparisons of this kind may be biased, though, because many immigrant populations are composed mainly of the type of persons responsible for crimes committed by native citizens, too (i.e., young males). Splitting overall statistics on criminal acts committed by native citizens and immigrants by the two most powerful determinants of deviance, age and gender, would often reveal that members of the immigrant population are no more deviant than members of the native population.

Apart from the potential of the SVP as a developer's tool in such applied settings, future research using the SVP should also consider describing deficits in the sensitivity to sampling bias at a higher resolution. Note that so far studies using the SVP have relied exclusively on global measures of subjective validity (and causal impact), supplemented at times by

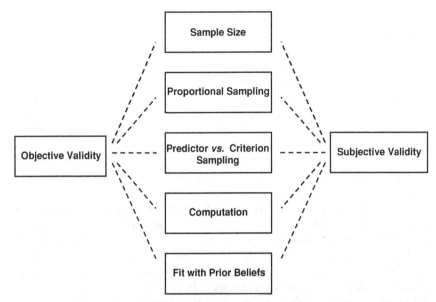

FIGURE 6.3. Schematic representation of a multiple-cue approach to sampling bias sensitivity. The validity of a scientific study cannot be assessed directly but has to be inferred from the presence or absence of validity cues. To the extent that diagnostic cues have no impact on validity judgments, associated sampling biases will carry over to biased judgments.

open-ended questions addressing participants' reasoning style or their recommendations regarding the methodology of a fictitious study. Although the use of such a global measure definitely helps obscure the ultimate goal of SVP studies, an undesirable side effect of global measures is that the specificity of the experimental manipulations is not matched by similarly specific measures. As long as we rely on global measures, all we know is that participants reject or accept a study as a whole, but we cannot tell which specific aspect of the fictitious research findings is responsible to what degree for acceptance versus rejection.

Thus, we would like to suggest that to understand the reasons for confidence in a study's validity we may have to manipulate valid as well as invalid cues to validity (see Figure 6.3) and to examine which features of a sampling procedure account for variation in subjective validity judgments. Of course, such an approach would be mainly descriptive. However, it would still constitute an improvement if we could not only say *that* people use cues to validity inappropriately but also describe *which* cues to validity are used to what degree. That is, our understanding of the insensitivity to sampling bias should benefit from an increase in the specificity of the origins of (biased) subjective validity judgments.

Author Note

The research underlying this manuscript was supported by a grant from the German Research Foundation (Deutsche Forschungsgemeinschaft, DFG) to the second author. The authors are grateful to Peter Juslin, Andreas Mojzisch, and Christian Unkelbach for valuable comments on drafts of this manuscript. Correspondence concerning this manuscript should be addressed to Peter Freytag, Department of Psychology, University of Heidelberg, Hauptstrasse 47–51, 69117 Heidelberg, Germany.

References

Bar-Hillel, M. (1980). The base-rate fallacy in probability judgments. *Acta Psychologica, 44,* 211–233.

Chomsky, N. (2002). *Media control.* New York: Seven Stories Press.

Crocker, J. (1981). Judgment of covariation by social perceivers. *Psychological Bulletin, 90,* 272–292.

Eddy, D. M. (1982). Probabilistic reasoning in clinical medicine: Problems and opportunities. In D. Kahneman, P. Slovic, & A. Tversky (Eds.), *Judgment under uncertainty: Heuristics and biases.* Cambridge, UK: Cambridge University Press.

Evans, J. St. B. T. (1989). *Bias in human reasoning: Causes and consequences.* Howe, UK: Lawrence Erlbaum.

Fiedler, K. (2000a). Beware of samples! A cognitive-ecological sampling approach to judgment biases. *Psychological Review, 107,* 659–676.

Fiedler, K. (2000b). Illusory correlations: a simple associative algorithm provides a convergent account of seemingly divergent paradigms. *Review of General Psychology, 4,* 25–58.

Fiedler, K., Brinkmann, B., Betsch, T., & Wild, B. (2000). A sampling approach to biases in conditional probability judgments: Beyond baserate neglect and statistical format. *Journal of Experimental Psychology: General, 129,* 399–418.

Fiedler, K., Freytag, P., & Unkelbach, C. (2004a). *Great oaks from giant acorns grow: A sampling approach to causal attribution.* Unpublished manuscript, University of Heidelberg.

Fiedler, K., Freytag, P., Unkelbach, C., Bayer, M., Schreiber, V., Wild, B., & Wilke, M. (2004b). *Subjective validity judgments of fictitious research findings: A paradigm for investigating sensitivity to sampling bias.* Unpublished manuscript, University of Heidelberg.

Fiedler, K., Walther, E., Freytag, P., & Nickel, S. (2003). Inductive reasoning and judgment interference: Experiments on Simpson's paradox. *Personality and Social Psychology Bulletin, 29,* 14–27.

Fiedler, K., Walther, E., Freytag, P., & Plessner, H. (2002). Judgment biases and pragmatic confusion in a simulated classroom – A cognitive-environmental approach. *Organizational Behavior and Human Decision Processes, 88,* 527–561.

Fiedler, K., Walther, E., Freytag, P., & Stryczek, E. (2002). Playing mating games in foreign cultures: A conceptual framework and an experimental paradigm for trivariate statistical inference. *Journal of Experimental Social Psychology, 38,* 14–30.

Fiedler, K., Walther, E., & Nickel, S. (1999). The auto-verification of social hypotheses: Stereotyping and the power of sample size. *Journal of Experimental Social Psychology, 77*, 5–18.

Fischhoff, B., Slovic, P., & Lichtenstein, S. (1979). Subjective sensitivity analysis. *Organizational Behavior and Human Performance, 23*, 339–359.

Gigerenzer, G., & Hoffrage, U. (1995). How to improve Bayesian reasoning without instructions: Frequency formats. *Psychological Review, 102*, 684–704.

Kahneman, D., & Tversky, A. (1972). Subjective probability: A judgment of representativeness. *Cognitive Psychology, 3*, 430–453.

Plessner, H., Hartmann, C., Hohmann, N., & Zimmermann, I. (2001). Achtung Stichprobe! Der Einfluss der Informationsgewinnung auf die Bewertung sportlicher Leistungen (Beware of samples! The influence of information gathering on the evaluation of performance in sports). *Psychologie und Sport, 8*, 91–100.

Schaller, M. (1992a). Sample size, aggregation, and statistical reasoning in social inference. *Journal of Experimental Social Psychology, 28*, 65–85.

Schaller, M. (1992b). In-group favoritism and statistical reasoning in social inference: Implications for formation and maintenance of group stereotypes. *Journal of Personality and Social Psychology, 63*, 61–74.

Schaller, M., Asp, C. H., Rosell, M. C., & Heim, S. J. (1996). Training in statistical reasoning inhibits the formation of erroneous group stereotypes. *Personality and Social Psychology Bulletin, 22*, 829–844.

Schwarz, N. (1994). Judgment in a social context: Biases, shortcomings, and the logic of conversation. In M. P. Zanna (Ed.), *Advances in experimental social psychology* (Vol. 26, pp. 123–162). San Diego: Academic Press.

Sedlmeier, P. (1999). *Improving statistical reasoning: Theoretical models and practical implications.* Mahwah, NJ: Lawrence Erlbaum.

Wells, G., & Windschitl, P. D. (1999). Stimulus sampling and social psychological experimentation. *Personality and Social Psychology Bulletin, 25*, 1115–1125.

7

An Analysis of Structural Availability Biases, and a Brief Study

Robyn M. Dawes

Benjamin Franklin once stated that "experience is a dear teacher" (in Poor Richard's Almanac, 1757/1793). I have often heard that misquoted as, "Experience is the best teacher." Clearly, however, Franklin meant something other than "good" or "best" by the word "dear," because he finished the thought with "and fools will learn from no other."[1]

What experience provides is a sample of instances. Those who maintain that we are very good at learning from experience are correct in that experience with a sample teaches us about the characteristics of the sample and a population if the sample is "representative" of it (assured, for example, by random sampling). For example, we are superb at producing intuitive estimates of frequencies and relative frequencies in a sample of materials to which we are actually exposed. (See Anderson and Schooler, 1991.) A problem arises, however, when a population about which we wish to generalize or learn is systematically different from the sample available to us. This systematic difference may provide a *structural availability bias* in that however accurate we may be about summarizing characteristics of our sample, we may fail to realize that it does not provide a good basis for making inferences that are important to us. Here, indeed, experience may be "dear" in the sense of "expensive." (Generally, the literature on availability biases involves those of selective memory or estimation based on subjective judgments of the *ease* of recall – often due to vividness of particular instances – but that type of availability bias is not the topic of this chapter.)

---

[1] I had heard this statement repeatedly from old New England farmers in the 1940s and 1950s whose vegetable gardens produced more rocks than vegetables year after year, but who often claimed that their experience about how to grow things was superior to any "book larning." In contrast, some of their sons who survived first-year calculus at the University of New Hampshire (where many of these sons went with the ambition of becoming engineers) learned enough about economics and land development through books to become wealthy in real estate some thirty years later.

Let me illustrate the structural availability problem with a (favorite) example that occurred when I was President of the Oregon Psychological Association in 1984. At that time, the rules involving conduct of psychotherapy were "sunsetted," and although most of the rules were then "sunrised" as they were, the fact that they were allegedly rewritten *de novo* provided an unusual opportunity for change. One change that my colleagues on the Board of Directors at the Oregon Psychological Association wished to make was in the rules about violating confidentiality to report suspected child sexual abuse. The rules previously mandated that confidentiality must be broken and suspicions reported to the child welfare authorities and police if there was any possibility that the child was in danger of continued abuse. If the children had been removed to a place distant from the abuser (e.g., sent to an aunt and uncle or a grandparent), or if the suspicions were about abuse that had occurred in the past and the potential victim was no longer a child, then there was no need to violate confidentiality to report the suspicion. My colleagues objected to the law as it was currently specified because "one thing we all know about child sexual abusers is that they never stop on their own without therapy." Thus, my colleagues on the board argued that *all* suspected child abuse should be reported, even suspicions about behavior that has occurred decades earlier, because there was likely to be (would be) a repeat. That also meant changing the law so that their suspicions did not have to involve a particular, identifiable child.

When I asked these colleagues how they *knew* child sexual abusers never stopped on their own without therapy, they answered that they – or a close colleague – had gained extensive experience with such abusers, so they "knew what they were like." I then asked how this experience came about. The standard answer was that these abusers had been brought to the attention of the court system, and then referred to clinical psychologists for treatment, perhaps as a condition for staying out of jail. "So then your sample of sexual abusers is *by definition* limited to those who did *not* stop on their own," I pointed out. Yes, they understood the logic. In fact, one of my closest colleagues began two subsequent discussions of the problems of changing the laws by mentioning that I had pointed out a "logical problem" with what everyone knew to be the truth. That did not, however, affect the opinion. The personal experience with such people was so compelling and vivid that my colleagues believed they knew what *all* such people were like.

This example illustrates three characteristics of a structural availability bias. First, a large proportion of judges may not be aware of its existence (in my example, virtually all my fellow board members). Second, even when people become aware of its existence, this awareness may have little or no effect on judgment, particularly if the context has great emotional salience to the judges. Third, an awareness of such biases may yield a judgment about their direction, but often their magnitude must remain unknown, again for structural reasons. In the example, my colleagues who accepted my logic acknowledged that their experience would be biased toward repeaters, but there was no way of deciding how much to "correct" this experience as a result of the bias. (Lack of such knowledge may be one factor in ignoring the bias, even if it is acknowledged.)

A better known example of a problem of this magnitude can be found in the work of Ross, Amabile, and Steinmetz (1977). Stanford students were randomly assigned to the role of questioner versus contestant in a "trivial knowledge" game, where the questioner was asked to make up questions that were difficult but not impossible to answer. Naturally, the questioner appeared more knowledgeable than the contestant (to both the contestants themselves and to observers, but *not* to the questioners, who had "privileged knowledge" of how the questions were developed). But how should an observer "debias" an impression of questioners and contestants in light of knowing the arbitrarily assigned social roles? One possibility is to ignore the interaction entirely and just judge each student to possess the trivia knowledge ascribed to the average Stanford student. This judgment would be a bit "heroic" given the pattern of available successes and failure. (It also raises the question of whether a contestant who does particularly well or poorly can really be ignored altogether.). What happens for sophisticated subjects is an amalgamation, which *cannot* be justified logically or statistically, of what is observed with knowledge of the bias.

The current little study is based on my hypothesis that understanding existence of a structural availability bias is a necessary condition yielding a directional correction for such a bias – and, consequently, that when people make a correction it is in the right direction. Whether there is a two-step procedure, by which subjects first understand the bias and only later incorporate the understanding into their judgment, or whether the understanding and incorporation are simultaneous, is a fascinating question, but one not addressed by the current study.

OVERVIEW OF THE SMALL STUDY

In a course for graduate social workers at the University of Pittsburgh, one of the participants volunteered to have a survey conducted about estimates of frequency, just so long as the students were given feedback about the purpose of the survey and its results. Subsequently, subjects were told that there were certain frequencies of characteristics of child abusers (not necessarily sexual abusers) in a sample of people referred to them for treatment, and they were then asked whether they thought that the relative frequency in the sample would be a good estimate, an overestimate, or an underestimate of the frequency in the general population of child abusers. Four items were chosen that on an a priori basis should be overrepresented

TABLE 7.1.

Characteristics	Sample Frequency	More /Less Common among All Abusers				
** The spouse of the abuser made a formal charge of abuse	4	–	–	0	+	++
The abuser was abused him/ herself as a child	19	–	–	0	+	++
0 The abuser is left-handed	8	–	–	0	+	++
The abused child has temper tantrums	18	–	–	0	+	++
* The abuser has not abused a child for 2 years	3	–	–	0	+	++
* The spouse of the abuser has tried to cover up the abuse	4	–	–	0	+	++
The abuser believes in God	22	–	–	0	+	++
** The abuser has abused a child in a public place	3	–	–	0	+	++
0 The abuser is over 50 years old	4	–	–	0	+	++
The abuser is less than 18 years old	2	–	–	0	+	++
The abuser attends church regularly	7	–	–	0	+	++
The abuser is a female	21	–	–	0	+	++

or underrepresented in their samples. For example, being turned in to the authorities by a spouse should lead to overrepresentation of abusers turned in, whereas having the spouse cover up for such abuse should result in underrepresentation (to the degree to which the spouse is successful and the abuse remains unknown). We also included two items that we thought would not lead to a structural availability bias; thus, there were two items that we thought would be biased high in the sample, two that we thought would be biased low, and two that we thought that should have no bias. Insofar as possible, we attempted to use realistic estimates from what is known in the literature for those characteristics (e.g., left-handedness) that we thought would not be biased by the fact of referral.

Materials

The items are presented in Table 7.1 together with our statement of the frequency with which they appeared in the sample of 25 hypothetical abusers. Those items that we thought would appear more frequently in the sample than in the unselected population are indicated by a double asterisk, those who we believed would appear less often in the sample are indicated by a single asterisk, and the two "control" items are indicated by zeros. Sixteen subjects taking the graduate class in social work agreed to make judgments of whether the characteristic should be "more or less common among all abusers." That judgment was indicated by simply circling the alternative at the right of the item.

Results

By averaging across subjects, it appeared that they were myopic about what could be changed and how; whereas 10 of the 16 subjects thought that the alleged fact that the spouse had made a formal charge of abuse would lead to overrepresentation in the sample of this characteristic, only 2 of 16 thought that people who did not abuse a child for two years would be underrepresented, 11 of 16 thought that people whose spouses had tried to cover up would be underrepresented in the sample relative to "all abusers," and 7 of 16 thought that those who abused the child in a public place would be overrepresented among those referred. Thus, there were 30 judgments consistent with the structural availability bias, and 34 inconsistent; it should be pointed out that the consistent judgments were equally split among those involving overrepresentation with those involving underrepresentation. Within-subject analysis was consistent with myopia. Scoring a correct movement as a $+1$ and a failure to move or an incorrect movement as a -1 gave an average subject score of .25, yielding a t value of .69.

Ignoring no movement at all, however, subjects were correct about the direction of adjustment. That is, ignoring all those responses of "zero" (that is, neither more nor less common among all abusers) the subjects gave an average score of .8125, for a t value of 2.25. That is, when we restrict our attention to the situations in which some adjustment had been made, the adjustment was significantly in the *correct direction* (again, defined by our a priori ideas). Including the "control" items in a parallel analysis yielded the same conclusions.

Discussion

To deal rationally with the structural availability bias, three criteria must be met: (1) The subject must understand that the bias exists, (2) the subject must understand the direction of the bias – given the understanding that it exists, and (3) the subject must act on the previous conclusions. This study does not involve action. For example, in the anecdote about reporting child sexual abuse in Oregon, it was clear that although several people understood the nature of the bias, they were unwilling to incorporate it into their recommendations about (*not*) changing the law. Whether the first two factors happened simultaneously or in sequence cannot be ascertained by the current data. What is important, however, is that *when* subjects implicitly acknowledge the existence of a structural availability bias – by suggesting that the characteristic should be either more or less true of all abusers – they are generally *correct* in the directional judgments of "more" versus "less." That is a characteristic of a structural availability bias hypothesized in the introduction of this chapter.

There exists much speculation about "two-system" processing underlying these cognitive biases. I suggest that the understanding that biases exist at all may be part of the automatic processing system, not the self-conscious "rational" one. It is only when this automatic system triggers the need to make an adjustment that an adjustment is made on a deliberative reasoning basis. For an explication of this distinction between automatic processing (System 1) and "more deliberative reasoning" (System 2) see Kahneman and Frederick, (2002).

Finally, this small study illustrates the importance of looking at individual responses in assessing bias. Had we just examined the overall average results, we would have concluded that our subjects were blissfully unaware of the bias, and we would have overlooked the phenomenon that *when* subjects modified their judgments, they did so in the correct direction. The importance of looking at individual responses has been emphasized by Bar-Hillel (1990) in her discussion of base-rate underutilization. When given "indicant" (witness recall) information about whether a hit-and-run cab was blue or green, most subjects attended only to the specific case. A sizable minority of Bar-Hillel's subjects, however, attempted to integrate this witness judgment with base-rate information about the accident rate of blue and green cabs. That attempt is hard to evaluate, given that it is neither Bayesian nor a simple averaging process. If both indicant and base-rate information were in the same direction, however, a proper Bayesian inference would be a probability judgment that is more extreme than either alone. Here, again, simple direction is critical.

References

Anderson, J. R., & Schooler, L. J. (1991). Reflections on the environment in memory. *Psychological Science, 2*, 396–408.

Bar-Hillel, M. (1990). Back to base rates. In R. M. Hogarth (Ed.), *Insights in decision making: A tribute to Hillel J. Einhorn* (pp. 200–216). Chicago: University of Chicago Press.

Franklin, B. (1757/1773) *Poor Richard's almanac.* New York: David McKay, Inc.

Kahneman, D., & Frederick, S. (2002.) Representativeness revisited: Attribute substitution in intuitive judgment. In T. Gilovich, D. Griffin, & D. Kahneman (Eds.) *Heuristics of intuitive judgment: Extensions and applications* (pp. 49–81). New York: Cambridge University Press.

Ross, L., Amabile, T. M., & Steinmetz, J. L. (1977). Social roles, social controls, and biases in the social perception process. *Journal of Personality and Social Psychology, 35*, 485–494.

8

Subjective Confidence and the Sampling of Knowledge

Joshua Klayman, Jack B. Soll, Peter Juslin,
and Anders Winman

INTRODUCTION

In many situations people must, at least implicitly, make subjective judgments about how certain they are. Such judgments are part of decisions about whether to collect more information, whether to undertake a risky course of action, which contingencies to plan for, and so on. Underlying such decisions are subjective judgments about the quality of the decision maker's information. Accordingly, many researchers have been interested in the mental processes underlying such judgments, which go under the general label of *confidence*.

There are many ways in which confidence can be expressed, both in the real world and in the research lab. For example, Yaniv and Foster (1995) present the concept of *grain size*. People communicate their confidence in an estimate via the precision (grain size) with which they express it. "I think it was during the last half of the nineteenth century" implies a different degree of confidence than "I think it was around 1875." Listeners expect speakers to choose a grain size appropriate to their level of knowledge. People also use a variety of verbal probability terms to describe confidence in their predictions, choices, and estimates (e.g., "I'm pretty sure ...," "It's definitely not ...," "I really don't know, but ..."), which people understand to imply different degrees of certainty (Wallsten & Budescu, 1983; Zimmer, 1984).

In the lab, most studies use one of three predominant paradigms. The two most often used involve providing a series of factual questions for which two alternative answers are given, one of which is correct. With the *half-range format*, questions take the following form: "Who lived longer – Ho Chi Minh or Claude Monet?" Judges select the answer they think is more likely to be correct and express their confidence using a numerical probability scale ranging from .50 to 1.0. With the *full-range format*, judges receive the statement "Ho Chi Minh lived longer than Claude Monet did" (or vice versa) and give the probability that the statement is true, from 0 to 1. The third paradigm, and the one on which we will focus in this chapter, involves subjective interval estimates. This *interval production format* requires judges to indicate a range of values that correspond to a given level of confidence. One example of a subjective interval would be, "I am 80% sure that Claude Monet was between the ages of __ and __ when he died." Here, too, there are variations in how such intervals are elicited from judges, as we will discuss later.

Judgments akin to subjective intervals play important roles in many real decisions that involve preparing for different possible outcomes. Take, for example, an Internet service provider considering offering an all-you-can-use pricing plan. In order to plan options for increasing capacity, the provider would need to know the chance of different degrees of change in average usage. Yaniv and Foster's (1995) work suggests that this basic judgment task is widespread in everyday life, albeit without numerical precision: Their concept of grain size can be seen as more or less equivalent to a rough expression of a subjective interval.

Decision analysts also rely on subjective intervals to capture judges' knowledge and opinions. This typically takes the form of fractile judgments. One form is like this: "There is a 90% chance that demand will be higher than __. There is a 70% chance it will be higher than __. There is a 50% chance..." Again, various other elicitations techniques are employed (see, e.g., Clemen, 1996; Spetzler & Staël von Holstein, 1975; von Winterfeldt & Edwards, 1986).

Most research on confidence has focused on what Hammond (1996) calls *correspondence*, that is, how well judgments correspond to reality. In the case of confidence, this usually entails comparing the expressed probability of being correct for each item to the actual proportion of correct answers across a set of questions. Such comparisons show that subjective confidence judgments are far from perfect. The estimated probability of being correct often agrees poorly with the actual proportion correct (poor calibration) and people do not distinguish very clearly between sets of items that have different success rates (poor resolution).

Are there also overall biases in confidence, and if so, are they in the direction of overconfidence or underconfidence? In the early days of research on confidence, the answer seemed clear: People were consistently and strongly biased toward overconfidence (see the review by Lichtenstein, Fischhoff, & Phillips, 1982). However, more recently it has become clear that biases vary greatly depending on the topic, the respondent, and the type of judgment (Gigerenzer, Hoffrage, & Kleinbölting, 1991; Juslin, Wennerholm, & Olsson, 1999; Klayman et al., 1999; Koehler, Brenner, & Griffin, 2002; Sniezek, Paese, & Switzer, 1990). In this chapter, we focus on differences among types of judgments – specifically, on the effect of different assessment formats for eliciting confidence judgments.

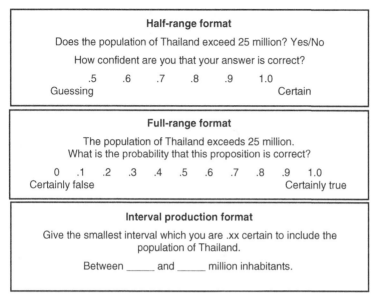

FIGURE 8.1. Application of the half-range, the full-range, and the interval production formats to elicit a participant's belief about the population of Thailand.

Figure 8.1 illustrates the half-range, full-range, and interval formats introduced earlier. A person might answer yes to the question of whether the population of Thailand exceeds 25 million in the half-range format and assign probability .9. To be consistent, we would expect this person to assign .9 in the full-range format. We would also expect a lower limit of 25 million in an 80% interval format, because an 80% interval is established by the 10th and 90th fractiles, and the answers to the previous questions imply that the 10th fractile is 25 million. The three formats in Figure 8.1 are merely different ways of probing a person's confidence in how much he or she knows about Thailand. It seems reasonable to assume that all the formats would generate similar levels of overconfidence.

Studies in which these formats are applied to the same items demonstrate a remarkable degree of format dependence: The realism of people's confidence in a knowledge domain varies profoundly depending on the assessment format. The least amount of overconfidence is exhibited in the half-range task. Averaging across multiple domains and participants, Juslin, Winman, and Olsson (2000) found that the mean estimated probability of choosing the correct answer approximately matched the actual percentage correct. In a similar multi-domain study using the half-range, Klayman et al. (1999, Experiment 1) reported a small but statistically reliable net overconfidence of 4.6%. Overconfidence is slightly higher with the full-range format, averaging around 10% (Juslin et al., 1999). Things are very different for the interval production format. Here, overconfidence is severe. The proportion of answers inside, say, 80% or 90% intervals is typically 15–45% lower than expected (Juslin & Persson 2002; Juslin et al., 1999; Klayman et al, 1999). This difference between two-alternative and interval formats is larger than the effects of most experimental manipulations and debiasing techniques applied to any one format. Indeed, it seems that, when the interval production format is used, the most effective debiasing method is to change to a different format.

In this chapter, we discuss why there is such a large difference among these types of judgments, especially between the interval format and the other formats, but also among different kinds of interval estimates. We believe that the explanation lies partly in the mechanisms by which information is pulled together to support an answer and partly in people's limited understanding of those mechanisms. We propose that when asked for an answer or an estimate, people "sample" the information available to them, from the environment and from inside their heads. We do not mean "sample" only in the strict, statistical sense of drawing a subset from a fixed population of items. Rather, our view of sampling includes drawing from all potential sources of information, including extracting facts from memory, but also accessing knowledge via mental simulations, deduction from causal models, reasoning by analogy, and so on.

If we are being so all-inclusive, then why refer to it as sampling? The information applied to forming an answer or estimate is a process of sampling in two senses. First, what one knows is only a sample of all the potentially available relevant information. Second, what one thinks of at the time is only a sampling from all the potentially relevant information in one's memory. We hypothesize that these processes of information sampling are the primary locus of errors in confidence.

We will next briefly review some of the relevant findings from studies of subjective confidence. Then we will present our hypotheses about how overconfidence can result from the limitations and biases of the information sampling process. These samples, we argue, tend to be small and nonrandom. Overconfidence results from the fact that judges do not fully anticipate, and thus do not adjust for, the effects of these properties. We will conclude with a discussion of ideas for future research to help understand the processes by which confidence is judged and how those processes might be improved.

PHENOMENA OF SUBJECTIVE CONFIDENCE

Like so many other aspects of psychology, subjective confidence turns out to be more complex than early findings suggested. Through the 1980s, there seemed to be a single, predominant finding: People are overconfident. In a review of previous two-choice studies, for example, Lichtenstein et al.

(1982) reported that when participants say they are about 70% sure they have the correct answer, they are right less than 60% of the time; when they say they are 90% sure, they are right about 75% of the time. The degree of overconfidence seemed to depend primarily on one variable: difficulty (Lichtenstein & Fischhoff, 1977). With very easy items (i.e., those for which people often chose the correct answer), confidence was about right; maybe even a little too low. For difficult items, overconfidence was extreme.

Most of the early work on confidence relied on two-choice questions in the half-range format. However, there was also some research using interval judgments. Overconfidence seemed to be worse with interval judgments (Lichtenstein et al., 1982). Russo and Schoemaker (1992), for example, found that business managers asked to provide 90% confidence ranges had the correct answer within the stated range between 42% and 62% of the time, depending on the domain and the participant group. Meanwhile, 50% ranges contained the correct answer about 20% of the time. So, prior to the mid-1980s most investigators concluded that there were three major phenomena of overconfidence that applied across the board. (1) People are poorly calibrated. Their chance of being correct is only in moderate agreement with their expressed confidence. (2) There is an effect of difficulty. The harder the questions (i.e., the lower the chance of being correct), the more overconfident people are. (3) Overall, people are overconfident, often by a wide margin.

The general explanation for these effects was cognitive bias. Here is a brief sketch of a typical explanation. People access some small, initial sample of evidence regarding the question. That evidence produces an initial impression. Subsequent retrieval and interpretation of evidence are biased such that information consistent with the initial impression is more accessible and more influential than information contradicting it. People treat this biased collection of information as though it were unbiased, and thus they overestimate the extent to which the evidence verifies the initial hypothesis or estimate (see Klayman, 1995; Koehler, 1991; Koriat, Lichtenstein, & Fischhoff, 1980; McKenzie, 1998).

In the past couple of decades, new research has been conducted using better methods to control for several potential confounds in earlier research. These recent studies demonstrate that overconfidence is not a universal phenomenon. It depends on the context in which confidence is elicited. Studies have found that (over)confidence varies with the domain of knowledge being tested, the gender of the judge, and from individual to individual (e.g., Juslin, Olsson, & Björkman, 1997; Juslin et al., 1999; Juslin, Winman, & Olsson, 2003; Klayman et al., 1999; Soll, 1996; Soll & Klayman, 2004). Recent research with the two-choice half-range format has confirmed that people are poorly calibrated, but it has called into question the last two of the conclusions from earlier research. It seems that, when possible confounding features of research methods are controlled, overconfidence and the effect of difficulty either disappear or are greatly diminished. Studies of subjective interval estimates using equivalent methodological controls produce different results. Those studies still find plenty of overconfidence. So, it seems that overconfidence with the two-choice half-range is either modest or nonexistent, whereas overconfidence with the interval format is robust and large. Why would that be?

What Happened to Overconfidence in Two-Choice Questions?

Recent research has engendered two major developments in the way researchers think about overconfidence in two-choice questions. These involve a greater appreciation of the roles of item selection by experimenters and of stochastic variability in judgments. With regard to the former, Gigerenzer, Hoffrage, and Kleinbölting (1991) raised the possibility that the bias is not in the mind of the judge, but rather in the actions of the experimenter. They suggested that experimenters traditionally have favored difficult questions, that is, questions that are less likely to be answered correctly than the average question in that domain. Why this is the case is unclear. Gigerenzer et al. suggest it might stem from a long history of preparing challenging class examinations. It may also be that researchers studying the variables that affect overconfidence may naturally gravitate toward tests that produce lots of overconfidence to study. In any case, a difficult set of questions will tend to include many questions for which the judge's information suggests the wrong answer. Unless special care is taken, these "contrary" questions may be overrepresented in the experimental stimuli.[1] An extreme version of this would be for the experimenter to select only contrary questions. In that case, the judge might reasonably be quite confident but would answer none of the questions correctly.

At a minimum, Gigerenzer et al.'s (1991) critique meant that studies in which questions were hand-picked by experimenters could no longer be relied upon as proof of overconfidence. Instead, studies should use questions that are randomly sampled from well-specified domains of knowledge. There is now a large and growing set of studies that meet this requirement (e.g., Budescu, Wallsten, & Au, 1997; Gigerenzer, et al., 1991; Griffin & Tversky, 1992; Juslin, 1993, 1994; Juslin et al., 1997;

---

[1] An example of a contrary question might be "Which city is farther west, Reno or Los Angeles?" Many people rely on the position of the state as a cue to answer this question. California is generally to the left of Nevada on the map, so people guess that Los Angeles is further west. In fact, the answer is Reno. People expect some contrary questions, so they are often not 100% confident on such questions. Still, the experimenter (intentionally or not) can make people look overconfident if contrary questions appear more frequently in the stimuli than they do naturally.

Juslin et al., 1999; Juslin, Winman, & Olsson, 2000; Klayman et al., 1999, Soll, 1996). Although some of these studies show pronounced overconfidence, others show substantial underconfidence. Across studies, the net bias is much smaller than had been demonstrated prior to 1991. Moreover, studies using multiple domains have found that under- or overconfidence varies considerably from one topic to another (Juslin et al., 1997; Klayman et al., 1999). This variation in overconfidence across topics helps explain why studies relying on only one or two topics have found such different results.

There have also been significant developments in understanding the role of stochastic variability as a cause of biases in confidence. The basic idea is that the judge has some store of information on which to base an answer. Across questions, the quality of information varies (how much information there is, how reliable it is, how strongly it favors one answer over another, etc.), such that the better the quality of information, the more likely the judge is to guess the right answer. In general, people are better than chance at identifying which questions have better quality information. Hence quality of information is positively correlated with the judge's confidence, and also with the probability of choosing the correct answer. Even in the absence of bias, these correlations are bound to be imperfect. Judgments about the appropriate confidence given the available information are subject to chance variations in one's experiences (Soll, 1996), variability in feelings of confidence given identical information (Erev, Wallsten, & Budescu, 1994), and unreliability in assigning a numerical value to one's feelings (Budescu, Erev, & Wallsten, 1997; Juslin et al., 1997). At the same time, the proportion of correct answers that actually obtains with a given quality of information is also susceptible to variation. For any quality level, there is some proportion of answers that run contrary to what the information suggests. In a finite sample of questions, there is inevitably some variability in how many of those contrary questions end up in the sample.

Since confidence and accuracy are both inexact manifestations of the underlying quality of information, mean-reversion effects are inevitable. That is, high confidence is usually too high and low confidence is usually too low – the typical pattern of miscalibration. Similarly, when accuracy is low for a set of questions it is likely to be lower than the available information would lead one to expect, and when accuracy is high it is likely to be higher than one would expect (Dawes & Mulford, 1996; Juslin et al., 2000; Klayman et al., 1999). This leads to overconfidence for "hard" question sets, and occasionally even underconfidence for "easy" sets – the so-called hard–easy effect. Soll (1996) has shown that other effects of stochastic variability may also account for any net overconfidence that remains in the half-range format. As far as we aware, the effects of stochastic variability are sufficient to explain the results of studies that have randomly sampled two-choice questions. However, given the large differences among domains, it is possible that cognitive bias will still be implicated in some situations.

One might reasonably have expected a similar pattern of results from other forms of confidence judgments, but instead some very sharp differences have emerged. In particular, the methodological improvements that changed our view of overconfidence in two-choice questions have not had a similar effect regarding intervals. Studies using dozens of randomly sampled questions from dozens of different domains still find very significant amounts of overall overconfidence bias with the interval production format (Juslin et al., 1999; Klayman et al., 1999). For example, Klayman et al. asked participants to complete range estimates, like "I am 90% sure that the answer is between __ and __," and they found that correct answers fell inside participants' 90% confidence intervals 43% of the time. Using a similar procedure with different domains, Soll and Klayman (2004) found 39% correct answers within 80% intervals. Soll and Klayman (2004) and Juslin et al. (2003) also demonstrated that overconfidence is sensitive to the method used to elicit the intervals. We discuss this in more detail later.

Soll and Klayman (2004) extended the examination of interval estimates to include consideration of possible effects of variability in interval sizes in the absence of bias. They showed that, indeed, variation in interval width that is unbiased on average in terms of the scale of measurement (e.g., years, meters, or kilograms) can produce a net bias toward overconfidence in terms of how many correct answers fall within the intervals. Perhaps judges simply fail to compensate for this rather obscure property. So, Soll and Klayman developed measures by which to estimate how the average interval size offered by judges compares to the size that would be well calibrated, ignoring the biasing effect of variability. Using range estimates with .8 confidence, they estimated that subjective intervals are in fact less than half the size they would need to be to produce appropriate confidence, excluding effects of variability.

These results provide the framework for our discussion of the processes underlying confidence judgments. We present hypotheses about how overconfidence, or the lack of it, can be understood in terms of the ways people process information. In so doing, we pay close attention to the differences that have emerged between different ways of asking for confidence, and what those differences may reveal about underlying processes. First, we consider the possibility that the large overconfidence bias with the interval production format is explained by a failure to understand the relationship between sample and population, where the sampling and processes applied to the sample are unbiased in other respects. Thereafter, we consider how systematic differences between judges' sampling and random sampling could also contribute to overconfidence. Both explanations provide insight into why overconfidence differs so dramatically across formats.

OVERCONFIDENCE IN INTERVAL ESTIMATES I: NAÏVE SAMPLING

To understand how biases might arise in interval estimates, it is helpful to think of people as naïve statisticians (Juslin et al., 2003). In this view, people behave like statisticians, drawing random samples and making inferences from them. Unlike real statisticians (good ones, that is), our naïve statistician fails to correct for sampling error. Other than this one, small difference, we assume for now that people otherwise behave as statisticians. This metaphor will allow us to generate and test new predictions about the conditions that lead to overconfidence. Conspicuously absent from the naïve sampling model is confirmation bias – the idea that people collect and interpret evidence in a way that is biased toward their beliefs. Although confirmation bias may play a role in producing overconfidence, in this section we explore how far we can get without it. Simply put, the naïve sampling model posits that people (1) have unbiased samples of their environments, (2) encounter only a small part of the environment, and (3) provide accurate and unbiased descriptions of these samples. The naïveté refers to an inclination to only represent the variability and error that are explicitly represented in the sample, in effect neglecting the corrections needed when sample properties are used to estimate population properties.

Recently, Juslin et al. (2003) applied the naïve sampling model to format dependence in confidence judgments. The key idea is that if people uncritically take sample properties as estimators of population properties (Kareev, Arnon, & Horwitz-Zeliger, 2002; see Chapter 2 of this volume), they will be overconfident with some formats but accurate with others. Consider the assessment of a subjective probability distribution (SPD) for an unknown quantity, for example, the population of Thailand. If the judge does not know the population exactly, a plausible range has to be inferred from other facts known about Thailand. Figure 8.2A illustrates the judge's uncertainty about Thailand, in the form of an SPD, assuming that all of the potentially available evidence is appropriately considered. Reading off this SPD, the judge assigns probability .25 to the event that Thailand has at least 35 million inhabitants and .75 to the event that it has no more than 55 million. The lower and upper boundaries jointly define a 50% interval such that the judge implicitly assigns a 50% chance to the event that the interval contains the true population of Thailand.

According to the naïve sampling model the key distinction is between probability judgment (either in terms of the half-range or the full-range format) and the production of intervals. Juslin et al. (2003) therefore concentrated on two formats that are comparable in all ways except that one format requires full-range assessment of a prestated interval whereas the other format requires the production of an interval of a prestated probability. To illustrate the distinction, consider two ways of eliciting the SPD

A

Population of evidence

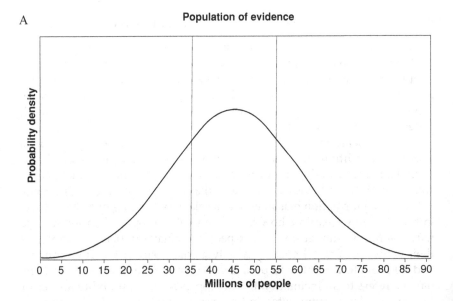

B

Sample of evidence, N = 4

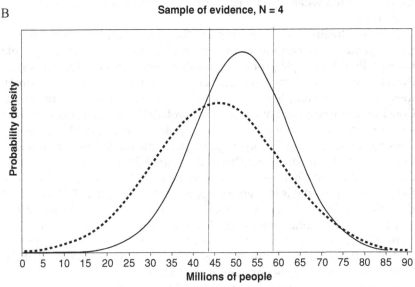

FIGURE 8.2. (A) Hypothetical probability density function that would result from consideration of all potentially available information about the target value (in this example, the population of Thailand). The 25th and 75th fractiles are shown, indicating that 50% of the probability lies between 35 million and 55 million. (B) Probability density function for a sample of four examplars drawn from the population, with an average amount of displacement from the population mean and average variance for samples of size 4. The 25th and 75th fractiles of the sample are shown, indicating a 50% interval between 43.5 and 58.5 million. Note that this interval includes only 36% of the population.

shown in Figure 8.2A. With *interval evaluation* the judge is provided with an interval and is required to assess the probability that a target quantity falls inside. For example, if asked for the probability that the population of Thailand falls between 35 million and 55 million the judge in Figure 8.2A would report .5. In contrast, *interval production* requires the decision maker to state the smallest central interval such that he or she is *.xx* confident that it includes the true value. For a 50% interval the answer should be "between 35 and 55 million." In summary, whereas interval evaluation supplies fractiles and requests a probability, interval production supplies a probability and requests fractiles.

The Naïve Sampling Model

The naïve sampling model assumes that the judge forms an SPD for an uncertain quantity associated with a target by retrieving from memory a sample of n observations that are similar to the target and for which the quantity is known. Similarity is determined by one or more cues that the observations must share with the target to be eligible for inclusion in the sample. The SPD is assumed to be a simple extrapolation of the values in the sample. For example, suppose that the target is the country Thailand and the uncertain quantity is population. The cue in this case might be that the observation is a country in Asia. So, imagine that the judge roughly knows the populations of five Asian countries and retrieves these into the sample: China (population 1.3 billion), South Korea (50 million), Japan (125 million), Syria (17 million), and Laos (6 million).

The mechanics for the interval evaluation task are straightforward. The assessed probability is simply the proportion of observations inside the range. For example, if the range 35 million to 55 million is provided, only the population of South Korea lies inside, so the assessed probability is .20. For the interval production task, the model assumes that people treat the sample as if it were perfectly representative of the population. The general idea that people report sample data without making the sophisticated corrections needed for unbiased estimates of the population distribution can be implemented in different ways, either by assuming that people report the sample percentiles as estimates of the population percentiles, or by assuming that they compute some measure of sample variance and use that directly to estimate population variance. Juslin et al. (2003) focused on a variation in which the judge interprets percentiles in the sample as reasonable estimates of percentiles in the population. For example, according to one standard method, in a sample of 5 observations the median of the upper half of the distribution is the 75th percentile and the median of the lower half of the distribution is the 25th. In the previous problem, the populations of Syria and Japan, 17 million and 125 million, respectively, would serve as the lower and upper limits of a 50% interval. When the exact percentiles are not available in the sample, the model provides an algorithm for interpolation (see Juslin et al., 2003).[2]

For both formats, a sample of relevant observations is retrieved from memory. Then, the judge reports the characteristic of the sample that corresponds to the required estimate: the proportion of observations within the specified interval (interval evaluation) or the values that correspond to the specified percentiles of the observations (interval production). The naïve statistician makes only one mistake, which is to treat the sample as if it were the actual population. Interestingly, the effect of this mistake plays out very differently for the two response formats.

Why Should Assessment Format Matter with Naïve Sampling?

Consider first the interval evaluation format. An important point here is that the interval and the sample are independent. The experimenter first chooses an interval of any size and location, and then the judge samples. The naïve sampling model assumes that the experimenter cannot refer to the sample to choose an interval and that the judge's sample is not influenced by the choice of interval. Given these constraints, the sample proportion is an unbiased estimator of the proportion in the population. This setup leads to the prediction that people will be appropriately confident, on average, with interval evaluation.

In contrast, in interval production the interval is constructed from the sample; interval and sample are no longer independent as in the evaluation task. This dependence implies that intervals constructed directly from samples will be biased. By definition, the sample interval will gravitate to wherever the data are most tightly clustered in the sample. The data in the underlying population are unlikely to be most tightly clustered in exactly the same spot as in the sample. Thus, the proportion of the sample that the interval contains is likely to be larger than the proportion of the population that falls into that same interval.

---

[2] There exist several methods for interpolating fractiles from finite samples (see http://www.maths.murdoch.edu.au/units/c503/unitnotes/boxhisto/quartilesmore.html for a discussion of the most common methods). The simulations relied on a standard method, sometimes referred to as the EXCEL method. It is calculated as follows: Let n be the number of cases, and p be the percentile value divided by 100. Then $(n-1)p$ is expressed as $(n-1)p = j + g$, where j is the integer part of $(n-1)p$, and g is the fractional part of $(n-1)p$. The fractile value is calculated as x_{j+1} when $g = 0$ and as $x_{j+1} + g(x_{j+2} - x_{j+1})$ when $g > 0$. This method captures the naïveté implied by the naïve sampling model in the sense that there is no correction for the fact that the sample dispersion is expected to be smaller than the population dispersion. Although there are other ways to implement the idea that people naïvely use sample dispersion to estimate population dispersion without the appropriate corrections, in general they will produce qualitatively similar results: intervals that are too tight, with extreme overconfidence.

It is instructive to decompose the bias in the interval production task into two types of sampling error. First, relative to the population the sample will exhibit an error of location – the mean of the sample will typically differ from the mean of the population. This is shown in Figure 8.2B. One reason why the .5 interval for the sample includes less than .5 of the population is that it is off-center. For any single-peaked distribution, even asymmetric ones, shifting an interval away from the peak will generally result in a loss of data.

The second type of sampling error has to do with dispersion. The distance between the 25th and 75th percentiles in the sample will differ from the distance between those percentiles in the population. There are two ways that this can lead to bias. One is that the distance between any two percentiles in a sample will on average be shorter than the distance between the same percentiles in the population, as shown in Figure 8.2B. This is related to the fact that the expected variance of a sample is smaller than the variance of a population by a factor of $(n - 1)/n$. This is reflected in the standard formula for estimating population variance from sample variance, but people's intuitions do not include this aspect of sampling (see Kareev, Chapter 2 of this volume). In addition, interval width will vary across samples, and that also contributes to bias. For single-peaked distributions, more data are lost when the interval is too narrow than are gained when it is too wide, and hence variability in width can lead to overconfidence. This effect and its implications are discussed in depth by Soll and Klayman (2004). For statistical confidence intervals, Student's *t* distribution takes this into account. Assuming that intuitive intervals fail to make similar corrections, they will be overly narrow, producing overconfidence. Figure 8.2B shows a sample that has the amount of displacement and the amount of variance that are average for samples of size 4. The 50% interval on the sample includes only 36% of the population.

Juslin et al. (2003) tested the hypothesis that people would fail to adjust for the factors that make sample intervals smaller than the corresponding population intervals. They predicted that people would be fairly well calibrated on the interval evaluation task and severely overconfident on the interval production task when both formats are applied to the same judgment content. They first verified the model by applying Monte Carlo simulation to a database of world countries. Estimates concerned a given country's population, and samples of data were drawn randomly from other countries on the same continent as the target country. The algorithms described here for the production and evaluation formats were executed numerous times. Figure 8.3A presents the proportion of true values falling within the intervals in an interval production task, as predicted by the naïve sampling model. At small sample sizes the proportions are far too low, as is commonly observed in data. Figure 8.3B shows how mean overconfidence bias changes with sample size when computed across the 50%,

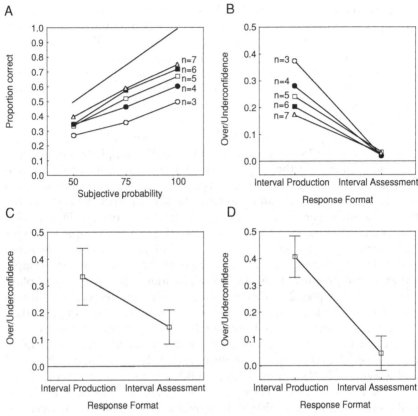

FIGURE 8.3. (A) Proportions of correct values included by the intervals plotted as a function of prestated probability of the interval as predicted by the naïve sampling model. (B) The format dependence predicted by the naïve sampling model for independent samples in conditions for interval production and interval assessment. (C) The observed overconfidence in a standard interval production task and a task with probability assessments for the same intervals, but made by other participants (error bars are 95% confidence intervals). (D) The observed overconfidence in a standard interval production task and in a task with probability assessment for independently defined intervals of the same size (error bars are 95% confidence intervals). Adapted from Juslin, Winman, & Hansson, 2003.

75%, and 100% intervals presented in Figure 8.3A. In the production task, overconfidence starts out at 40% when $n = 3$, but drops below 20% for $n = 7$. The intervals generated by the production task were then treated as inputs for the evaluation task. In other words, each interval produced in the first simulation, between x and y, defined an event to be evaluated (in terms of the probability of the target falling between x and y). The evaluation of each interval was determined by retrieving a new random sample of n observations and ascertaining the proportion of observations in this new sample that fell between x and y. As shown on the right-hand side of Figure 8.3B, overconfidence is negligible for the interval evaluation task at all sample sizes, as anticipated by the naïve sampling model.

As reviewed earlier, there is plenty of evidence showing that people are severely overconfident on the interval production task. In fact, extrapolating from the simulation, it appears that people assess uncertainty as if they were operating with samples of $n = 3$, since overconfidence in empirical studies is often near 40%. It should be emphasized that this sample size refers to the ideal case of random sampling from the environment. In reality, samples are likely to deviate systematically from random selection. These deviations may include confirmation biases, which we discuss in the next section. Even in the absence of such cognitive biases, there are likely to be biases inherent in the way information reaches the judge (Fiedler, 2000). For example, in one judgment domain it may be that the larger the observation, the more likely one is to be aware of it (e.g., populations of countries). In another domain, the information may be biased toward one's nearest environment, in yet another domain dramatic events may be overrepresented, and so on. For a variety of idiosyncratic reasons the samples we encounter are likely not to be perfectly independent random samples from the relevant reference classes. To the extent that such effects apply across formats, they may produce additional overconfidence, even at larger sample sizes. However, we would still expect to see the predicted difference between interval production and interval evaluation.

Juslin et al. (2003) compared interval evaluation and production formats to see how well the model would hold up to data from human judges. Some participants produced 50%, 80%, and 100% intervals for the populations of 40 randomly selected countries (e.g., "Assess the smallest interval for the population of Bangladesh for which you are .8 (80%) confident that it includes the true value: Between ___ and ___ with probability .8"). The results shown on the left-hand side of Figure 8.3C replicate the extreme overconfidence bias with the interval production format. Other participants assessed the probability that the intervals provided by the first group included the target value (e.g., "What is the probability that the population of Bangladesh falls between 55 million and 85 million?"). Although they are the same events, the overconfidence bias is significantly reduced.

One reason why some overconfidence remained in the evaluation task was that the samples retrieved by the two sets of participants are not perfectly independent, as presumed in the simulations. Because of the shared environment (effects of media coverage, salience of neighboring countries, etc.) different persons are likely to retrieve overlapping samples when they judge a target. Therefore, the sample of evidence used to evaluate the interval is not a new and independent random sample but in part involves retrieval of the same observations that served to misplace the interval in the first place. Near-zero overconfidence in interval evaluation is only predicted if the evaluation is based on a new, random sample of observations that is completely independent of the sample used to produce the interval. To more closely approximate independence, a new experiment replicated the production experiment, with one critical variation. As before, those in the evaluation condition received intervals of the same size as those generated by the production group. However, this time, before being presented, each interval was shifted so that the midpoint of the interval was the true population figure of another, randomly assigned, country. Figure 8.3D shows that under these conditions, when interval and sample are independent, the overconfidence bias in the evaluation task is nearly eliminated. These results show that naïve estimation from small samples can produce the format differences of the magnitude observed empirically.

How should we interpret the naïve sampling model? Do people literally infer probabilities from a retrieved pool of similar exemplars? Perhaps they do in some circumstances. For example, a marathon runner may remember race times and use a sample of past times to assess confidence. The model can also be applied when people store a perceptual impression of a quantity, such as how long it takes to drive to work. In this case, a sample of past experiences may be retrieved, but these will still need to be translated somehow to numerical responses. Even if there is no bias in how people interpret their samples, the naïve sampling model says that people will be overconfident unless they correct for sampling error. Even if people do not literally retrieve quantities and report percentiles off a distribution, the model is still valid at another level. A person's sample of experiences, whatever they are and however they are represented, will tend to deviate from the population of possible experiences. If people neglect to correct for this when producing intervals, they will be overconfident.

OVERCONFIDENCE IN INTERVAL ESTIMATES II: BIASED
SAMPLING AND INTERPRETATION

In the previous section, we suggested that a misunderstanding of the relation between samples of evidence and the population can produce overconfidence. This is true even if the sampling process and the interpretation of the evidence collected are otherwise unbiased. To complement this explanation, there is a great deal of evidence that people's processing of information is biased in ways that also contribute to overconfidence. Information that is consistent with one's hypothesis is more easily retrieved, more comprehensible, more credible, and less likely to be distorted (Koehler, 1991; see also Klayman, 1995). Accordingly, once some initial information has been retrieved to favor an answer or an estimate, the information subsequently retrieved, and the interpretation of that information, will likely be biased toward favoring the initial impression.

How does this apply to subjective intervals? Again, we turn to the notion that the variability in the implications of the sample of retrieved evidence is one important cue by which people judge appropriate confidence. (Other cues may include the total amount of information collected and the ease or speed of collecting it.) Now, suppose, for example, in the year 2005 Judge J is asked to estimate the age of the current British prime minister. The first thing that comes to mind is a recent television image, which gives J the impression that Tony Blair is not likely to be less than forty-five years of age, is maybe in his early fifties, and is very unlikely to be older than sixty. Now, looking for further information, J is more likely to access additional facts that are consistent with this general impression, such as a description of Blair as having "boyish" looks, along with general knowledge that that term is often used to describe certain good-looking, middle-aged men – say, in their early fifties. Information that was less consistent with the prevailing impression would face a retrieval disadvantage. For example, J might be less likely to retrieve general knowledge about the typical age range for leaders of major democracies, which might suggest that Blair was unlikely to be younger than fifty.

This associative process would produce a collection of evidence that was more consistent than it would be if it had been sampled randomly from J's available store. Now, a biased sample of information poses no particular problem for judgment *if* the judge recognizes the bias and takes it into account, or at least has learned from experience to compensate for it. In general, people seem not to have this metacognitive insight (Fiedler, 2000; Klayman, 1995). If judges act as though the consistency and coherence of their sample of evidence is an unbiased estimate of the evidence as a whole, this retrieval bias will contribute to their underestimation of the variance in the overall body of evidence.

An additional potential source of bias in information sampling involves how the retrieved information is used in forming a judgment. A variety of confirmation biases have been documented, in which an initial or focal hypothesis gains an advantage in the way that information is weighed and interpreted (see Klayman, 1995). This seems to be the case even when there is little or no commitment to the hypothesis. The main conditions for confirmation bias are that one of two or more alternatives becomes the focal or frontrunner hypothesis and that there is some potential vagueness or ambiguity in the evidence and its implications (e.g., Hoch & Ha, 1986; Russo, Meloy & Medvec, 1998). Thus, it is not necessary that judges make any explicit decisions, even to themselves, for confirmation biases to begin. Judge J in our example took the "boyish looks" remark to be consistent with J's initial impression that Tony Blair was roughly in his early fifties. Now consider Judge K. K's initial impression was that Blair was at least in his mid-fifties, probably around sixty, and possibly older. Now, K also recalls seeing Blair on television and hearing that remark about "boyish looks."

K concludes that the "boyish" remark implies that Blair is older than he looks, confirming K's original impression. Again, this is not inherently a problem for forming judgments of confidence, if J and K compensate for this natural biasing by discounting confirming evidence and putting extra weight on the disconfirming evidence that does get through. Absent such correction, however, confirmation biases also will contribute to the underestimation of the variability of evidence.

Biased sampling of evidence can affect other likely cues to confidence as well. For example, judges may monitor the flow of evidence favoring the current impression versus evidence suggesting alternatives. Positive evidence is likely to be more easily and rapidly retrieved (Hoch, 1984), and that in turn may be taken as evidence that it is also more plentiful (consistent with the idea of availability as a cue to probability; Tversky & Kahneman, 1974).

The mechanisms proposed here are by no means novel; they are very similar to those that were prevalent in the 1980s (Hoch, 1985; Koriat et al., 1980). Those arguments were developed mainly in the context of two-choice questions and are now largely being abandoned in that domain. However, as we have attempted to show, those mechanisms potentially apply to judgments about intervals. Unlike with two-choice questions, interval production shows pervasive overconfidence sufficient to suggest that cognitive biases are indeed functioning.

Why Should Assessment Format Matter with Biased Sampling?

In a previous section we showed that naïve sampling provides one reason why assigning a probability to a given event (i.e., a given interval) produces little overconfidence, whereas generating a subjective interval produces plenty. In addition, we hypothesize that, when it comes to confidence, an important variable is the presence or absence of specific alternatives. That is, when asked to provide a range corresponding to a given degree of certainty, the judge attempts to form a single impression, along with some sense of the reliability, strength, or vagueness of the impression. So, for example, when asked to estimate Tony Blair's age, the judge collects the available evidence and evaluates the collection as to its implication for a plausible range. In contrast, consider the two-choice format. A two-choice question, we hypothesize, prompts *two* collections of evidence – one supporting or refuting the first alternative, and another supporting or refuting the second alternative. Which country has a higher population, Thailand or Pakistan? There are potentially four types of evidence here: for and against Thailand and for and against Pakistan. Evidence pro and evidence con may not be weighted equally for either hypothesis, and evidence for one may not be automatically encoded as evidence against the other (see McKenzie, 1998). Nevertheless, Thailand and Pakistan each serve as foci for evidence gathering, resulting in two collections of evidence that, though not independent, are not completely redundant. Other formats also lend themselves to consideration of multiple hypotheses. For example, judging the statement "The population of Thailand is greater than 25 million" (see Figure 8.1) automatically suggests its alternative, namely that it is less. Evaluating the statement "The population of Thailand is between 25 million and 75 million" suggests two other alternatives, namely that it is less than 25 million and that it is more than 75 million. For now, though, we will explain our hypotheses by contrasting interval production with two-alternative, half-range questions. If range estimates are, psychologically, a single hypothesis, whereas two-choice questions are a comparison of two hypotheses, then the latter will suffer less from the effects of limited sampling and less from biased sampling. Suppose, for example, that the judge thinks about the choice between Thailand and Pakistan by forming a subjective impression about each one's population. Each will be based on its own sample of evidence, although the two may have some common members. For example, both sets of evidence may include knowledge of India's large population, suggesting that Asian countries are highly populous. The comparison of the two samples is less biased by naïve sampling inferences than is the estimate of each small sample itself.

Similarly, several lines of research suggest that explicit alternatives are also less prone to biased evaluation than is a single hypothesis. Klayman (1995) reviews a number of studies suggesting that an explicit contrast between alternatives reduces confirmation biases. For example, people have a preference for positive hypothesis testing, that is, for searching for evidence that would, if present, support the hypothesis. Such a strategy is not intrinsically confirmatory, because failure to find the evidence sought is taken as disconfirmation. However, when combined with other elements of hypothesis testing, positive testing can open the door to confirmation bias. For example, people are too little influenced by the probability of observing the evidence given the alternative (see also Dawes, 2001). Thus, they give too much weight to evidence that is expected under the current hypothesis and also fairly likely under the alternative. This is referred to as "pseudodiagnosticity" (Doherty et al., 1979). Instead of following the Bayesian principle that the impact of evidence on belief (its diagnosticity) is a function of the ratio of the probability of the evidence given the hypothesis to the probability of the evidence given the alternative(s), people are influenced largely by the numerator alone (see also Slowiaczek et al., 1992). However, in comparing two explicit alternatives, people may be prompted to consider both sides of the equation. Alternatively, they may accomplish a similar effect if they form a final confidence judgment by comparing the strength of evidence supporting each alternative. Pseudodiagnostic data may be erroneously counted in favor of each alternative when considered separately, but those two errors will roughly balance out in the comparison, producing a more nearly Bayesian result (Klayman and Brown, 1993; McKenzie, 1998). Tversky and Koehler (1994) find that, given two explicit, exhaustive, and mutually exclusive alternatives, the probability assigned to the two of them, in separate consideration, sums to just about the normative 1.0. In contrast, when each of several hypotheses is considered with the alternatives unspecified, the sum of probabilities is greater than 1.0, indicating that judgments are biased in favor of focal hypotheses.

Results such as these suggest why two-choice questions may be relatively immune to cognitive biases whereas range estimates remain susceptible. The former present two, clearly specified alternatives; the latter seems to ask for a single, best judgment (albeit a fuzzy one.)

Differences among Methods of Eliciting Subjective Intervals

Difference among different methods of eliciting subjective intervals provides further evidence concerning the psychology of subjective confidence. Soll and Klayman (2004) tried three different procedures for obtaining subjective interval estimates across a variety of domains. One method was to ask for ranges corresponding to a specific certainty – in this case, 80%. Another way to delineate an 80% interval is to ask separately about upper and lower bounds. For example,

In what year was [item or process] invented?
I am 90% sure that this happened after ＿＿.
I am 90% sure that this happened before ＿＿.

Thirdly, Soll and Klayman asked some participants for an additional, median judgment in conjunction with separate high and low estimates, such as

I think it's equally likely that this happened after or before ＿＿.

Soll and Klayman found considerable differences among these three ways of eliciting subjective intervals of equal certainty. Ranges that were supposed to capture the correct answer 80% of the time captured about 40%, with intervals that were less than half the well-calibrated size. Separate estimates of high and low boundaries captured about 55%, with intervals about two-thirds the well-calibrated size. With high, low, and median judgments, the hit rate was about 65%, and intervals were about three-quarters well-calibrated size. Juslin et al. (1999) found similar results when they asked participants for a best-guess point estimate with range estimates (see also Block & Harper, 1991; Clemen, 2001; Selvidge, 1980).[3]

---

[3] Note that the absolute amount of overconfidence observed depends on the set of domains chosen for study, since overconfidence varies considerably from domain to domain. Some

These results are consistent with our hypothesis that range estimates are processed as a single judgment and that consideration of explicit alternatives reduces bias by increasing the sample of evidence and by neutralizing some of the effects of biased sampling. By asking for separate high and low boundaries, judges focus their attention on two different judgments, namely, what is the lowest plausible number, and what is the highest plausible number. It is likely that these two judgments share some evidence. In both cases, the judge may conjure up an image of Tony Blair from memory of a television newscast. However, it is also likely that the two questions will not access identical evidence. Positive testing suggests, for example, that in estimating the lowest plausible age for Tony Blair judges will especially search for evidence of youth, whereas the upper estimate would encourage more search for evidence of maturity.

So, asking for the two bounds separately produces a larger total sample of evidence. This should lead to greater accuracy, which in this context translates to intervals that are better centered on the truth. This indeed turns out to be the case (Soll & Klayman, 2004). It should also reduce biases due to differences between sample variance and population variance, as the sample becomes larger. Moreover, confirmation biases now tend to pull the two estimates apart, rather than together. The search for youngest plausible will be biased toward youth, and that for the oldest plausible will be biased toward age. Thus, we would expect to find that, even as estimates become more accurate, the size of intervals becomes larger. Soll and Klayman also find this effect. Hence separating the range judgment into two parts reduces overconfidence by simultaneously increasing accuracy and increasing subjective interval widths.

Now, suppose we add a best-guess estimate in addition to high and low boundaries (Soll & Klayman, 2004) or in addition to ranges (Juslin et al., 1999). In the former case, it is easy to see the parallel to our earlier arguments. Now there are three judgments: lowest plausible, highest plausible, and plausible middle. These subjective fractiles are closer together than just a high and low estimate, so there is bound to be more redundancy in the information accessed. However, such redundancy will not be total; additional focal hypotheses still produce more information and thus greater accuracy combined with larger intervals.

When we combine a point estimate with a range estimate (as in Juslin et al., 1999), the story must be modified slightly. We hypothesize that the range estimate is already treated as a kind of fuzzy best-guess estimate, so why would adding an explicit best guess reduce overconfidence? One possibility has to do with how retrieved evidence is used to support

studies show significant differences between male and female judges as well. However, differences between elicitation formats are fairly consistent across domains and genders (Soll & Klayman, 2004).

different answers. When only the boundaries are requested, some evidence supporting a best guess or intermediate value will likely be retrieved as well. This evidence might tend to counterbalance evidence favoring the extremes, pulling in the boundaries. When a best guess is requested in addition to the boundaries, the judge collects additional evidence favoring intermediate values. Why does this not pull the boundaries further in? We speculate that, given three different estimates (low, middle, and high), judges attempt to collect different evidence for each. Thus, intermediate evidence is "reserved" for the middle case and tends to be excluded from consideration in the boundary cases. As an example of how this might work, suppose that in the range format a judge uses Tony Blair's boyish looks as a reason why he is unlikely to be older than fifty-five. If an explicit median were requested, the same judge might think of Blair's boyish looks as an argument for why he must be around forty-five. Could the judge then use "boyish looks" again in setting the upper boundary? Certainly, but using the same argument multiple times might feel inappropriate. We would suggest, tentatively, that the impact of "boyish looks" on the boundary would diminish once it has been used to develop a best guess.

DISCUSSION

The history of research on overconfidence is a bit of an odd one. For a few decades, results seemed to point consistently to one basic conclusion: Overconfidence was a pervasive and fundamental characteristic of human judgment, resulting from biases in information processing. Now, it is hard to resist saying "we shouldn't have been so sure." During the 1990s, evidence mounted against the presumption of overconfidence. Recent research on two-choice questions has deemphasized explanations based on selective memory, confirmation biases, and motivational factors. Instead, attention has focused on regression effects, the use of bounded response scales, and the relationship between the task sample and the natural environment about which inferences are made. Judgments of confidence are indeed prone to error, but overall those errors are only slightly biased toward overconfidence. Whether or not this is a sufficient explanation for observed overconfidence remains a topic of discussion, but it is clear that investigators had not fully understood the relations among item selection, stochastic variability, and apparent bias.

   Then, in the past few years, several extensive studies have shown that although a cognitive bias toward overconfidence may have been a mirage in some judgment tasks, it seems to be very material in others. In particular, when judges express confidence by producing subjective intervals, they are indeed biased. The intervals they produce are too narrow given their knowledge level, often considerably so. It is of course possible that we have, yet again, failed to recognize some subtle, unintended processes by which biased responses can emerge from noisy but unbiased judgments.

However, this time, researchers have looked much more closely at such matters, and we must, at least for now, conclude that there are important cognitive biases for overconfidence in intervals.

So, the question that seemed interesting in the 1980s and irrelevant in the 1990s is back, in a revised form, in the 2000s: Why are people prone to overconfidence bias? Now, though, we must reframe the question as, "Why are people prone to overconfidence in some contexts and not in others?" The gender of the judge, the domain of questions, and the format of elicitation have all been shown to be important elements in determining whether confidence is too high, too low, or about right.

There is not yet much solid evidence concerning the processes by which people arrive at subjective judgments of confidence. Instead, what we have to work with at this point is primarily evidence about how confidence judgments fare across a variety of different modes of elicitation. These include two-choice questions in which judges select the more likely answer and provide confidence on the 50–100% scale, two-choice questions in which they express confidence between 0% and 100% that a proposed answer is correct, quantitative estimates for which judges give the probability that the answer is within a specified interval, and quantitative estimates in which people set an interval corresponding to a given probability by providing a range or two endpoints, with or without an additional median or best-guess estimate.

Based on the results from these different formats, we hypothesize two classes of explanations for the presence or absence of overconfidence. These are (a) limited sampling of evidence and (b) biased sampling and interpretation of evidence. These explanations are by no means mutually exclusive, and at present we do not have sufficient evidence to say whether one or another dominates. It may in fact be the case that one is more prominent a factor in some environments, and the other in others. In any case, the two types of explanations share an important feature: They both presume that confidence arises from a process of reaching into one's store of knowledge and then extrapolating from what one retrieves to a larger, hypothetical population. The retrieved information is a *sample* in several ways. It represents only a portion of the relevant information potentially available in the environment. It is even only a portion of the relevant information potentially available in the judge's own head. Moreover, it is prone to error: We may misremember what we read or heard or experienced, and we may do so differently on different occasions. So, we argue, the process of inferring confidence from retrieved evidence is analogous to estimating the mean and variance of a population from a sample. Uncertainty is of course inherent in such estimates, and that uncertainty determines an appropriate confidence interval.

The sample the judge has to go on is limited, and it is also not a random sample. Both of those characteristics limit the accuracy of the judge's estimates, but they would not interfere with appropriate confidence *if the judge correctly anticipated their effects.* We hypothesize that judges do not in fact have a good, intuitive understanding of the effects of either the limited size of their samples or their deviations from random or representative selection.

In the former category, we propose that judges act as though the dispersion of implications within their sample mirrors the dispersion of implications in the hypothetical population of all evidence. A judge may think of five facts, images, principles, or experiences that suggest answers of 40, 45, 55, 57, and 62. Confidence, we presume, is based on an intuitive reading of the amount of evidence and the extent to which it is consistent. This is roughly parallel to considering the size of a sample and the within-sample variance. The interval that captures a given portion of the sample is, on average, smaller than the interval that captures that proportion of the population, and more so as the sample gets smaller. It is also located in a different spot. Thus, although 60% of the sample falls between 44 and 58, it is likely that less than 60% of the population does. Judges lack insight into this feature of small samples. Thus, they tend to overestimate the extent to which agreement within retrieved evidence indicates certainty in the world.

Moreover, items of evidence are more likely to be retrieved when they are more similar to items already retrieved. The perceived implications of a given piece of evidence are also likely to be pulled toward prior inferences. Thus, the judge tends to work from a sample with dispersion less than would be obtained by random sampling. Again, it is unlikely that judges anticipate this and correct for it appropriately.

Although we so far lack direct process data, the effects of limited and biased information samples can be mapped to observed differences among ways of eliciting confidence judgments. The most overconfidence-prone method is to ask judges to generate a range of values corresponding to a given confidence level. In that format, we argue, judges collect a single sample of evidence and base their interval on the range that applies to the sample. They do not compensate adequately for the effects of the limited size of the sample nor for the likelihood that the sample is biased toward consistency.

Less overconfidence is observed when judges separately set each end-point of an interval. In that case, we argue, judges gather two, somewhat independent samples of evidence – one to determine the lower limit, and one to determine the upper. This produces more evidence, reducing the biasing effects of small samples. It also reduces the tendency for evidence to be self-confirming, because the two collections of evidence will cluster around very different foci. Adding a third, median estimate further reduces overconfidence by increasing the amount of information gathered and by implicitly demanding that each of the three samples of evidence be at least somewhat independent.

Three formats seem to produce little overconfidence. One is to ask judges to estimate the probability that the correct answer lies within a given interval. Here, the location of the interval is not a function of the evidence retrieved to evaluate it. With this independence, random samples provide an unbiased (though unreliable) estimate of the proportion of the population that will fall within the interval. Lack of bias in this format might be taken as evidence that the sample itself is not biased relative to random sampling. However, the effects of bias here are not straightforward. Suppose the judge's initial impression of the answer falls within the proposed interval, and subsequent evidence is biased toward consistency. The result will be an overestimate of the probability that the answer lies within the interval. However, if the initial impression falls far from the center of the proposed interval, biased sampling will exaggerate the probability that the answer lies outside the interval, producing an underestimate of the within-interval probability. Moreover, one can argue that the effects of sampling bias may be reduced if judges regard the proposed interval as essentially defining a set of alternative hypotheses (the amount is lower than, within, or higher than the specified range), rather than accepting it as an initial estimate.

The other formats that show little overconfidence are the ones using two-alternative questions, either 0–100% confidence about a particular answer or 50–100% confidence about one's preferred answer. We argue that little overconfidence is to be expected because judges treat the two possible answers as separate and equal hypotheses. From the point of view of limited sampling, this corresponds to evaluating the probability within a given interval, because the alternative answers are not determined by the evidence collected. From the point of view of biased sampling, evidence is collected with regard to each alternative, and although each collection may be biased in favor of the alternative being tested, these biases more or less balance out when the two are compared.

In addition to our favored explanations, we acknowledge that there are a number of alternatives worth considering. One that comes easily to mind is anchoring and insufficient adjustment. Although this does indeed seem to be a widespread phenomenon, we question its usefulness in this context. It does not appear that formats that engender less overconfidence are those in which one would expect less anchoring. For example, asking people to explicitly name a median value within their interval produces wider intervals (Juslin et al., 1999; Soll & Klayman, 2004.) Many of the best-known demonstrations of anchoring involve anchors provided by others (Tversky & Kahneman, 1974), but we find little overconfidence when evaluating an interval proposed by the experimenter. Moreover, recent work (e.g., Mussweiler & Strack, 1999) suggests that much of what has been labeled anchoring and insufficient adjustment may be better understood as the result of confirmation biases. Nonetheless, there may be some tasks in which there is, indeed, a process of anchoring on a particular point estimate and then an adjustment from there (see Epley & Gilovich, 2001), and it will be valuable to consider the role of such processes in confidence.

Another possibility is that judges' problems stem from their unfamiliarity with the task of producing numerical equivalents for feelings of confidence. We agree that this likely detracts from the accuracy of confidence judgments, but we are less clear that it can explain the findings. All of the formats we discussed ask for similar, numerical equivalents of confidence, but some show very large overconfidence and others none. Substantial overconfidence in range estimates has been demonstrated at various confidence levels (e.g., 50% intervals in studies by Russo & Schoemaker, 1992). Besides, we doubt that confusion about numerical equivalents can explain why, for example, participants asked to provide 80% or 90% ranges in fact have hit rates of less than 50%. Nevertheless, it will be worthwhile to study how intuitive understandings of numerical probabilities may affect confidence judgments.

Another model that challenges our sampling approach is that of Gigerenzer and his associates (e.g., Gigerenzer & Goldstein, 1999; Gigerenzer et al., 1991). They propose that judges tend to use only a single diagnostic cue for choosing among alternatives. So far, the evidence of the plausibility of this mechanism is greater than the evidence that it is actually the predominant strategy used by judges. However, if people rely on a single cue in producing intervals, then overconfidence presumably reflects an overestimate of the validity of that cue. Soll (1996) provides evidence that people do not generally overestimate the validity of the available cues in two-choice questions, but there may be other mechanisms that apply in translating a perception of cue validity into an interval size, and we would need to shift our focus to understanding the processes by which people estimate cue validities. Gigerenzer et al. (1991) tend to imagine cue validity as determined by retrieval of cases in which the cue did or did not point to the correct conclusion. We question whether one often receives the necessary feedback to keep such tallies, and we suspect that many of the same sampling-related processes we propose would also apply to intuitions about the validity of individual cues.

We believe the sampling model of confidence judgments provides a plausible and intriguing proposal. Clearly, there will need to be much more research into the processes underlying confidence judgments to determine whether this is an appropriate metaphor and to learn more about how limitations and biases affect the process. Studying different formats for eliciting confidence has proven useful, and there are plenty more available for study. Some interesting possibilities include the following: multiple-choice questions, such as "Which of these four universities is the oldest?"; confidence in the truth of propositions with no easily specified alternatives, such as "'Nimbus' is a kind of cloud"; and interval judgments concerning objective populations, such as "80% of students at this university are between ___ and ___ in height." The large differences in over- or underconfidence observed across different content domains provide another potential source of clues. As yet, the question of what accounts for domain differences is still an open one, as it is for the gender differences and individual differences that have been observed (Soll & Klayman, 2004). We could also learn a great deal from simple process data such as verbal protocols, reaction times, or priming effects.

Recent research on interval estimates has certainly not settled the question of why people are overconfident when they are, or why they are not when they are not. However, it is clear that there are still interesting phenomena to explain. The trend in the 1990s was to attribute overconfidence to stochastic variability in judgments and experimenters' misguided analyses. Those are indeed important element, and thanks to that research we now understand much more clearly how to define and measure over- and underconfidence. With interval estimates, though, there is something psychologically interesting going on; reports of the death of overconfidence were greatly exaggerated. Now we need to get on with figuring out what is keeping it alive.

References

Block, R. A., & Harper, D. R. (1991). Overconfidence in estimation: Testing the anchoring-and-adjustment hypothesis. *Organizational Behavior and Human Decision Processes, 49,* 188–207.

Budescu, D. V., Erev, I., & Wallsten, T. S. (1997). On the importance of random error in the study of probability judgment. Part I: New theoretical developments. *Journal of Behavioral Decision Making, 10,* 157–171.

Budescu, D. V., Wallsten, T. S., & Au, W. T. (1997). On the importance of random error in the study of probability judgment. Part II: Applying the stochastic judgment model to detect systematic trends. *Journal of Behavioral Decision Making, 10,* 172–188.

Clemen, R. T. (1996). *Making hard decisions: An introduction to decision analysis* (2nd ed.). Boston: PWS-Kent Publishing.

Clemen, R. T. (2001). Assessing 10–50–90s: A surprise. *Decision Analysis Newsletter, 20*(1), 2, 15.

Dawes, R. M. (2001). *Everyday irrationality.* Cambridge, MA: Westview Press.

Dawes, R. M., & Mulford, M. (1996). The false consensus effect and overconfidence: Flaws in judgment or flaws in how we study judgment? *Organizational Behavior and Human Decision Processes, 65,* 201–211.

Doherty, M. E., Mynatt, C. R., Tweeney, R. D., & Schiavo, M. D. (1979). Pseudo-diagnosticity. *Acta Psychologica, 43,* 111–121.

Epley, N., & Gilovich, T. (2001). Putting adjustment back in the anchoring and adjustment heuristic: Differential processing of self-generated and experimenter-provided anchors. *Psychological Science, 12*, 391–396.

Erev, I., Wallsten, T. S., & Budescu, D. V. (1994). Simultaneous over- and underconfidence: The role of error in judgment processes. *Psychological Review, 101*, 519–527.

Fiedler, K. (2000). Beware of samples! A cognitive-ecological sampling approach to judgment biases. *Psychological Review, 107*, 659–676.

Gigerenzer, G., & Goldstein, D. G. (1999). Betting on one good reason: The take the best heuristic. In G. Gigerenzer, P. M. Todd, & the ABC Research Group (Eds.), *Simple heuristics that make us smart* (pp. 75–96). New York: Oxford University Press.

Gigerenzer, G., Hoffrage, U., & Kleinbölting, H. (1991). Probabilistic mental models: A Brunswikian theory of confidence. *Psychological Review, 98*, 506–528.

Griffin, D., & Tversky, A. (1992). The weighing of evidence and the determinants of confidence. *Cognitive Psychology, 24*, 411–435.

Hammond, K. R. (1996). *Human judgment and social policy: Irreducible uncertainty, inevitable error, unavoidable injustice.* New York: Oxford University Press.

Hoch, S. J. (1984). Availability and interference in predictive judgment. *Journal of Experimental Psychology: Learning, Memory, and Cognition, 10*, 649–662.

Hoch, S. J. (1985). Counterfactual reasoning and accuracy in predicting personal events. *Journal of Experimental Psychology: Learning, Memory, and Cognition, 11*, 719–731.

Hoch, S. J., & Ha, Y.-W. (1986). Consumer learning: Advertising and the ambiguity of product experience. *Journal of Consumer Research, 13*, 221–233.

Juslin, P. (1993). An explanation of the hard-easy effect in studies of realism of confidence in one's general knowledge. *European Journal of Cognitive Psychology, 5*, 55–71.

Juslin, P. (1994). The overconfidence phenomenon as a consequence of informal experimenter-guided selection of almanac items. *Organizational Behavior and Human Decision Processes, 57*, 226–246.

Juslin, P., Olsson, H., & Björkman, M. (1997). Brunswikian and Thurstonian origins of bias in probability assessment: On the interpretation of stochastic components of judgment. *Journal of Behavioral Decision Making, 10*, 189–209.

Juslin, P., & Persson, M. (2002). PROBabilities from EXemplars (PROBEX): A "lazy" algorithm for probabilistic inference from generic knowledge. *Cognitive Science, 26*, 563–607.

Juslin, P., Wennerholm, P., & Olsson, H. (1999). Format dependence in subjective probability calibration. *Journal of Experimental Psychology: Learning, Memory and Cognition, 28*, 1038–1052.

Juslin, P., Winman, A., & Hansson, P. (2003). *The naïve intuitive statistician: A sampling model of format dependence in probability judgment.* Manuscript, Department of Psychology, Uppsala University, Sweden.

Juslin, P., Winman, A., & Olsson, H. (2000). Naive empiricism and dogmatism in confidence research: A critical examination of the hard-easy effect. *Psychological Review, 107*, 384–396.

Juslin, P., Winman, A., & Olsson, H. (2003). Calibration, additivity, and source independence of probability judgments in general knowledge and sensory discrimination tasks. *Organizational Behavior and Human Decision Processes, 92,* 34–51.

Kareev, Y., Arnon, S., & Horwitz-Zeliger, R. (2002). On the misperception of variability. *Journal of Experimental Psychology: General, 131,* 287–297.

Klayman, J. (1995). Varieties of confirmation bias. In J. R. Busemeyer, R. Hastie, & D. L. Medin (Eds.), *Decision making from the perspective of cognitive psychology* (pp. 385–418). New York: Academic Press.

Klayman, J., & Brown, K. (1993). Debias the environment instead of the judge: An alternative approach to reducing error in diagnostic (and other) judgment. *Cognition, 49,* 97–122.

Klayman, J., Soll, J. B., González-Vallejo, C., & Barlas, S. (1999). Overconfidence: It depends on how, what, and whom you ask. *Organizational Behavior and Human Decision Processes, 79,* 216–247.

Koehler, D. J. (1991). Explanation, imagination, and confidence in judgment. *Psychological Bulletin, 110,* 499–519.

Koehler, D. J., Brenner, L., & Griffin, D. (2002). In T. Gilovich, D. Griffin, & D. Kahneman (Eds.), *Heuristics and biases: The psychology of intuitive judgment* (pp. 686–715). New York: Cambridge University Press.

Koriat, A., Lichtenstein, S., & Fischhoff, B. (1980). Reasons for confidence. *Journal of Experimental Psychology: Human Learning and Memory, 6,* 107–118.

Lichtenstein, S., & Fischhoff, B. (1977). Do those who know more also know more about how much they know? *Organizational Behavior and Human Performance, 20,* 159–183.

Lichtenstein, S., Fischhoff, B., & Phillips, L. D. (1982). Calibration of subjective probabilities: The state of the art up to 1980. In D. Kahneman, P. Slovic, & A. Tversky (Eds.), *Judgment under uncertainty: Heuristics and biases* (pp. 306–334). New York: Cambridge University Press.

McKenzie, C. R. M. (1998). Taking into account the strength of an alternative hypothesis. *Journal of Experimental Psychology: Learning, Memory, and Cognition, 24,* 771–792.

Mussweiler, T., & Strack, F. (1999). Hypothesis-consistent testing and semantic priming in the anchoring paradigm: A selective accessibility model. *Journal of Experimental Social Psychology, 35,* 136–164.

Russo, J. E., Meloy, M. G., & Medvec, V. H. (1998). Predecisional distortion of product information. *Journal of Marketing Research, 35,* 438–452.

Russo, J. E., & Schoemaker, P. J. H. (1992). Managing overconfidence. *Sloan Management Review, 33,* 7–17.

Selvidge, J. E. (1980). Assessing the extremes of probability distributions by the fractile method. *Decision Sciences, 11,* 493–502.

Slowiaczek, L. M., Klayman, J., Sherman, S. J., & Skov, R. B. (1992). Information selection and use in hypothesis testing: What is a good question, and what is a good answer? *Memory & Cognition, 20,* 392–405.

Sniezek, J. A., Paese, P. W., & Switzer, F. S. C. (1990). The effects of choosing on confidence in choice. *Organizational Behavior and Human Decision Processes, 46,* 264–282.

Soll, J. B. (1996). Determinants of overconfidence and miscalibration: The roles of random error and ecological structure. *Organizational Behavior and Human Decision Processes, 65,* 117–137.

Soll, J. B., & Klayman, J. (2004). Overconfidence in interval estimates. *Journal of Experimental Psychology: Learning, Memory, and Cognition, 30*, 299–314.

Spetzler, C. S., & Staël von Holstein, C.-A. S. (1975). Probability encoding in decision analysis. *Management Science, 22*, 340–358.

Tversky, A., & Kahneman, D. (1974). Judgment under uncertainty: Heuristics and biases. *Science, 185*, 1124–1131.

Tversky, A., & Koehler, D. J. (1994). Support theory: A nonextensional representation of subjective probability. *Psychological Review, 101*, 547–567.

von Winterfeldt D., & Edwards, W. (1986). *Decision analysis and behavioral research.* Cambridge, UK: Cambridge University Press.

Wallsten, T. S., & Budescu, D. V. (1983). Encoding subjective probabilities: A psychological and psychometric review. *Management Science, 29*, 151–173.

Yaniv, I., & Foster, D. P. (1995). Graininess of judgment under uncertainty: An accuracy informativeness trade-off. *Journal of Experimental Psychology: General, 124*, 424–432.

Zimmer, A. C. (1984). A model for the interpretation of verbal predictions. *International Journal of Man–Machine Studies, 20*, 121–134.

9

Contingency Learning and Biased Group Impressions

Thorsten Meiser

Impression formation about social groups relies on the sampling and integration of group-related information from the environment and from memory. As a consequence, biased group impressions may arise from increased prevalence of certain kinds of social information in the world, from differential retrieval of social information from memory, and from an uncritical use of environmental or stored information in the judgment process of the "naïve intuitive statistician." The present chapter reframes well-known judgment biases in group impression formation, such as illusory correlations and spurious correlations, as sampling phenomena. In particular, it is argued that group-related probabilistic information is extracted quite accurately from the environment and that unwarranted inferences on the basis of correctly learned proportions and correlations can cause systematic judgment biases. Whereas traditional explanations of biased group impressions focus on distorted retrieval or simplistic representation of the co-occurrence of group membership with positive or negative behaviors, the theoretical perspective of pseudo-contingencies (Fiedler, 2000) implies that erroneous impressions may reflect inferences from true frequency relations and from true covariations in complex social scenarios. The present chapter reviews empirical evidence for the pseudo-contingency account from previous studies on biased group impression formation, and it presents two new experiments that test specific predictions of the novel account. The final section addresses the question of whether pseudo-contingencies merely mirror a metacognitive deficit in inductive inference or whether they result from a rational decision strategy.

INTRODUCTION

Stereotypes about social groups can be regarded as contingencies between group membership and other attributes in the social environment (Tajfel, 1969; van Knippenberg & Dijksterhuis, 2000). Stereotype formation, as a

TABLE 9.1. *Stimulus Distribution in the Illusory Correlation Paradigm*

	Group A	Group B
Desirable behaviors	18	9
	(69%)	(69%)
Undesirable behaviors	8	4
	(31%)	(31%)

special case of contingency learning, can be based on different kinds of information, such as shared social knowledge about groups in a society, incidental exposure to group-related information in the media, or actual encounters with group members. Inasmuch as stereotypes rest on various pieces of group-related information experienced during an individual's learning history, and inasmuch as group judgments rely on the activation of such information in memory, stereotype formation is a case in point of the interplay among initial information sampling from the environment, later information sampling from long-term memory, and the integration of available information to overall judgments. Biases can occur at each of these stages, and many findings of distorted group impressions and erroneous stereotypes can be interpreted as sampling biases according to the general sampling framework introduced by Fiedler (2000).

Hitherto, theoretical accounts of biased group stereotypes have mainly concentrated on the memory and processing of co-occurrences of group membership with behavioral outcomes. A prominent example is stereotype formation in the illusory correlation paradigm (Hamilton & Gifford, 1976). In this paradigm, twice as many pieces of information are presented concerning target Group A as target Group B (see Table 9.1). Furthermore, about twice as many pieces of information contain desirable behaviors as opposed to undesirable behaviors. Importantly, the proportion of desirable behaviors is identical for the two groups (i.e., 69% in Table 9.1), corresponding to a zero correlation between group membership and desirability. Despite the factual zero correlation, perceivers tend to judge Group A more positively than Group B.

The traditional explanation of illusory correlations rests on the distinctiveness of infrequent events and on the paired distinctiveness of co-occurring infrequent events (Hamilton, 1981; Hamilton & Gifford, 1976; Stroessner & Plaks, 2001). It is assumed that the combination of Group B with undesirable behaviors is especially salient because both events are infrequent in the stimulus set. Therefore, undesirable behaviors performed by members of Group B attract more attention during encoding, and they are more easily available at the judgment stage than are the other kinds of stimuli. As a consequence, the co-occurrence of minority Group B with infrequent negative behaviors is overestimated, which results in less positive judgments of Group B than of Group A.

In terms of the sampling framework, the distinctiveness account of illusory correlations claims a biased selection, or weighting, of information from the social environment and from memory. First, it is presumed that combinations of events with a low marginal probability are processed more intensively than other events during perception and encoding. Second, in an extension of the original distinctiveness account, it was posited that paired infrequent events may be subject to more elaborate processing at a postencoding stage even if they were not perceived as distinctive during encoding (McConnell, Sherman, & Hamilton, 1994). Finally, because of the enhanced processing of paired distinctive stimuli at the encoding or postencoding stage, these events are more easily available from memory at judgment than are other events. Thus, according to the distinctiveness account, group judgments are based on a biased mental sample that overrepresents paired infrequent events because of their enhanced encoding, processing, and retrieval.

The illusory correlation effect corresponds to a general tendency to overestimate correlations in finite stimulus sequences (Kareev, 1995; Kareev, Lieberman, & Lev, 1997). Moreover, alternative explanations have been suggested in terms of associative learning principles that do not rely on differential encoding and retrieval processes (Fiedler, 1996; Sanbonmatsu, Shavitt, & Gibson, 1994; Smith, 1991). Associative learning models imply that information loss, caused by environmental uncertainty and the limited capacity of the human cognitive system, results in more accurate extraction of the preponderance of desirable behaviors from the larger sample of Group A than from the smaller sample of Group B. Following this account, the mental sample that forms the basis for group judgments is not biased per se, but the actual predominance of desirable information is discerned more reliably for the larger group than for the smaller group. This account gains support from the finding that differential judgments of the two target groups are due to a rather indeterminate evaluation of Group B together with a positive evaluation of Group A, rather than to a negative evaluation of Group B (Fiedler, 1991; Meiser & Hewstone, 2001).

In this chapter, a theoretical perspective is pursued to explain biased group impressions in terms of unwarranted inferences about correlations on the basis of largely accurate mental samples. In particular, it is argued that *pseudo-contingencies* (Fiedler, 2000) can play a pivotal role in stereotype formation. The concept of pseudo-contingency denotes the inference of a correlation between two variables from their marginal distributions or from their pairwise correlations with a third variable. In simple scenarios with only two variables, the alignment of two skewed distributions may give rise to a pseudo-contingency between the two variables even if no information is provided about their combination. Applied to the illusory correlation paradigm, the skewed marginal distributions of group membership and desirability may lead to the simple, yet fallacious conclusion that the two frequent events and the two infrequent events belong together. Thereby, associations of Group A with desirable behaviors and Group B with undesirable behaviors may be inferred on the basis of well-learned frequency information concerning both group membership and desirability. In contrast to the explanations by distinctiveness or associative learning, the pseudo-contingency account does not require information about the co-occurrence of group membership and desirability, neither in the environment nor in the mental sample, because pseudo-contingencies can be inferred from mere knowledge about marginal distributions. In line with this assumption, a substantial illusory correlation was found in an experiment in which participants were verbally informed of the fact that Group A is twice as large as Group B and then read the list of frequent desirable and infrequent undesirable behaviors without group labels (McGarty et al., 1993). Because the desirable and undesirable behaviors were not linked to the group labels in this experiment, the observed illusory correlation reflects an inference from the marginal distributions of group and desirability.

In the remainder of this chapter, the pseudo-contingency approach is advanced as an explanation of biased group impressions in more complex social environments, which contain information about group membership, behavioral outcome, and a confounding context variable. In those scenarios, pseudo-contingencies emerge if correlations of the context variable with both group membership and behavior are taken to imply a contingency between group and behavior. For example, stereotypic associations between gender and job-related behavior may be strengthened by unequal distributions of men and women across different positions (Eagly & Steffen, 1984). Imagine that men are overrepresented in manager positions and that women are overrepresented in administrative positions. Furthermore, let manager positions be associated with self-assertive behavior and administrative positions with more defensive behavior. Then the impression may arise that men are more assertive than women, even though women may actually be more assertive than men within both manager and administrative positions. The misleading impression can result from a simplistic reasoning process that fails to consider the context variable of position. However, it can also result from a pseudo-contingency that utilizes existing covariations with the context variable for gender judgments. That is, if the pairwise correlations of gender with position and position with assertiveness are taken to imply that men are more assertive than women, then a pseudo-contingency is formed.

As a precondition for pseudo-contingencies from covarying context factors, correlations of context variables with group membership and behavior

TABLE 9.2. *Trivariate Stimulus Distribution Used in Previous Research on Biased Group Impression Formation*

	Town X		Town Y	
	Group A	Group B	Group A	Group B
Desirable behaviors	16	6	0	6
	(73%)	(100%)	(0%)	(27%)
Undesirable behaviors	6	0	6	16
	(27%)	(0%)	(100%)	(73%)

have to be extracted from the environment and represented in one's social knowledge. The following sections will review previous studies that indicated that such correlations are actually learned and that accurate extraction of existing correlations with a confounding factor may lead to judgment biases reflecting pseudo-contingencies. Following the review of previous studies, two new experiments will be presented to provide further evidence that pseudo-contingencies can form a source of biased group judgments in complex stimulus environments. Because the reviewed studies and new experiments followed a similar procedure and employed a specific model of statistical analysis to measure sensitivity to pairwise correlations in complex stimulus designs, an overview of the experimental procedure and analysis is given first.

OVERVIEW OF THE PROCEDURE AND ANALYSIS

Table 9.2 displays a typical trivariate stimulus distribution used in previous experiments on group impression formation (Meiser, 2003; Meiser & Hewstone, 2004). As is obvious from the table, there are two strong bivariate correlations, one between town of residence and group membership, and the other between town of residence and desirability. That is, there are more members of Group A than of Group B in Town X, whereas there are more members of Group B than of Group A in Town Y. Likewise, there are more desirable than undesirable behaviors in Town X, whereas there are more undesirable than desirable behaviors in Town Y. Numerically, the correlations of town of residence with group membership and desirability amount to $\varphi = .57$. Two important research questions are, therefore, whether these correlations are extracted from the trivariate stimulus set and whether they are taken into account when forming an impression about Groups A and B. According to the pseudo-contingency account pursued herein, the pairwise correlations of town of residence with group membership and desirability should be learned, and they should constitute the basis for inferred associations of Group A with desirability and of Group B with undesirability.

The true relation between group membership and desirability is rather ambiguous in Table 9.2. Within each town, the proportion of desirable behaviors is larger for Group B than for Group A. In fact, there is a correlation of $\varphi = -.27$ between group membership and desirability in each of the subtables for Town X and Town Y. If the table is collapsed across the context factor of town, however, a larger proportion of desirable behaviors results for Group A (i.e., 57%) than for Group B (i.e., 43%), yielding a positive correlation of $\varphi = .14$ between group membership and desirability. Thus, group judgments favoring Group B seem warranted on the basis of the subtable information conditional on town of residence, whereas group judgments favoring Group A seem warranted if the factor of town of residence is ignored.

In the experiments summarized in the following, it is shown that participants judge Group A more favorably than Group B after encountering the kind of trivariate information displayed in Table 9.2. Moreover, the experiments tested whether the observed preference of Group A is due to a neglect of the factor of town of residence or to a pseudo-contingency that makes use of well-learned bivariate contingencies of town of residence with group membership and desirability.

General Procedure

Presentation Phase. The experiments began with a presentation phase, in which a series of stimulus sentences were displayed on a computer screen. Each stimulus sentence contained the first name of a fictitious male target person, information about his membership in either Group A or Group B, his residence in either Town X or Town Y, and a behavior description that was either moderately desirable or moderately undesirable. Examples of the stimulus sentences are "Michael, a member of Group B living in Town X, helps colleagues to solve software problems" or "Thomas, a member of Group A living in Town Y, spreads rumours about colleagues." The stimulus sentences were presented in random order. Participants were instructed to read each of the sentences carefully.

Dependent Variables. Following the presentation phase, different tasks were administered to assess group impressions and perceived correlations. Herein, we focus on trait ratings, which indicate general likeability, and on assignments of desirable and undesirable behaviors to the target groups and towns, which reveal episodic memory and perceived correlations.

For the trait ratings, five positive trait adjectives (e.g., good-natured or sociable) and five negative trait adjectives (e.g., unpopular or irresponsible) were presented in a random order. For each trait, participants had to judge the degree to which the trait applies to the four combinations of target Groups A and B with Towns X and Y. Judgments were made on rating scales ranging from 0 ("Does not apply at all") to 9 ("Applies completely"). In the statistical analysis, responses to negative traits were recoded so that higher ratings reflect more positive evaluations. The ratings were then averaged across the ten traits.

In the assignment task, desirable and undesirable target behaviors from the presentation phase and desirable and undesirable distractor behaviors, which had not been presented before, were displayed as test items in random sequence. For each test item, participants had to decide in a first step whether the item had occurred in the presentation phase (i.e., response "old") or not (i.e., response "new"). If the response was "new," the next test item was presented. If the response was "old," source attributions were required with respect to the groups and towns. That is, participants had to decide in a second step whether an "old" item referred to Town X or Town Y and, in a third step, whether it referred to Group A or Group B. The assignment data were analyzed with a multinomial model of source monitoring.

Source Monitoring Analysis

Source monitoring denotes the cognitive processes that are involved in attributions of memories to their origin (Johnson, Hashtroudi, & Lindsay, 1993). These processes comprise actual recollection of context information from the encoding episode as well as reconstructive guessing. Multinomial measurement models of source monitoring (Batchelder & Riefer, 1990; Bayen, Murnane, & Erdfelder, 1996) are suitable means to decompose the latent memory and guessing processes that are intermingled in observed source attributions. In particular, multinomial models disentangle recognition memory for the test items, source memory for the source of target items, a guessing tendency to judge that items are old if item memory fails, and guessing tendencies for attributing items to a specific source if source memory fails.

In the assignment task for the trivariate stimulus distribution in Table 9.2, there were two binary source dimensions for desirable and undesirable test items. That is, items judged "old" had to be assigned to a source category on the dimension of Town X versus Town Y and on the dimension of Group A versus Group B. Therefore, an extended multinomial source monitoring model was used that allows the simultaneous measurement of source memory and guessing for crossed dimensions of source information. The model is a special case of a family of multinomial memory models that was formally specified and empirically validated in previous research (Meiser, 2005; Meiser & Bröder, 2002).

Multinomial Memory Model for Crossed-Source Information. Figure 9.1 illustrates the multinomial memory model for crossed-source information for the dimensions of town of residence and group membership. The model

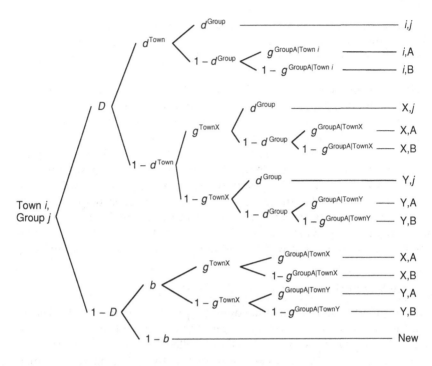

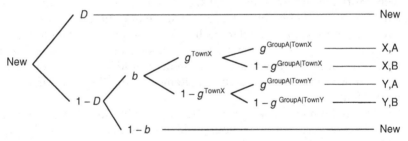

FIGURE 9.1. Processing tree representation of the multinomial memory model for crossed-source information. Test items labeled "Town i, Group j" denote target items from Group j in Town i, $i \in \{X,Y\}$ and $j \in \{A,B\}$. Test items labeled "New" denote distractor items. D, probability of recognizing target items as old and distractor items as new; b, probability of guessing that an unrecognized item is old; d^{Town}, probability of recollecting the town origin of recognized target items; d^{Group}, probability of recollecting the group origin of recognized target items; g^{TownX}, probability of guessing that an item originated in Town X; $g^{GroupA|TownX}$, probability of guessing that an item originated in Group A given assignment to Town X; $g^{GroupA|TownY}$, probability of guessing that an item originated in Group A given assignment to Town Y. Adapted from Meiser and Bröder (2002), copyright © 2002 by the American Psychological Association, with permission.

specifies the probabilities of various cognitive states that jointly lead to the selection of a response for target items from Town i and Group j, with $i \in \{X,Y\}$ and $j \in \{A,B\}$, and for new distractor items. First, the parameters D and b concern the processes involved in "old"/"new" recognition judgments. The parameter D reflects the probability of recognizing target items as old and distractor items as new, and b denotes the probability of guessing "old" if recognition memory fails. As D captures both the recognition of target items and the identification of distractors, the model is based on a two-high threshold assumption of "old"/"new" discrimination (Macmillan & Creelman, 1991). Second, the parameters d^{Town} and g^{TownX} specify the processes underlying source attributions with respect to town of residence. The source memory parameter d^{Town} reflects the probability of recollecting the actual town context of recognized target items. Provided that town recollection fails, items are attributed to Town X with guessing probability g^{TownX} and to Town Y with complementary probability $1 - g^{\text{TownX}}$. Third, the parameters d^{Group}, $g^{\text{GroupA}|\text{TownX}}$, and $g^{\text{GroupA}|\text{TownY}}$ refer to source attributions with respect to group membership. The source memory parameter d^{Group} specifies the probability of recollecting the actual group origin of recognized target items. The parameters $g^{\text{GroupA}|\text{TownX}}$ and $g^{\text{GroupA}|\text{TownY}}$ represent the guessing probabilities to attribute items to Group A given assignment to Town X and Town Y, respectively. Accordingly, items are attributed to Group B with complementary guessing probabilities $1 - g^{\text{GroupA}|\text{TownX}}$ and $1 - g^{\text{GroupA}|\text{TownY}}$. Source assignments of nonrecognized target items and of distractors that elicit false alarms are solely guided by the guessing processes g^{TownX}, $g^{\text{GroupA}|\text{TownX}}$, and $g^{\text{GroupA}|\text{TownY}}$.

Assessment of Contingencies in Source Attributions. Apart from providing process-pure measures of item recognition and source memory, multinomial source monitoring models yield unobtrusive indicators of perceived, expected, or inferred contingencies. Guessing processes in source attributions reflect schematic knowledge about correlations between item contents and source categories, which is shown by overproportional assignments of items to schematically associated source categories (Bayen et al., 2000). Analogously, social stereotypes, as special cases of schematic knowledge, guide source assignments under memory uncertainty (Klauer, Wegener, & Ehrenberg, 2002; Mather, Johnson, & DeLeonardis, 1999). Biased group judgments in the illusory correlation paradigm, for example, are reflected by overproportional assignments of desirable behaviors to the favored Group A and of undesirable behaviors to the less favored Group B, as qualified by a significant difference in the guessing parameter g for desirable versus undesirable test items (Klauer & Meiser, 2000; Meiser, 2003; Meiser & Hewstone, 2001). Applied to the multinomial model in Figure 9.1, a comparison of the guessing parameter g^{TownX} between desirable and undesirable test items can reveal perceived contingencies between town of residence and desirability. Likewise, comparisons of the guessing parameters $g^{\text{GroupA}|\text{TownX}}$ and $g^{\text{GroupA}|\text{TownY}}$ between desirable and undesirable test items may indicate subjective contingencies between group membership and desirability.

In addition to contingencies between item contents and source categories, the multinomial memory model for crossed-source information in Figure 9.1 allows one to test for perceived contingencies between different source dimensions. Because the guessing process concerning source attributions to Group A versus B is specified conditional on the source attribution to Town X or Town Y, the parameters $g^{\text{GroupA}|\text{TownX}}$ and $g^{\text{GroupA}|\text{TownY}}$ may reveal dependencies in the selection of source categories on the two dimensions. For instance, $g^{\text{GroupA}|\text{TownX}} > g^{\text{GroupA}|\text{TownY}}$ indicates that an item is more likely to be attributed to Group A if Town X is chosen than if Town Y is chosen. More generally, $g^{\text{GroupA}|\text{TownX}} \neq g^{\text{GroupA}|\text{TownY}}$ reflects a perceived contingency between the crossed-source dimensions of group membership and town of residence.

To recapitulate, the multinomial source memory model for crossed-source dimensions in Figure 9.1 allows one to test for perceived contingencies between

1. town of residence and desirability in terms of a difference in g^{TownX} for desirable versus undesirable test items,
2. group membership and desirability in terms of differences in $g^{\text{GroupA}|\text{TownX}}$ and $g^{\text{GroupA}|\text{TownY}}$ for desirable versus undesirable test items, and
3. town of residence and group membership in terms of $g^{\text{GroupA}|\text{TownX}} \neq g^{\text{GroupA}|\text{TownY}}$.

The analysis of guessing parameters thereby provides a rather complete picture of the contingencies that are extracted or inferred by the "naïve intuitive statistician." With respect to the stimulus distribution in Table 9.2, the multinomial analysis sheds light on our naïve statistician's sensitivity to the actual correlations of town of residence with desirability and group membership, and on the accompanying subjective contingency between group membership and desirability.

BIASED GROUP IMPRESSIONS FROM TRIVARIATE SAMPLES

The stimulus distribution in Table 9.2 has been employed in several previous experiments (Meiser, 2003; Meiser & Hewstone, 2004). Figure 9.2 displays a typical observed pattern of mean trait ratings, with more positive ratings for Group A than Group B and more positive ratings for Town X than Town Y.

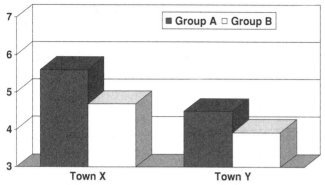

FIGURE 9.2. Mean trait ratings in an experiment on memory-based group impression formation with the stimulus distribution shown in Table 9.2 (Meiser & Hewstone, 2004, Study 1). Higher scores indicate more positive ratings.

Simplistic Reasoning or Pseudo-Contingency

More positive judgments of Group A than of Group B on the basis of stimulus distributions like the one in Table 9.2 were originally explained by a simplistic reasoning process in which the confounding third factor is neglected (Schaller, 1994; Schaller & O'Brien, 1992). According to this account, the differential impressions of the two target groups can be regarded as an immediate result of the positive correlation in the marginal table of group membership and desirability collapsed across town of residence. Stated in the sampling framework, the simplistic reasoning account assumes that group judgments are based on a mental sample of reduced dimensionality, that is, on a two-way representation of group membership and behavioral outcome aggregated across town. As a consequence, the actual negative contingency between group membership and desirability within each subtable of the trivariate distribution has to go unnoticed in judgment formation.

For several reasons, it appears unlikely that the simplistic reasoning account provides a viable explanation for the group stereotype observed in the present experiments. First, the trait ratings explicitly required ratings for the four combinations of Group A and Group B with Town X and Town Y. Therefore, town of residence can hardly be ignored at the judgment stage. Second, the significant main effect of town of residence on the trait ratings in Figure 9.2 shows that participants were not blind to the moderating role of town of residence with respect to desirability. Third, the guessing parameters in the multinomial source monitoring model, which will be discussed next, corroborate that participants were rather sensitive to the actual correlations of town of residence with both desirability and group membership. These observations are in line with research on causal learning, which brought pervasive evidence that third factors can be taken into account in causal judgments and that trivariate stimulus information is not generally reduced to two-way marginals (Fiedler et al., 2002, 2003; Spellman, Price, & Logan, 2001).

Figure 9.3 illustrates the guessing parameters for source attributions to Town X versus Town Y (upper panel) and the conditional guessing parameters for source attributions to Group A versus Group B given assignment to Town X or Town Y (lower panel). Separate parameters were estimated for source attributions of desirable and undesirable behaviors. As illustrated in the upper panel of Figure 9.3, desirable behaviors were more likely to be attributed to Town X than were undesirable behaviors. More precisely, desirable behaviors were assigned with higher probability to Town X than Town Y, whereas undesirable behaviors were assigned with higher probability to Town Y than Town X. Thus, the guessing parameter for town attributions, g^{TownX}, mirrors the actual correlation between town of residence and desirability in the stimulus set. Furthermore, the lower panel of Figure 9.3 shows that desirable and undesirable behaviors were more likely to be attributed to Group A following assignment to Town X than following assignment to Town Y. That is, the inequality $g^{\text{GroupA}|\text{TownX}} > g^{\text{GroupA}|\text{TownY}}$ held consistently across desirable and undesirable test items, reflecting the actual correlation between town of residence and group membership.

Together, the guessing parameters in Figure 9.3 indicate that the correlations of town of residence with desirability and group membership were extracted from the trivariate stimulus distribution and that they were used for source attributions in the case of memory uncertainty. At the same time, the guessing parameters $g^{\text{GroupA}|\text{TownX}}$ and $g^{\text{GroupA}|\text{TownY}}$ also reflected group judgments in favor of Group A. As can be seen in the lower panel of Figure 9.3, desirable behaviors were more likely to be attributed to Group A than were undesirable behaviors, and this difference proved significant irrespective of assignment to Town X or Town Y. Complementarily, undesirable behaviors were more likely to be assigned to Group B than desirable behaviors irrespective of town assignment. Hence, aside from indicating sensitivity to the context factor, the guessing parameters reflect a group stereotype in favor of Group A that resembles the correlation in the two-way table collapsed across the context factor.

The findings of biased group stereotype formation together with sensitivity to the covariations of the confounding context factor do not fit into the simplistic reasoning account. The results are readily accommodated, however, by an explanation in terms of pseudo-contingencies. According to the pseudo-contingency account, extraction of the correlations of town of residence with both group membership and behavioral outcome is the prerequisite for forming a stereotype in favor of Group A. In particular, it is assumed that participants represent the apparent correlations of town of residence in their mental sample, rather than reducing the trivariate stimulus

Guessing Parameters for Town Attributions

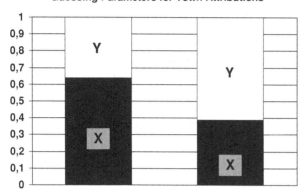

Desirable Behaviors   Undesirable Behaviors

**Guessing Parameters for Group Attributions
Conditional on Town Assignments**

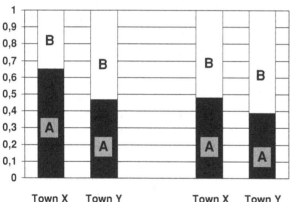

Town X   Town Y   Town X   Town Y
Desirable Behaviors   Undesirable Behaviors

FIGURE 9.3. Guessing parameters of source attributions in the source monitoring model for the dimensions of town of residence and group membership (Meiser & Hewstone, 2004, Study 1). In the upper panel, black bars show the guessing probabilities of attributing desirable and undesirable behaviors to Town X (i.e., g^{TownX}). White bars indicate the complementary guessing probabilities of attributing items to Town Y (i.e., $1 - g^{TownX}$). In the lower panel, black bars show the guessing probabilities of attributing desirable and undesirable behaviors to Group A given assignment to Town X (i.e., $g^{GroupA|TownX}$) or Town Y (i.e., $g^{GroupA|TownY}$). White bars indicate the complementary guessing probabilities of attributing items to Group B (i.e., $1 - g^{GroupA|TownX}$ and $1 - g^{GroupA|TownY}$).

information to a two-way mental sample. On the basis of the extracted correlations of town of residence with group membership and desirability, participants may then draw an invalid conclusion about the correlation between group and desirability, because this correlation is less clearly discernible from the stimulus information. Biased stereotype formation is, thus, explained by an unwarranted inference from a rather accurate mental sample. In other words, pseudo-contingencies do not originate in an overly simplistic mental representation of multidimensional information but in a metacognitive deficit to evaluate the validity of inferences from well-learned correlations.

Analysis of Interindividual Differences

Apart from the observed sensitivity to the covariations with the confounding factor in the average trait ratings and guessing parameters, an analysis of interindividual differences may reveal whether the simplistic reasoning account or the pseudo-contingency account provides a better explanation of the biased group judgments. Following the simplistic reasoning account, participants who do not pay attention to the covariations with the confounding factor should be highly susceptible to group judgments in favor of Group A, because they are likely to misrepresent the trivariate stimulus information in terms of an aggregate two-way table (Schaller et al., 1995). Participants who are well aware of the role of the confounding factor, however, should use a more complex reasoning strategy and, thus, be less prone to biased group judgments. Applied to the trait ratings in Figure 9.2, the simplistic reasoning account implies a negative correlation between the evaluative differentiation of Town X versus Town Y, which reflects the true correlation of town of residence with desirability, and the stereotype in favor of Group A. In contrast, a positive correlation between more favorable ratings of Town X than Y and more favorable ratings of Group A than B follows from the pseudo-contingency account. In this account, the stereotype in favor of Group A is interpreted as an inference on the basis of learned covariations with town of residence. Therefore, participants are expected to make the fallacious inference of a pseudo-contingency concerning group membership and desirability only to the extent to which they have acquired sufficient knowledge about the pairwise correlations of town of residence with desirability and group membership.

To test the competing predictions, the empirical correlation between the effect of town of residence (i.e., ratings of Town X minus ratings of Town Y averaged across target group) and the effect of target group (i.e., ratings of Group A minus ratings of Group B averaged across town) was computed for the trait rating data in Figure 9.2. The correlation was positive with $r = .35$, $p = .029$, supporting the pseudo-contingency account.

A similar result was obtained in a quasi-experiment in which participants were split into two groups according to their trait ratings for the two target groups (Meiser & Hewstone, 2004, Study 3). One quasi-experimental condition contained those participants who gave more favorable ratings to Group A than to B, and the other condition contained participants who gave either equal ratings to the two groups or more positive ratings to Group B. After the sample of participants was split, the multinomial source monitoring model in Figure 9.1 was applied to the assignment data of each quasi-experimental condition. The guessing parameters for source attributions to the towns and groups confirmed that biased stereotype formation was not related to a neglect of the moderating role of town of residence in the trivariate stimulus distribution. On the contrary, the crucial parameter differences in Figure 9.3 that revealed sensitivity to the correlations of town of residence with desirability and group membership were replicated in both quasi-experimental conditions. Contradicting the simplistic reasoning account, but supporting the pseudo-contingency account, some of the differences were even more pronounced for those participants who gave group ratings in favor of Group A.

TESTING NEW PREDICTIONS OF THE PSEUDO-CONTINGENCY ACCOUNT

Two new experiments addressed specific predictions of the pseudo-contingency account that provide critical tests of the competing explanations for biased group stereotypes from trivariate stimulus distributions. First, the conclusiveness of the empirical evidence for the simplistic reasoning account may be limited owing to a sampling bias in experimental research. That is, inasmuch as all experiments conducted in the simplistic reasoning tradition used stimulus distributions that entailed an actual spurious correlation in the aggregate two-way table of group membership and behavioral outcome (e.g., Schaller, 1994; Schaller & O'Brien, 1992), it has not been shown that reliance on this correlation is the causal agent of biased group judgments. Therefore, the first experiment employed a trivariate stimulus distribution that rendered no correlation whatsoever in the aggregate table of group membership and desirability. At the same time, the stimuli contained strong bivariate correlations of group membership and desirability with town of residence. Thus, if participants use a simplistic reasoning strategy and ignore the confounding factor, an undifferentiated judgment of the two groups should result. If, however, participants learn the two bivariate correlations and form a pseudo-contingency concerning group membership and desirability on that basis, a biased stereotype should occur.

Second, pseudo-contingencies do not require complete trivariate stimulus information. Instead, pseudo-contingencies between group

TABLE 9.3. *Trivariate Stimulus Distribution in the First Experiment*

	Town X		Town Y	
	Group A	Group B	Group A	Group B
Desirable behaviors	16	8	0	8
	(67%)	(100%)	(0%)	(33%)
Undesirable behaviors	8	0	8	16
	(33%)	(0%)	(100%)	(67%)

membership and desirability can also be formed on the basis of two separate kinds of bivariate information, one concerning the confounding factor and group membership and the other concerning the confounding factor and desirability. Following this rationale, participants in the second experiment were presented with stimulus sentences that provided either information about group membership and town of residence or about town of residence and desirability, but not about the co-occurrence of group membership and desirability. It was predicted that a pseudo-contingency between group membership and desirability is inferred from this incomplete stimulus information.

Group Judgments on the Basis of a New Stimulus Distribution

Table 9.3 shows the new trivariate stimulus distribution used in the first experiment. As in Table 9.2, there are considerable bivariate correlations of town of residence with group membership and desirability, with both $\varphi = .50$. Moreover, the proportion of desirable behaviors is larger for Group B than Group A within each of the two towns, yielding a negative correlation of $\varphi = -.33$ between group membership and desirability in the subtables of Town X and Town Y. Unlike Table 9.2, however, collapsing Table 9.3 across town of residence results in a uniform two-way distribution of group membership and desirability with sixteen stimuli in each cell. Because the uniform two-way table shows a correlation of $\varphi = .00$, it does not provide a basis for differential judgments of the two target groups. As a consequence, the account of stereotype formation by simplistic reasoning predicts no group stereotype in favor of Group A. The pseudo-contingency account, in contrast, predicts that the apparent covariations of town of residence with group membership and desirability are discerned and that these covariations form the basis for the fallacious inference that Group A is more likeable than Group B. The analysis of group impressions emerging from the presentation of Table 9.3 therefore provides a critical test of the two theoretical accounts.

As argued in the foregoing, pseudo-contingencies presumably result from a metacognitive deficit to monitor and control the use of probabilistic information in judgment formation (see Fiedler, 2000). Accordingly, the first experiment aimed at showing the role of metacognitive operations in stereotype formation by manipulating the availability of metacognitive cues that guide information integration. For this purpose, different instructions were used that induced either memory-based impression formation on the basis of stored group exemplars or online impression formation during the encoding of the stimuli (Hastie & Park, 1986). In the memory condition ($n = 36$), the initial instructions explained that the experiment investigates information processing and memory for visually presented information concerning people and their behaviors, and participants were asked to read each of the stimulus sentences carefully. In the online condition ($n = 37$), the instructions explained that the experiment investigates impression formation about different groups, and participants were asked to form thorough impressions of Group A and Group B while reading the sentences.

Recent research showed that relative frequencies and contingencies are extracted more accurately after online impression formation instructions than after memory instructions (Meiser, 2003). For the present experiment, it was therefore expected that the covariations of town of residence with group membership and desirability are discerned especially well in the condition of online impression formation, setting the stage for pseudo-contingencies. To show that participants can nevertheless avoid the inference of a fallacious pseudo-contingency if they are equipped with appropriate metacognitive guidance, the online instruction was enriched by cues that should channel the process of impression formation and, thus, help to overcome the metacognitive deficit in inductive inference. That is, the instruction contained a priori information that the behaviors can differ between Town X and Town Y, which should disambiguate the causal status of the confounding factor and encourage the use of subtable information conditional on town of residence (Fiedler et al., 2002; Waldmann & Hagmayer, 2001). Moreover, participants in the online condition were explicitly asked to take notice of the town information in the stimulus sentences and to form impressions of the target Groups A and B within Town X and within Town Y. Thereby, participants were guided to base their judgments on the subtable information, rather than on overall contingencies in the trivariate stimulus set. Similar instructions had been found to reduce susceptibility to biased judgments in previous studies (Schaller & O'Brien, 1992, Experiment 2). Taken together, the online instruction stressed awareness of the town factor and provided metacognitve guidance to conditionalize group judgments on town of residence.

To recapitulate, the experiment tested three predictions. First, a biased group impression in favor of Group A was expected in the condition of memory-based impression formation, indicating a pseudo-contingency.

TABLE 9.4. *Mean Trait Ratings in the First Experiment*

	Town X				Town Y			
	Group A		Group B		Group A		Group B	
	M	*SD*	*M*	*SD*	*M*	*SD*	*M*	*SD*
Memory-based	5.85	1.21	4.78	1.33	4.43	0.98	4.14	1.29
Online	6.07	1.40	5.78	1.29	3.77	1.10	3.42	1.42

Note: The rating scale ranged from 0 to 9. Higher scores indicate more positive ratings.

Second, the pairwise correlations of town of residence with group membership and desirability should be extracted better in the online condition than in the memory condition, replicating earlier results (Meiser, 2003) and reflecting the a priori information given to the participants. Third, the biased group stereotype should be eliminated in the online condition because of the metacognitive cues.

Evidence of a Pseudo-Contingency from the Trait Ratings. As shown in Table 9.4, Group A received more positive trait ratings than Group B, with $F(1, 71) = 12.61$, $p = .001$, which reflects the predicted pseudo-contingency. Furthermore, Town X received more positive ratings than Town Y, with $F(1, 71) = 71.82$, $p < .001$, which matches the existing correlation of town of residence with desirability. There was no main effect of processing condition, but the evaluative differentiation between Town X and Town Y was stronger in the online condition ($M = 5.92$ versus $M = 3.59$) than in the memory condition ($M = 5.31$ versus $M = 4.28$). This effect was substantiated by a significant interaction of processing condition with town, $F(1, 71) = 10.72$, $p = .002$, and it indicates increased sensitivity to actual contingencies in the online condition. The interaction between processing condition and target group fell short of significance, with $F(1, 71) = 1.59$, $p = .211$. Nonetheless, simple effect analyses within each processing condition showed that the judgment bias in favor of Group A was significant in the memory condition, $F(1, 35) = 14.97$, $p < .001$, but not in the online condition, $F(1, 36) = 2.16$, $p = .150$.

Taken together, the trait ratings lent empirical support to the predictions that were tested in the first experiment. A substantial stereotype in favor of Group A was obtained in the memory condition, online impression formation strengthened awareness of the correlation between town of residence and desirability, and the biased group stereotype did not emerge in the online condition that guided the process of impression formation by explicit cues. As delineated in the foregoing, the observed group stereotype in the memory condition cannot be accommodated by the simplistic reasoning account, because ignoring town of residence in Table 9.3 would

TABLE 9.5. *Parameter Estimates and Asymptotic 95% Confidence Intervals (CI) in the Multinomial Memory Model for Crossed-Source Information in the First Experiment*

	Desirable Behaviors		Undesirable Behaviors	
	Estimate	**CI**	**Estimate**	**CI**
Memory-based				
D	$.47_a$	(.42, .52)	$.55_b{}^*$	(.50, .60)
d^{Town}	$.05_a$	(.00, .21)	$.10_a$	(.00, .24)
d^{Group}	$.12_a$	(.00, .25)	$.00_a$	(.00, .12)
g^{TownX}	$.63_a{}^*$	(.58, .67)	$.39_b$	(.34, .43)
$g^{GroupA\,\vert\,TownX}$	$.75_a{}^*$	(.70, .81)	$.42_b{}^*$	(.35, .49)
$g^{GroupA\,\vert\,TownY}$	$.37_a$	(.30, .45)	$.45_a$	(.40, .51)
b	$.39_a$	(.34, .43)	$.37_a$	(.32, .43)
Online				
D	$.49_a$	(.45, .54)	$.64_b{}^*$	(.59, .68)
d^{Town}	$.18_a$	(.04, .32)	$.18_a$	(.05, .30)
d^{Group}	$.13_a$	(.00, .25)	$.10_a$	(.00, .21)
g^{TownX}	$.70_a{}^*$	(.66, .75)	$.33_b$	(.29, .38)
$g^{GroupA\,\vert\,TownX}$	$.64_a{}^*$	(.58, .69)	$.57_a{}^*$	(.49, .65)
$g^{GroupA\,\vert\,TownY}$	$.42_a$	(.33, .51)	$.45_a$	(.40, .51)
b	$.33_a$	(.28, .37)	$.37_a$	(.31, .43)

Note: The multinomial model showed a good overall fit to the assignment data of the first experiment, $G^2(36) = 28.90$, $p = .794$. Different subscripts within a row indicate significant differences in model parameters between desirable and undesirable behaviors, $p < .05$. Superscript asterisks indicate significant differences in model parameters between the memory condition and the online condition, $p < .05$. Confidence intervals with a lower limit of .00 were bounded by the parameter space.

not result in more positive judgments of Group A than Group B. The pseudo-contingency account, in contrast, predicts the group stereotype on the basis of the two correlations of town of residence with group membership and desirability.

Evidence of a Pseudo-Contingency from Behavior Assignments. The data from the assignment task were analyzed with the source monitoring model in Figure 9.1. Separate parameter sets were estimated for the assignment of desirable and undesirable behaviors in each processing condition. The parameter estimates are displayed in Table 9.5. The table also shows the results of parameter comparisons between responses to desirable and undesirable behaviors (subscripts) and between the two conditions (superscripts).

The memory parameters of item recognition, D, source memory for town, d^{Town}, and source memory for group, d^{Group}, replicated earlier findings. Item memory was better for undesirable behaviors than for desirable behaviors (see Klauer & Meiser, 2000; Meiser & Hewstone, 2001), and memory was somewhat more accurate in the online condition than in the memory condition (see Meiser, 2003). Of greater importance in the present context, the guessing parameters g^{TownX}, $g^{\text{GroupA|TownX}}$, and $g^{\text{GroupA|TownY}}$ indicated perceived correlations of town of residence with both desirability and group membership.

Replicating the pattern in Figure 9.3, the guessing parameter for attributing behaviors to Town X, g^{TownX}, in Table 9.5 shows that desirable behaviors were more likely to be attributed to Town X than to Town Y (i.e., $g^{\text{TownX}} >$ .5), whereas undesirable behaviors were more likely to be attributed to Town Y than to Town X (i.e., $g^{\text{TownX}} < .5$). The difference in g^{TownX} for desirable and undesirable behaviors was significant in the memory condition and in the online condition, indicating that knowledge about the correlation between town of residence and desirability was acquired in both conditions. Nonetheless, the tendency to attribute desirable behaviors to Town X was significantly stronger in the online condition, which suggests that sensitivity to the correlation was higher under online processing than under memory processing.

Likewise, the guessing parameters for attributing behaviors to Group A given assignment to either Town X or Town Y, $g^{\text{GroupA|TownX}}$ and $g^{\text{GroupA|TownY}}$, in Table 9.5 indicate that the actual correlation between group membership and town of residence was discerned. For desirable behaviors, the inequality $g^{\text{GroupA|TownX}} > g^{\text{GroupA|TownY}}$ was significant in the memory-based condition and in the online condition, with $\Delta G^2(1) = 63.16$ and $\Delta G^2(1) = 16.88$ (both $p < .001$). For undesirable behaviors, the difference was significant in the online condition [$\Delta G^2(1) = 5.55$, $p = .019$] but not in the memory condition [$\Delta G^2(1) < 1$]. This finding provides further support for more accurate extraction of contingency information in the online condition compared with the memory condition.

Finally, comparisons of $g^{\text{GroupA|TownX}}$ and $g^{\text{GroupA|TownY}}$ between desirable and undesirable behaviors yielded an indication of a biased group judgment in favor of Group A in the memory condition, but not in the online condition. In the memory condition, $g^{\text{GroupA|TownX}}$ was significantly larger for desirable behaviors than for undesirable ones. More specifically, desirable items assigned to Town X were more likely to be attributed to Group A (i.e., $g^{\text{GroupA|TownX}} > .5$), whereas undesirable items assigned to Town X were more likely to be attributed to Group B (i.e., $g^{\text{GroupA|TownX}} < .5$). In the online condition, neither $g^{\text{GroupA|TownX}}$ nor $g^{\text{GroupA|TownY}}$ varied as a function of desirability. Moreover, for items assigned to Town X, the tendencies to attribute desirable behaviors to Group A and undesirable behaviors to Group B were significantly stronger in the memory condition than in the online condition. We may thus conclude that evidence of a biased group stereotype favoring Group A was found in the case of memory-based impression formation and that this effect was significantly reduced, and in fact eliminated, in the online condition.

In summary, the guessing processes in behavior assignments corroborated the predictions that were derived from the pseudo-contingency account. A stereotype in favor of Group A was expressed by the behavior assignments in the memory condition, but not in the online condition that contained metacognitive cues to use conditional subtable information for group judgments. Moreover, behavior assignments in the online condition provided a better reflection of the actual contingencies of town of residence with group membership and desirability than did the behavior assignments in the memory condition.

Group Judgments on the Basis of Incomplete Trivariate Information

Building on the results of the first experiment, the second experiment aimed at a more direct test of pseudo-contingencies in biased group stereotype formation. The experiment tested the specific prediction that biased stereotypes arise on the basis of pairwise correlations even if group membership and behavioral outcome do not co-occur in the stimuli. For this purpose, an incomplete trivariate stimulus distribution was created by deleting the information about group membership from half of the stimulus sentences in each cell of Table 9.3 and by deleting the behavior description from the other half. As a consequence, each stimulus carried only bivariate information either about town of residence and behavioral outcome or about town of residence and group membership. The stimulus distribution is displayed in Table 9.6. Sentences about town of residence and group membership and sentences about town of residence and behavioral outcome were presented in random order. All participants ($N = 37$) received memory instructions before the presentation phase.

The pairwise correlations of town of residence with group membership and desirability in Table 9.6 equal the correlations of $\varphi = .50$ from the first experiment. However, Table 9.6 does not provide any information about the contingency between group membership and desirability, neither conditional on town of residence nor in an aggregate two-way table. According to the pseudo-contingency account, it was predicted that social perceivers draw an inference concerning the missing relation between group membership and desirability on the basis of the their pairwise correlations with

TABLE 9.6. *Stimulus Distribution in the Second Experiment*

	Town X	Town Y		Town X	Town Y
Desirable	12	4	Group A	12	4
Undesirable	4	12	Group B	4	12

TABLE 9.7. *Mean Trait Ratings in the Second Experiment*

Town X				Town Y			
Group A		Group B		Group A		Group B	
M	SD	M	SD	M	SD	M	SD
5.51	1.33	4.92	1.01	4.31	1.13	3.97	1.43

Note: The rating scale ranged from 0 to 9. Higher scores indicate more positive ratings.

town of residence. Hence, a pseudo-contingency was expected in terms of a stereotype favoring Group A.

Evidence of a Pseudo-Contingency from the Trait Ratings. The mean trait ratings are displayed in Table 9.7. As predicted, Group A received more positive ratings than Group B, with $F(1, 36) = 4.07$, $p = .051$, which reflects the expected pseudo-contingency. Moreover, Town X was judged more positively than Town Y, with $F(1, 36) = 14.73$, $p < .001$, matching the correlation between town of residence and desirability in the stimulus set. Thus, a biased group impression in favor of Group A was formed on the basis of incomplete stimulus information that provided only information about the bivariate relations of town of residence with group membership and desirability. An explanation in terms of simplistic reasoning seems precluded, because ignoring town of residence in Table 9.6 would not allow any differentiated group judgments. The pseudo-contingency account, in contrast, readily accommodates the group stereotype as a result of an unwarranted inference from learned correlations.

Evidence of a Pseudo-Contingency from Behavior Assignments. The assignment data were analyzed with the model in Figure 9.1. Deviating from the previous model applications, however, the source memory parameter for group, d^{Group}, was fixed to zero. This restriction was imposed because the stimulus sentences provided no information about the group origin of target behaviors (see Table 9.6). As a consequence, it was impossible to remember the group origin of a given behavior, so that group attributions had to be based on guessing processes only. Table 9.8 displays the parameter estimates and the results of parameter comparisons between desirable and undesirable behaviors.

Again, the memory parameters largely paralleled earlier findings. Item recognition D was better for undesirable behaviors than for desirable ones (see Klauer & Meiser, 2000; Meiser & Hewstone, 2001). Furthermore, source memory for town, d^{Town}, did not reliably differ from zero (see Meiser, 2003; Meiser & Hewstone, 2004). More interestingly, the guessing parameters provided evidence of true contingency learning and biased

TABLE 9.8. *Parameter Estimates and Asymptotic 95% Confidence Intervals (CI) in the Multinomial Memory Model for Crossed-Source Information in the Second Experiment*

	Desirable Behaviors		Undesirable Behaviors	
	Estimate	CI	Estimate	CI
D	$.54_a$	(.49, .58)	$.61_b$	(.56, .65)
d^{Town}	$.04_a$	(.00, .17)	$.07_a$	(.00, .20)
d^{Group}	.00	–	.00	–
g^{TownX}	$.70_a$	(.65, .74)	$.37_b$	(.33, .42)
$g^{GroupA \mid TownX}$	$.68_a$	(.63, .73)	$.47_b$	(.39, .54)
$g^{GroupA \mid TownY}$	$.47_a$	(.39, .56)	$.43_a$	(.38, .49)
b	$.26_a$	(.21, .30)	$.28_a$	(.22, .33)

Note: The multinomial model showed a good overall fit to the assignment data of the second experiment, $G^2(20) = 20.20$, $p = .445$. Different subscripts within a row indicate significant differences in model parameters between desirable and undesirable behaviors, $p < .05$. Confidence intervals with a lower limit of .00 were bounded by the parameter space. The parameter d^{Group} was fixed to .00 because no group membership was presented for the target behaviors in the second experiment.

stereotype formation that resembled the memory condition in the first experiment.

As revealed by the guessing parameter for source attributions to Town X, g^{TownX}, in Table 9.8, desirable items were more likely to be attributed to Town X than to Town Y (i.e., $g^{TownX} > .5$), and undesirable items were more likely to be attributed to Town Y than to Town X (i.e., $g^{TownX} < .5$). The significant difference in g^{TownX} for desirable and undesirable behaviors indicates that the correlation between town of residence and desirability had been learned.

Similarly, the guessing parameters for attributing behaviors to Group A given assignment to either Town X or Town Y, $g^{GroupA \mid TownX}$ and $g^{GroupA \mid TownY}$, reflect sensitivity to the correlation between town of residence and group membership. The inequality $g^{GroupA \mid TownX} > g^{GroupA \mid TownY}$ was significant for desirable behaviors, with $\triangle G^2(1) = 17.24$, $p < .001$. As in the memory condition of the first experiment, the difference between $g^{GroupA \mid TownX}$ and $g^{GroupA \mid TownY}$ failed to reach significance for undesirable items $[\triangle G^2(1) < 1]$.

Finally, the guessing parameter for attributing behaviors to Group A given assignment to Town X, $g^{GroupA \mid TownX}$, was larger for desirable behaviors than for undesirable ones. That is, desirable items assigned to Town X were more likely to be attributed to Group A than were undesirable behaviors. Just like in the memory condition of the first experiment, the differential guessing tendency reflects a biased group impression in favor of Group A.

In summary, in the second experiment, the trait ratings and the behavior assignments brought evidence of a biased group stereotype in favor of Group A. This stereotype emerged despite the fact that no information was given about the co-occurrence of group membership and behavioral outcome. The results support the notion that pseudo-contingencies can play an important role in biased group impression formation. That is, the correlations of town of residence with group membership and desirability were learned from the incomplete stimulus information, and these correlations constituted the basis for an unwarranted inference concerning the missing relation of group membership with desirability.

DISCUSSION

The empirical findings of the earlier studies (Meiser & Hewstone, 2004) and of the new experiments reported herein indicate that biased group impressions can result from the inference of a pseudo-contingency between group membership and behavioral outcome on the basis of learned correlations between these two variables with a context factor. The pseudo-contingency approach to biased judgments takes into account that existing correlations can be detected even in complex environments (see Fiedler et al., 2002; Spellman et al., 2001) and that correlations may be amplified, rather than concealed, by the limited capacity of the human cognitive system (see Kareev, 1995; Kareev et al., 1997). Thus, the pseudo-contingency approach integrates true contingency learning from finite samples of the environment and fallacious inferences from those contingencies resulting from a deficit of metacognitive skills in inductive reasoning.

As argued throughout this chapter, pseudo-contingencies are invalid inferences that may lead to systematically biased judgments. Nonetheless, pseudo-contingencies can be pragmatically adaptive inasmuch as they may increase the hit rate in contingency detection in many situations. As an illustration, take three random variables X_1, X_2, and X_3 with pairwise correlations r_{12}, r_{13}, and r_{23}. Then the range of r_{12} is restricted by the values of r_{13} and r_{23}. With $r_{13} = r_{23} = .50$, for example, the range of r_{12} is limited to the interval $[-.5, 1.0]$. More drastically, whenever r_{13} and r_{23} exceed .71, r_{12} is necessarily positive. In an environment in which two target variables X_1 and X_2 are strongly correlated with a third variable (i.e., $r_{13} \gg 0$ and $r_{23} \gg 0$), the probability that the two target variables also exhibit a positive correlation (i.e., $r_{12} > 0$) may therefore be well above 50%. From this reasoning, it seems that pseudo-contingencies reflect a rational inference strategy in situations in which direct information about the relation between two target variables is obscured or ambiguous.

The rationality of pseudo-contingencies is limited, however, by several conditions. First, observed correlations r_{13} and r_{23} narrow the range of r_{12}

within a given sample, but not necessarily in other samples of X_1, X_2, and X_3. Thus, if a contingency r_{12} is inferred from r_{13} and r_{23}, it may only be generalized if r_{13} and r_{23} can be regarded as representative of the entire population of trivariate samples. Applied to Tables 9.2 and 9.3, the unequal distribution of the Groups A and B across the Towns X and Y must reflect a stable feature of the environment, rather than the consequence of an arbitrarily unbalanced sampling scheme. If r_{13} and r_{23} are uncritically used to infer a relationship between X_1 and X_2 beyond the given sample, then the naïve intuitive statistician falls prey to metacognitive blindness. This caveat is especially relevant for pseudo-contingencies in group stereotype formation, because generalization of stereotypic beliefs to new group exemplars is an inherent mechanism of social stereotyping (e.g., van Knippenberg & Dijksterhuis, 2000).

Second, the total correlation between two variables X_1 and X_2 is not indicative of their conditional correlation within certain contexts even in the same sample. In Table 9.2, for instance, the overall correlation between group and desirability is opposite in sign from the conditional correlation within each of the two towns. Generalization of an inferred overall correlation r_{12} to concrete contexts may therefore result in serious misjudgments (Fiedler et al., 2003).

Taken together, the caveats concerning the representativity of observed correlations in a sample and concerning the generalization of inferred contingencies beyond the given sample or to specific contexts within a sample emphasize that rationality is not primarily a matter of the cognitive inference process, such as the inference of a pseudo-contingency. Instead, rationality is a matter of fit among the inference process, the sample to which it is applied, and the concrete problem that has to be solved.

SUMMARY

A review of previous research showed that pseudo-contingencies (Fiedler, 2000) may play an important role in biased group impression formation. Two new experiments tested specific hypotheses that were derived from the pseudo-contingency account. The first experiment used a stimulus distribution that gives rise to biased group judgments if a pseudo-contingency is inferred, but not if group impressions rely on a simplistic reasoning process that ignores confounding context factors. The second experiment presented only bivariate information concerning the relationship of group membership with a context factor and concerning the relationship of behavioral outcome with a context factor. The findings from both experiments supported the notion that pseudo-contingencies form a source of biased judgments. Although pseudo-contingencies reflect invalid inferences from a normative point of view, they may be pragmatically adaptive in many environments.

References

Batchelder, W. H., & Riefer, D. M. (1990). Multinomial processing models of source monitoring. *Psychological Review, 97*, 548–564.

Bayen, U. J., Murnane, K., & Erdfelder, E. (1996). Source discrimination, item detection, and multinomial models of source monitoring. *Journal of Experimental Psychology: Learning, Memory, and Cognition, 22*, 197–215.

Bayen, U. J., Nakamura, G. V., Dupuis, S. E., & Yang, C.-L. (2000). The use of schematic knowledge about sources in source monitoring. *Memory & Cognition, 28*, 480–500.

Eagly, A. H., & Steffen, V. J. (1984). Gender stereotypes stem from the distribution of women and men into social roles. *Journal of Personality and Social Psychology, 46*, 735–754.

Fiedler, K. (1991). The tricky nature of skewed frequency tables: An information loss account of distinctiveness-based illusory correlations. *Journal of Personality and Social Psychology, 60*, 24–36.

Fiedler, K. (1996). Explaining and simulating judgment biases as an aggregation phenomenon in probabilistic, multiple-cue environments. *Psychological Review, 103*, 193–214.

Fiedler, K. (2000). Beware of samples! A cognitive-ecological sampling approach to judgment biases. *Psychological Review, 107*, 659–676.

Fiedler, K., Walther, E., Freytag, P., & Nickel, S. (2003). Inductive reasoning and judgment interference: Experiments on Simpson's paradox. *Personality and Social Psychology Bulletin, 29*, 14–27.

Fiedler, K., Walther, E., Freytag, P., & Stryczek, E. (2002). Playing mating games in foreign cultures: A conceptual framework and an experimental paradigm for inductive trivariate inference. *Journal of Experimental Social Psychology, 38*, 14–30.

Hamilton, D. L. (1981). Illusory correlation as a basis for stereotyping. In D. L. Hamilton (Ed.), *Cognitive processes in stereotyping and inter-group behavior* (pp. 115–144). Hillsdale, NJ: Lawrence Erlbaum.

Hamilton, D. L., & Gifford, R. K. (1976). Illusory correlation in interpersonal perception: A cognitive basis of stereotypic judgments. *Journal of Experimental Social Psychology, 12*, 392–407.

Hastie, R., & Park, B. (1986). The relationship between memory and judgment depends on whether the judgment task is memory-based or on-line. *Psychological Review, 93*, 258–268.

Johnson, M. K., Hashtroudi, S., & Lindsay, D. S. (1993). Source monitoring. *Psychological Bulletin, 114*, 3–28.

Kareev, Y. (1995). Positive bias in the perception of covariation. *Psychological Review, 102*, 490–502.

Kareev, Y., Lieberman, I., & Lev, M. (1997). Through a narrow window: Sample size and the perception of correlation. *Journal of Experimental Psychology: General, 126*, 278–287.

Klauer, K. C., & Meiser, T. (2000). A source-monitoring analysis of illusory correlations. *Personality and Social Psychology Bulletin, 26*, 1074–1093.

Klauer, K. C., Wegener, I., & Ehrenberg, K. (2002). Perceiving minority members as individuals: The effects of relative group size in social categorization. *European Journal of Social Psychology, 32*, 223–245.

Macmillan, N. A., & Creelman, C. D. (1991). *Detection theory: A user's guide.* Cambridge, UK: Cambridge University Press.

Mather, M., Johnson, M. K., & DeLeonardis, D. M. (1999). Stereotype reliance in source monitoring: Age differences and neuropsychological test correlates. *Cognitive Neuropsychology, 16,* 437–458.

McConnell, A. R., Sherman, S. J., & Hamilton, D. L. (1994). Illusory correlation in the perception of groups: An extension of the distinctiveness-based account. *Journal of Personality and Social Psychology, 67,* 414–429.

McGarty, C., Haslam, S. A., Turner, J. C., & Oakes, P. J. (1993). Illusory correlation as accentuation of actual intercategory difference: Evidence for the effect with minimal stimulus information. *European Journal of Social Psychology, 23,* 391–410.

Meiser, T. (2003). Effects of processing strategy on episodic memory and contingency learning in group stereotype formation. *Social Cognition, 21,* 121–156.

Meiser, T. (2005). A hierarchy of multinomial models for multidimensional source monitoring. *Methodology, 1,* 2–17.

Meiser, T., & Bröder, A. (2002). Memory for multidimensional source information. *Journal of Experimental Psychology: Learning, Memory, and Cognition, 28,* 116–137.

Meiser, T., & Hewstone, M. (2001). Crossed categorization effects on the formation of illusory correlations. *European Journal of Social Psychology, 31,* 443–466.

Meiser, T., & Hewstone, M. (2004). Cognitive processes in stereotype formation: The role of correct contingency learning for biased group judgments. *Journal of Personality and Social Psychology, 87,* 599–614.

Sanbonmatsu, D. M., Shavitt, S., & Gibson, B. D. (1994). Salience, set size, and illusory correlation: Making moderate assumptions about extreme targets. *Journal of Personality and Social Psychology, 66,* 1020–1033.

Schaller, M. (1994). The role of statistical reasoning in the formation, preservation and prevention of group stereotypes. *British Journal of Social Psychology, 33,* 47–61.

Schaller, M., Boyd, C., Yohannes, J., & O'Brien, M. (1995). The prejudiced personality revisited: Personal need for structure and formation of erroneous group stereotypes. *Journal of Personality and Social Psychology, 68,* 544–555.

Schaller, M., & O'Brien, M. (1992). "Intuitive analysis of covariance" and group stereotype formation. *Personality and Social Psychology Bulletin, 18,* 776–785.

Smith, E. R. (1991). Illusory correlation in a simulated exemplar-based memory. *Journal of Experimental Social Psychology, 27,* 107–123.

Spellman, B. A., Price, C. M., & Logan, J. M. (2001). How two causes are different from one: The use of (un)conditional information in Simpson's Paradox. *Memory & Cognition, 29,* 193–208.

Stroessner, S. J., & Plaks, J. E. (2001). Illusory correlation and stereotype formation: Tracing the arc of research over a quarter century. In G. B. Moskowitz (Ed.), *Cognitive social psychology: The Princeton symposium on the legacy and future of social cognition* (pp. 247–259). Mahwah, NJ: Lawrence Erlbaum.

Tajfel, H. (1969). Cognitive aspects of prejudice. *Journal of Social Issues, 25,* 79–97.

van Knippenberg, A., & Dijksterhuis, A. (2000). Social categorization and stereotyping: A functional perspective. In W. Stroebe & M. Hewstone (Eds.), *European Review of Social Psychology* (Vol. 11, pp. 105–144). Chichester, UK: Wiley.

Waldmann, M. R., & Hagmayer, Y. (2001). Estimating causal strength: The role of structural knowledge and processing effort. *Cognition, 82,* 27–58.

10

Mental Mechanisms

Speculations on Human Causal Learning and Reasoning

Nick Chater and Mike Oaksford

A fundamental goal of cognition is to reason about the causal properties of the physical and social worlds. However, as Hume (2004/1748) observed, knowledge of causality is puzzling because although events are directly observable, causal connections between them are not. Hume's puzzle has both philosophical and psychological aspects. The philosophical puzzle is how causal knowledge can be *justified* – that is, when *should* people infer causality? Hume argued that this problem is simply unsolvable – causality can never justifiably be inferred. But this leaves the psychological puzzle. Whether defensibly or not, people routinely *do* infer causality from experience; the puzzle is to understand what principles underlie these causal inferences.

Hume believed that this psychological problem was solvable: He suggested, in essence, that people infer causality from constant association or, in statistical terms, correlation. However, inferring causality from correlation is fraught with peril. One particularly serious difficulty concerns the theme of this book: sampling. Biased samples can induce numerous correlations that are spurious from a causal point of view; and, conversely, can lead to no correlation, or anticorrelation, where there is a causal link between events.

From Hume's skeptical perspective, this difficulty might not seem important. Indeed, if causal knowledge is unjustified and unjustifiable, there is really no question of whether causality is inferred correctly, or incorrectly: The associations between events are all there is. From a modern perspective, however, such skepticism seems untenable. This is because causal inferences are not merely a theoretical veneer on a bedrock of associations; causal inferences are critical to determining how agents should *act* in the world.

In deciding how to act, an agent's causal knowledge becomes central. Suppose we notice that A and B are associated; and we want to bring about B. We wonder whether bringing about A will serve as means to bring about our end, B. If A causes B, and we make A happen, then we can predict that B will result; yet if A is merely associated with B, but does not cause it, changing A to bring about B will be futile. Thus, means-end reasoning presupposes causal assumptions about the world.

For example, suppose that we notice an association between specific types of animal and specific types of track. It is crucial that we realize that animals cause their tracks, and not the other way around. Otherwise, instead of using tracks to find our prey and avoid our predators, we will become engaged in futilely making tracks, to summon up our prey; and wiping out tracks, to ward off our predators.

We shall argue in this chapter that correcting for biases due to sampling, such as those discussed in the section *Biased and Unbiased Judgments from Biased Samples* in the present volume, may only be possible in situations in which people have substantial prior knowledge of the causal structure of the domain. This prior knowledge might include the knowledge that animals cause their tracks, by physical contact of their feet with muddy ground. Where this knowledge is not available, people will frequently find causal inference extremely difficult. But how is this causal knowledge represented? We suggest that the causal knowledge is represented in "mental mechanisms," which provide a causally "live" model of the causal structure of the aspect of the world being considered.

In a mental mechanism, the "internal" causal interventions can be defined over the mechanism, mirroring assumptions about the causal impact of interventions in the world. To indicate what this might involve, we draw directly on a recent, and we believe very important, theory of causation developed by Pearl (2000), which specifies, at least for simple classes of model, a representational and computational machinery that can carry out such inferences.

The term "mental mechanisms" is, to some degree, a placeholder. Apart from adverting to Pearl's work, and possible ways in which it might be extended, we put forward no detailed psychological theory of how such mechanisms might operate. Rather, we stress that most theories of mental representation and inference would not appear plausible candidates for a theory of mental mechanism, because they do not contain the appropriate computational and representational resources to deal appropriately with the distinction between causality and correlation. That is, their internal representations do not mirror the causal structure of the world. Only in cases where people have appropriate preexisting mental mechanisms are they able, we argue, to make appropriate causal inferences when faced with the problem of biased samples and other difficulties of inferring causation from correlation.

Causal learning is, however, also possible in domains in which we have relatively little prior knowledge – otherwise, we would face an insuperable bootstrapping problem in developing our knowledge of the world.

We suspect, here, that the problems of dealing with samples are substantial and that the cognitive system is able, nonetheless, to learn the causal structure of the world without using samples at all. Instead, we suggest that, by exploiting the richness of information shared in many cause–effect relationships, it is possible to infer causality using one-shot learning – in sampling terms, by using a sample of one.

Our argument divides into three main sections. In the first section, *Mental Mechanism versus Logical Representations*, we argue that knowledge is represented in causal, rather than in logical, terms; and we consider recent empirical results backing up our account. The second section, *Reasoning with and without Mental Mechanisms: Sampling in the Monty Hall Problem*, takes up the theme that reasoning is straightforward to the extent that reasoners can make reference to an underlying mental mechanism of the relevant causal structure in the world; but where there is no appropriate representation of this structure, reasoning breaks down. The third section, *Learning Mental Mechanisms from Data*, puts forward our speculations concerning the richness of the information available in causal learning about the everyday physical world and suggests that much learning is one-shot. We also set out the beginnings of a theoretical framework for understanding such learning, broadly in terms of "similarity" between cause and effect (or, more precisely, information sharing between cause and effect), an idea that casts a different light on the role of similarity in underpinning causal relations than has been traditional in philosophy. Finally, the *Discussion* briefly considers the implications of this viewpoint, along with the picture of the mind as an intuitive statistician or scientist.

MENTAL MECHANISMS VERSUS LOGICAL REPRESENTATION

In the framework that we outline here, causality is central to knowledge representation and reasoning. Yet in the psychology of reasoning, the focus has been on logical, rather than causal, structure. More specifically, according to most conventional theories in the psychology of reasoning, reasoning about verbal statements (e.g., a conditional statement, such as *If A then B*) depends on a mental representation that captures the logical form of that statement (although see, for example, Cheng & Holyoak, 1985, 1989; Cosmides, 1989; Poletiek, 2001, for a range of nonlogical viewpoints). If the causal viewpoint is right, then statements with identical logical structure may invoke very different patterns of reasoning; and people might affirm arguments that are causally justified, yet logically invalid. Consider, for example, the following argument:

Background:	*If A1 then B*
	If A2 then B
New Knowledge:	B
Conclusion:	A2

Here, for reasons that will become clear later, we have separated background knowledge from "new" knowledge – the issue of interest is the degree to which the new knowledge increases the degree of belief in the conclusion. From a logical point of view, this argument is invalid; that is, the premises can be true, but the conclusion false. Knowing $A1$ appears to allow us to infer B, but then knowing B does not allow us to apply the second premise and infer $A2$ – because there may be other ways in which B can hold, other than in cases in which $A2$ holds.

So, on the face of it, logic-based psychological theories of reasoning which will assume that this kind of inference is not typically drawn. Now, of course, psychological theories of reasoning must explain why people make logical mistakes. For example, do the mistakes result from comprehension errors or mistakes in applying rules of reasoning (from the point of view of a mental logic view of reasoning, e.g., Henle, 1962; Rips, 1994), or can we attribute them to the incorrect or incomplete construction of the relevant set of mental models (in a mental models framework, Johnson-Laird & Byrne, 1991)?

Still, these approaches should not be able to explain any impact of the *causal* interpretation of the *if ... then ...* form, in the aforementioned premises. This is because both approaches translate the form *if ... then ...* by standard logical material implication (Braine, 1978; Johnson-Laird, 1983; Johnson-Laird & Byrne, 1991; Johnson-Laird & Byrne, 2002; Rips, 1994; see Edgington, 1995, for review of different alternative viewpoints on the conditional), according to which *If A then B* is true just in case either the antecedent (A) is false and/or the consequent (B) is true.

Suppose, now, that we consider two parallel version of the previous argument, which have a very different causal structure. In the first, there are two antecedents, $C1$ and $C2$, and these are potential causes of the consequent, the effect E:

Background:	*If C1 then E*
	If C2 then E
New Knowledge:	C1
Conclusion:	C2

Using Pearl's (2000) notion, we can draw this situation as in Figure 10.1a, where arrows indicate causal direction.

Now how strong is this argument? It is, of course, logically invalid, as we have already noted. Moreover, the causal structure provides no indication that there is any inferential relationship between the alternative causes $C1$ and $C2$. Roughly, this is, in Pearl's theory of causality (see also Glymour & Cooper, 1999; Spirtes, Glymour, & Scheines, 1993), a result of the fact that the only path between the two in Figure 10.1a passes through a "collider" – two arrows that meet at a single node. We do not go into the rather subtle underlying formal theory of how causal and inferential relations can be propagated around these networks, provided by Pearl,

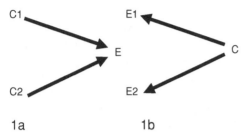

1a   1b

FIGURE 10.1. Different patterns of causal relation in a conditional inference problem. 1a shows two causes that lead to the same effect. 1b show two effects that can arise from the same cause. The graph structure shown, and a much larger class of graphs, can be used to propagate probabilistic information in a computationally local way, using some elegant results from probability (e.g., Lauritzen & Speigelhalter, 1988; Pearl, 1988; Spirtes, Glymour & Scheines, 1993). Moreover, Pearl (2000) has shown how to distinguish how information is propagated in such graphs, where variables have been modified by an "experimental" interventional from outside the system. The direction of the arrows indicates causal directionality – thus when a variable is modified "experimentally" the resulting modifications to other variables only propagate causally downstream of that variable. Such formal graph structures and their propagation rules provide an elegant and computationally efficient method of reasoning about causality and hence provide an initial candidate example of "mental mechanisms," which could underpin human causal reasoning.

but just use results of the approach here. To appreciate this case intuitively, consider the following setup: *If the battery is flat, the car won't start, if the electrical cables have been severed, the car won't start.* If we learn that the battery is flat, this does not, of course, tell us anything about whether the cables have been severed (or vice versa).

So we have no reason to believe that *C1* and *C2* are anything other than independent. Now suppose that we add a further premise, *E*, to our background knowledge:

Background:	*If C1 then E*
	If C2 then E
	E
New Knowledge:	*C1*
Conclusion:	*C2*

Now, in Figure 10.1a, we can see that *E* "blocks" the path between *C1* and *C2*. Interestingly, though, *C1* and *C2* are, according to Pearl's account, no longer independent – indeed, in the context of *E*, *C1* makes *C2 less* likely. In our example, suppose that we know that the car does not start. Then we wonder whether the battery is flat. Suppose this has some probability (e.g., .3) – it is, after all, a pretty plausible explanation of the nonstarting of the car. However, then we learn that the cables have been severed;

this immediately "explains away" the nonstarting of the car. Although it is possible that, additionally, the battery is also flat, there is now no particular reason to believe this – so the probability that the battery is flat will be revised downward (say to .001, or whatever is a reasonable default value for the probability that a randomly chosen car has a flat battery). The phenomenon of "explaining away" is well known in statistics and artificial intelligence (e.g., Wellman & Henrion, 1993), and "discounting" in the psychological literature on causal learning and attribution (e.g., Kelley, 1972). Other things being equal, learning that one causal explanation for some outcome is true reduces the evidence for alternative explanations.

Now let us contrast this pattern of inference with that for a logically identical reasoning problem in which the direction of causality is reversed; that is, instead of two causes of the same effect, we have two effects that result from the same cause:

Background:	*If E1 then C*
	If E2 then C
New Knowledge:	*E1*
Conclusion:	*E2*

This setup is graphically represented in Figure 10.1b. Whereas in the previous case, the "initial" configuration involved no dependence between the two antecedents (*C1* and *C2*), now the initial configuration indicates that there is a positive dependence between the two antecedents (*E1* and *E2*). Specifically, if we learn that *E1* is true, this raises the probability of *E2*, to some degree, because, if *E1* is true, then this raises the probability that *C* is true (after all, *C* is a potential cause of *E*); and this in turn raises the probability that other effects of *C* (including *E2*) hold.

This positive relationship is, however, undermined when the consequent (in this case *C*) is part of background knowledge. Once *C* is known *E1* carries no further information about *E2*. In diagrammatic terms (Figure 10.1b), the path between *E1* and *E2* is "blocked" by our knowledge of *C*.

Background:	*If E1 then C*
	If E2 then C
	C
New Knowledge:	*E1*
Conclusion:	*E2*

Overall, the key point is that the inferential relationship between the two antecedents depends, in a complex way, on causal direction that relates antecedents and consequent. Moreover, this pattern is modified in a complex way, by whether the consequent is part of background knowledge.

Does human reasoning with conditionals follow this pattern, indicating that causality, rather than logic, may be central to knowledge

TABLE 10.1. *How Causal Structure Affects Patterns of Conditional Inference*

	Consequent not Known	Consequent Known
Antecedents as Causes If C1 then E If C2 then E	*Antecedents independent* C1 independent of C2	*Antecedents negatively related* C1 and C2 count as evidence against each other, given E
Antecedents as Effects If E1 then C If E2 then C	*Antecedents positively related* E1 and E2 count as evidence against each other	*Antecedents independent* E1 independent of E2, given C

representation? When tested directly with arguments of this type, people show strong effects of the causal direction in arguments, just as would be expected if people were reasoning about causality in line with Pearl's theory (Ali, Chater, & Oaksford, 2004). This finding is not easy to account for from the point of view of a logic-based theory of reasoning, such as mental logic or mental models.

The fact that causal direction so radically affects how people perform with logically identical reasoning problems raises a further concern about conventional experimental work on human reasoning (see Evans, Newstead, & Byrne, 1993; Evans & Over, 1996; Oaksford & Chater, 1998). The problem is that experiments frequently use materials for which the causal direction is ambiguous; and when averaging over behavior using different materials, experimenters may inadvertently mix together materials that suggest different causal directions.

The viewpoint developed here is a special case of a broader tradition of disquiet with logic-based views of everyday reasoning. A range of theorists in psychology, philosophy, and artificial intelligence have argued that deductive logic is much too narrow to account for most everyday human reasoning. According to this viewpoint, almost all human reasoning is plausible rather than certain – conclusions are plausible, or perhaps probable, given a set of premises; but they do not follow from them with deductive certainty (e.g., Fodor, 1983; Oaksford & Chater, 2002; Putnam, 1974; Quine, 1951). For example, in artificial intelligence, there has been a particular focus of interest on default reasoning, that is, reasoning in which a conclusion follows as a default, but can be overturned by further information. Initial attempts to capture such inferences using extensions of logical reasoning (so-called nonmonotonic logics) have gradually be abandoned (e.g., McDermott, 1987; Oaksford & Chater, 1991) and replaced by probabilistic models of uncertain reasoning (e.g., Pearl, 1988).

We suggest that much of why people find verbal reasoning problems difficult is that they typically find it hard to reason when they are unable to construct a "mechanistic" model that captures the causal relationships in the scenario about which they are reasoning. Pearl's (2000) work on the mathematical and computational basis for causal inference has indicated how such mechanisms can be constructed, in an important class of cases.[1] According to this viewpoint, the basis for knowledge representation and reasoning is mental simulation of causal mechanisms in the world. We now turn to the question of how such mental mechanisms may be critical for understanding how we make inferences from sampled data, focusing as a case study on a celebrated, and puzzling, reasoning problem.

REASONING WITH AND WITHOUT MENTAL MECHANISMS: SAMPLING IN THE MONTY HALL PROBLEM

In this section, we consider variations on a reasoning problem that critically depends on the reasoner understanding the way in which data are *sampled*. Yet, understanding how the data are sampled requires, we argue, having a causal simulation (a mental mechanism) of the process by which data are sampled. We hope that these illustrations may have the force of a "psychophysical demonstration," which may be appreciated intuitively merely by attempting to solve the problems oneself, rather as many perceptual effects can be illustrated effectively by examples. Clearly, though, attempts to provide such demonstrations are merely a first step in research. We hope that this discussion encourages such experimental investigation.

Specifically, then, we wish to consider a set of problems of the following form. (In the spirit of first-hand psychophysical enquiry, readers might want to consider the problems in Figures 10.2–10.7, before proceeding further.)

Suppose you have to guess the location of a target, which may be in any of three positions (see Figures 10.2, 10.3, and 10.4). You are asked to make an initial guess (which you can only do at random, as you have no relevant information). Then, we choose to reveal one of the other two locations (i.e., to inspect whether the target is there or not). Crucially, we sample from these two locations in the following peculiar way: If the target is at neither location (i.e., we initially guessed the location of the target correctly), then we choose one of these two locations at random, *but if the target is at one of the two locations, we choose the other location.* This ensures that we will definitely

---

[1] The properties of such "graphical models" of probabilistic reasoning are now reasonably well understood. Only recently, though, has Pearl (2000) shown how to differentiate between the inferential impacts obtained when an observed variable has been changed by an "intervention" into a causally closed system, rather than that same variable (with the same value) being merely observed. The critical difference is that the impacts of a variable that has been modified by intervention should only probabilistically modify variables that are causally "downstream" of the modified variable; but if a variable is observed, inferences may be drawn about other variables that are causally upstream or downstream.

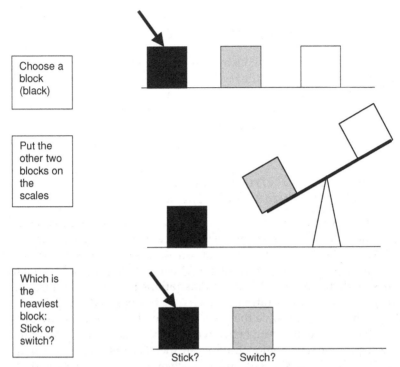

| Choose a block (black) |
| Put the other two blocks on the scales |
| Which is the heaviest block: Stick or switch? |

Stick?   Switch?

FIGURE 10.2. The balance-beam problem. The task, as before, is to guess which of three colored blocks is heavy. (The other two are roughly, but not exactly, equally light.) You are asked to choose a block; the other two blocks are put in the balance. You can now stick to your original choice or switch to the heavier block that was weighed. In this version, it seems intuitively clear that the best guess is to switch – there is only a 1/3 probability that you guessed right the first time; and if you initially guessed wrong, the alternative block must be the heaviest.

not reveal the target. The location that has been revealed to be empty can now be ignored, leaving just two possible locations for the target – the location that you choose to start with; and the other remaining location, which was not revealed in the rather peculiar sampling process. Now you are asked, again, to guess the location of the target. You can either stick to your original choice or you can switch to the other "live" location – the one that has not been revealed to be empty. Which should you do?

This problem has the structure of the famous "Monty Hall" problem, brought to prominence by an American TV game show and the subject of much subsequent psychological research (e.g., Gilovich, Medvec, & Chen, 1995; Krauss & Wang, 2003). [A formally identical, and also difficult, problem is the three prisoners problems (Shimojo & Ichikawa, 1989), although we do not consider this variant here.] The Monty Hall problem has proved

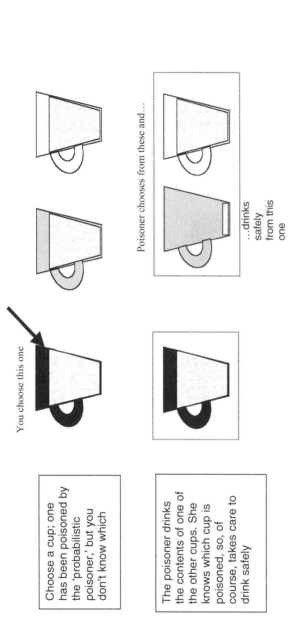

You choose this one

Choose a cup; one has been poisoned by the 'probabilistic poisoner,' but you don't know which

Poisoner chooses from these and...

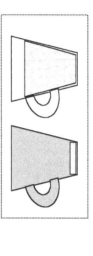

...drinks safely from this one

The poisoner drinks the contents of one of the other cups. She knows which cup is poisoned, so, of course, takes care to drink safely

Stick?

Switch?

Now you have to drink from one of the remaining cups

FIGURE 10.3. The probabilistic poisoner. In this problem, a poisoner has put poison in one cup. You choose a cup. The poisoner drinks from one of the other cups. You must then stick or switch.

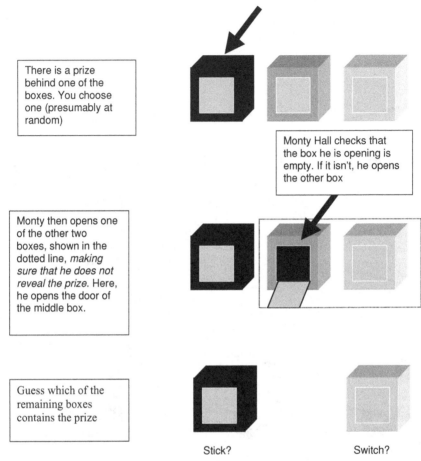

There is a prize behind one of the boxes. You choose one (presumably at random)

Monty Hall checks that the box he is opening is empty. If it isn't, he opens the other box

Monty then opens one of the other two boxes, shown in the dotted line, *making sure that he does not reveal the prize.* Here, he opens the door of the middle box.

Guess which of the remaining boxes contains the prize

Stick?   Switch?

FIGURE 10.4. The standard version of the Monty Hall problem. A prize is inside one the boxes. One box is chosen by the player. Monty then chooses to open another box, ensuring that he does not reveal the prize. Players must then stick to their original guess, or switch to the other unopened box. If they are correct, they receive the prize.

puzzling, even for some mathematicians. There are essentially two lines of reasoning, both of which seem attractive.

The first (correct) line of reasoning is that there is a 1/3 chance of guessing correctly with one's first guess. Then, because of the procedure for choosing the location that will be revealed, we know for sure that, if the target is among the other two items, it will not be the rejected one (i.e., the item to which one can switch). There is therefore a 2/3 chance that switching will correctly locate the target. Therefore you should switch.

The second (incorrect but intuitively attractive) line of reasoning is that, when considering the two "stick" and "switch" choices, one is simply choosing between two boxes; and there is an equal probability that the target is located within each. So the probability of success for sticking or switching is 1/2, and either option is equally good.

The difference between the two lines of reasoning concerns their assumptions concerning *sampling*. The first line of argument takes account of the peculiar way in which the "switch" location is chosen: If one of two locations from which it is chosen contains the target, then this is the location that will be retained – the other location will be revealed as empty. This crucially means that there are two ways in which the "switch" item can come to contain the target – as the target could have been in either of those two locations. There are three possible locations for the target, and hence, we see how there is a 2/3 probability that the target is at the switch location.

The second (incorrect) line of argument is, though, appealing. This is because the assumption that, other things being equal, if some kind of random process assigns a target to one of two locations, it is natural to presume that, in the absence of further information, the target is equally likely to be at either location. The rather complex sampling procedure that defines this class of problem is difficult to understand, or to trace the consequences of; in such problems it is not clear how a sampling bias might have crept in.

Indeed, the second line of argument would be correct if sampling of the "switch" item were indeed random. That is, you pick a locations, a coin is then flipped to decide which of the other two location is the "switch," and the other is revealed to see whether or not the target is present. As it happens, and purely by chance, the location that is revealed does not contain the target – of course it could have – and the probability that it did so is 1/3, because the target is equally likely to be in any of the three locations in the initial setup. (Of course, had the target been revealed at this stage, then the game is over.) Then the question is asked whether we should stick or switch – and now the answer is that the options are equally good: Each has a 1/2 chance of being correct. Now the sampling procedure is not biased. All that is happening, in essence, is that we ponder the three locations, check one of them, and find it to be empty. We then ask what is the probability that the target is in either of the two remaining locations (the stick and switch locations). With no additional information, we are forced to assume that either location is equally probable. The problems in Figures 10.5, 10.6, and 10.7 are analogues of the problems in Figures 10.2, 10.3 and 10.4 but have been altered to have this structure.

From our perspective, the reason that the Monty Hall problem is so difficult is that we have great difficulty in reasoning about sampling and that potential biases are induced by sampling. This point is consistent with arguments in many of the chapters in this book, with the power of demonstrations such as Simpson's paradox (1951), and with elegant explanations of a wide range of biases in social judgment in terms of inability to take

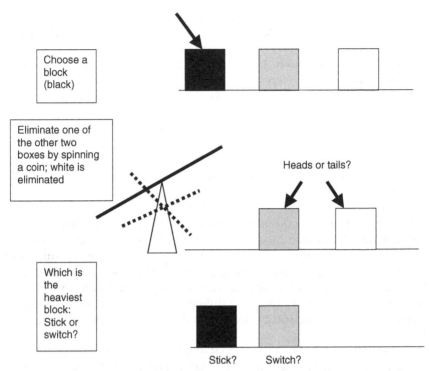

FIGURE 10.5. Random sampling version of the balance-beam problem. Here, the problem is as before, but the balance-beam scales are broken. So, instead, after you have chosen a particular block initially, one of the other blocks is eliminated by flipping a coin. If the eliminated item happens to be the heaviest one, this is explained to the player, and the game is over. However, in the example shown this does not happen; the result of this random process is to eliminate the white block, just as occurred in the previous version of the problem with a functioning balance beam, when the scales were used to show that the white block could not be the heavy block, because it is lighter than the gray block. Switching or sticking are equally good options, in this version of the problem.

account of sampling procedures effectively (e.g., Fiedler, 2000; Fiedler et al., 2000).

We suggest, though, that it is not that the cognitive system has intrinsic problems with dealing with sampling. It is rather that reasoning about sampling, just as with the conditional reasoning described in the previous section, is very difficult without a "mental mechanism," that is, without a mental representation of the underlying causal structure that gives rise to the data.

According to this line of thinking, the traditional version of the Monty Hall problem is that there is no causal process that underpins the peculiar sampling rule that Monty is supposed to use. When Monty reveals an empty location, it is not apparent whether this location has been chosen randomly (but just happens not to contain the prize) or whether Monty has followed his rather complex sampling rule.

However, in logically analogous problems, in which the sampling rule emerges naturally from a causally comprehensible scenario, reasoning is much simpler. Figure 10.2 shows a balance-beam version of the problem, in which the target is a heavy block (with the other two blocks being light, though not identical). Here, one block is chosen, and the other two are weighed. The question is whether one should stick to one's original block or switch to the block that was the heavier of the other two. Now it seems intuitively natural to switch (correctly) – because the causal setup makes it clear that the "switch" block has not been sampled randomly but has instead been chosen for heaviness (in relation to the other block).[2]

If, though, we change the problem so that it is clear that one of the blocks is rejected at random (without use of the balance; Figure 10.5), then this causal mechanism for producing a biased sample of the "switch" item is removed, and it is intuitively apparent that the stick and switch items are equally plausible.

In the balance-beam task, simple physical principles govern how blocks are selected (i.e., how the weight of blocks determines how the scales tilt). However, other, apparently intuitive, straightforward versions of the problem can be generated that require "intentional" or "folk-psychological" explanation (e.g., Fodor, 1968), that is, explanation of people's behavior by reference to their beliefs and desires. Thus, in this regard, such problems are analogous to the classic Monty Hall problem – but instead of having to assimilate an apparently arbitrary rule that Monty uses to make his choices, we instead can see the choice regularity as emerging naturally from our understanding of human behavior.

To illustrate, consider what we call the "probabilistic poisoner" problem (Figure 10.3). You are confronted by a poisoner, who demands, at gunpoint, that you drink from one of three cups (black, gray, and white). One of these cups has been laced with a fatal dose of cyanide. The poisoner allows you to choose a cup – you choose, say, the white cup. She then notes that she knows, of course, which cups and poisoned and which are safe; she then

---

[2] It might seem disanalogous with the original problem that the two nontargets differ here. This does not make any formal difference to the problem. However, we eliminate this by an alternative cover story, in which the two lighter weights are identical, but the balance has almost no friction at its bearing, and hence will come down one side or the other, even when comparing identical weights (e.g., determined by currents of wind). The important point is simply that the balance is not level with the two light blocks. Otherwise, of course, the target could trivially be located in a single weighing – if the balance is level, then target must be the weight that is not on the balance; if the balance tilts, the heavier item on the balance would be the target.

You choose this

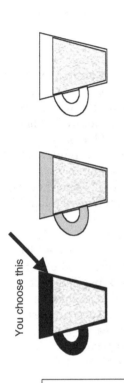

Choose a cup; one has been poisoned by the 'probabilistic poisoner,' but you don't know which

Other victim randomly chooses from these...

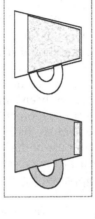

...drinks safely from this one

The poisoner then forces a second victim to choose and drink from one of the other cups. The victim does not know which cup is poisoned, so can only choose randomly. Happily, the victim is not poisoned.

Switch?

Stick?

Now you have to drink from one of the remaining non-empty cups

224

drinks from one of the others, say the gray cup; the cup is not, of course, poisoned. You then must stick or switch.

Here, it seems there is a strong intuition that it makes sense to stick. There is a 1/3 chance you were unlucky enough to pick the poisoned cup initially; if not (i.e., with probability 2/3), then the remaining cup contains the poison (after all, you know that the poisoner would not drink the poisoned cup). That is, the intentional rationale behind the poisoner's sampling of cups (i.e., that this sampling is systematic rather than random) is transparent to the player.

Now, consider a slightly different setup, where there are two hapless participants, both of whom have no idea which cup is poisoned, who are held at gunpoint by the poisoner (Figure 10.6). The first participant chooses a cup. The second is then told to choose *and drink* a cup. Thankfully, the participant survives. So that cup was not poisoned. Now, the first player must choose which cup to drink – the question, as ever, is whether to stick or switch. Here, the sampling of the cups is clearly random. Moreover, there is no reason to prefer to stick or switch. Essentially, the situation is just as if one of the cups happened to be knocked over by chance, and the poisoner were to comment that this was not the poisoned cup.

The point here is that where the problem has a transparent causal structure from which differences in modes of sampling are derived, these differences in sampling may be reflected in people's reasoning; and this may be true, whether the causal structure of the situation is underwritten by physical causality (as in the balance-beam problem) or intentional explanation (as in the probabilistic poisoner). When people invoke a mental mechanism that mirrors the causal structure in the problem under consideration, reasoning is straightforward. When causal structure is opaque (as in the Monty Hall problem), then reasoning is poor.

This conclusion may play some role in explaining what Oaksford and Chater (1998) have called the paradox of human reasoning: that human

---

FIGURE 10.6. In this version of the probabilistic poisoner problem (Figure 10.3), the poisoner has put poison in one of three drinks, and has two unlucky victims who will be forced to choose and drink from a cup. As one of the victims, you are first asked to provisionally choose a cup. The other victim is then told to choose *and drink from* one of the other cups. Happily, the victim drinks from a cup that does not happen to be poisoned and are unharmed. You are then asked whether you would like to stick with your original choice, or drink from the other cup. Given that the other victim's choice of cup was random (i.e., the victim had no idea which cup is poisoned), each of the remaining cups is equally likely to contain the poison. The probability of being poisoned is $^{1}/_{2}$, whether you choose to stick with your original choice, or whether you switch. This case is crucially unlike the case where the poisoner drinks from a cup – because the poisoner's selection of cup will be non-random, as the poisoner will presumably be careful to avoid drinking from the poisoned cup.

Choose a box

Heads or tails?

Monty then opens one
of the other two
boxes, shown in the
dotted line, by tossing
a coin. As it happens,
the prize is not
revealed

Guess which of the
remaining boxes
contains the prize

Stick?   Switch?

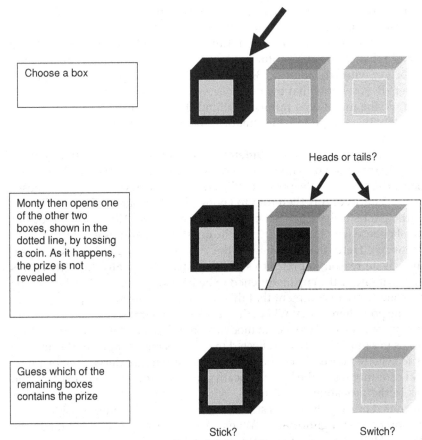

FIGURE 10.7. Random sampling version of the Monty Hall problem. In this version
of the problem, a box is chosen; and Monty randomly choses to open one of the
other two boxes. If he happens to choose the box with the prize, the game is void.
Here, he happens to choose a box (the middle box) that does not reveal the prize.
Thus, from the player's point of view, the outcome is precisely the same as in the
standard version of the game. The change in the way that Monty samples which
box to open is crucial – if the box is opened randomly, then there is merely a half
chance that the prize is in either of the other two boxes. We suggest that the problem
with the standard Monty Hall problem is that people find it difficult to appreciate
the difference between this setup and the standard version. The parallel versions
of the balance-beam and probabilistic poisoner problems indicate that this may not
be because people are unable to take account of the difference between different
sampling methods; instead, we suggest that sampling is difficult to take account
of without a clear causal model of the situation.

reasoning is spectacularly impressive in dealing with the richness of common-sense reasoning about the physical and social world and far exceeds the capacity of any mathematical or computational models of reasoning; yet when faced with apparently simple logical tasks, which are trivially handled by mathematical or computational models, people appear to reason very poorly. The explanation may be that human reasoning is built around causal models of the world – around what we have called mental mechanisms; and hence reasoning about these models is easy for the cognitive system (although how, at a computational level, this is achieved remains, of course, mysterious). Yet, in typical laboratory problems in the psychology of reasoning, the causal structure of the problem is either not specified or presumed to be irrelevant to the solution of the problem (essentially, because logic does not have the resources to express causality). Hence, when people reason about conditionals, syllogisms, or, for that matter, probabilistic relationships, it is likely that they will attempt to fill out the problems, perhaps in a rather haphazard way, to allow them to employ their standard reasoning methods. However, this will often lead to conclusions that depart from the causality-neutral dictates of logic.

LEARNING MENTAL MECHANISMS FROM DATA

We have used the term "mental mechanisms" to label the models of the causal structure of the world that we presume to underpin everyday human reasoning, and we have illustrated how the idea of a mental mechanism can be related to a standard reasoning task, the Monty Hall problem. This illustrates our general idea that the interpretation of samples of data, and in particular the ability to correct for biased sampling, depends on the learner possessing an appropriate mental mechanism, which appropriately represents the causal structure of the environment.

Yet if interpreting samples of new information depends on the pre-existence of appropriate mental mechanisms, then we face the possibility of an infinite regress in considering how mental mechanisms can themselves be learned. One viewpoint is, of course, that they are not – that, instead, causal knowledge is innately specified. This viewpoint is bolstered by studies suggesting that causal knowledge emerges surprisingly early (see, e.g., Leslie, 1987; Leslie & Keeble, 1987; Spelke, 1998; Sperber, Premack, & Premack, 1995).

Theorists who, by contrast, assume that causal relations are, at least in some part, learned typically focus on associative or statistical mechanisms of causal learning (for reviews, see de Houwer & Beckers, 2002; Dickinson, 2001), which are rooted in theories of animal learning (e.g., Rescorla & Wagner, 1972; Wagner, 1981). Thus, they follow in the tradition tracing back through behaviorism, and back to Hume's view (2004/1748) that causality is inferred from associations between events.

In practice, this associative point of view is typically spelled out in a rather specific, and, we will suggest, narrow, way (e.g., see Cheng, 1997; Shanks, 1995). For a particular event, the cause and effect are regarded as either present or absent. A two-by-two contingency table is then constructed, from which the learner must infer a causal relationship. Thus, in Figure 10.8, the learner is presumed to attempt to infer whether there is a causal relationship between throwing a stone and a window breaking, using frequencies of the four different possible event types (of course, the situation is more complicated where there are several possible causes, but the additional complications are irrelevant here). With such a simple representation of the raw data, the learner needs ample amounts of data; thus, learning is viewed as a matter of association, or statistical inference, over many trials.

By contrast, we suggest that by using a much richer encoding of the data, a single event may be sufficient to establish a causal connection. For example, when a stone hits a window, there is a very precise temporal coincidence between the landing of the stones and the breaking of the window and there is a precise spatial relationship between the point where the stone strikes and the center of the break; and, for that matter, there is a precise relationship between the pattern of cracks and the impact, such that the cracks emanate from the point of impact (see Figure 10.8).

From this perspective, the probability of this particular effect spontaneously arising, in this particular way, in this particular point in space and time, is extremely remote. Therefore, a single instance of such an event provides strong evidence of a causal connection. That is, a single observation may be enough to establish a causal link.

One difference between the one-shot approach and the standard associative viewpoint is that the associative viewpoint has been formalized in a range of specific models, whose quantitative predictions have been assessed against empirical data. For the one-shot alternative to be more than an interesting speculation, the same must, of course, be attempted with this approach. This project lies beyond the scope of this chapter; we suggest, though, that a core formal notion might be the amount of shared information between cause and effect. Intuitively, this means that amount of information that can be saved by specifying cause and effect together, rather than specifying them separately. More formally, this can be captured by the notion of *mutual information* from classical information theory, or, more generally, by an analogue of this notion, in algorithmic information theory (Li & Vitányi, 1997).[3] This notion captures the degree of coincidence that relates potential causes and effects – and hence the evidence that the

---

[3] Specifically, for cause C and effect E, the quantity is $K(C) + K(E) - K(C \& E)$, where $K(X)$ measure the complexity of X, where this is measured by the length of the shortest computer program that can generate X. See Li & Vitányi (1997) for a discussion.

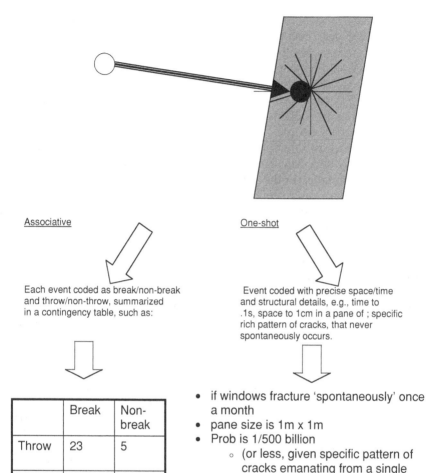

Associative

One-shot

Each event coded as break/non-break and throw/non-throw, summarized in a contingency table, such as:

Event coded with precise space/time and structural details, e.g., time to .1s, space to 1cm in a pane of ; specific rich pattern of cracks, that never spontaneously occurs.

	Break	Non-break
Throw	23	5
Non-Throw	1	263

- if windows fracture 'spontaneously' once a month
- pane size is 1m x 1m
- Prob is 1/500 billion
  - (or less, given specific pattern of cracks emanating from a single point)

FIGURE 10.8. Associative and one-shot views of causal learning. The associative viewpoint considers events "atomically," that is, with regard to whether they did or did not occur at a particular time. The one-shot view relies on the rich structure of the pattern between events.

events are causally connected (although this measure says nothing about the direction or nature of that causal connection, of course; in particular, many such links will be via third causes, which causally influence both events of interest, where those events may not causally influence each other).

This measure is particularly interesting, because it is closely related to one account of similarity between objects (e.g., Chater & Vitányi, 2003;

Hahn, Chater, & Richardson, 2003). Thus, it may be that evidence for a causal link between events may depend on their *similarity*.[4] For example, consider the causal link between a hand and its shadow. The hand and shadow have similar shape; and if the hand is moved, the shape of the shadow changes also, in close spatio-temporal synchrony. Thus, where similarity is so apparent between cause and effect, we might suggest that very rich information is available to the learner – and that substantial learning might, at least in principle, occur in one shot. After a century of psychology focusing on spatio-temporal contiguity as the driving force for causal learning, it may be worth seeing how far learning can get, by considering a richer representation of the nature of events that may reveal similarities and hence perhaps causal connections between them.

This use of similarity here may appear to be rather loose. Is it really the case that causes are typically similar to their effects? For example, is the ball that strikes a window really similar to the broken window? We argue, though, the relevant similarity comparison is not between the *objects*, but rather a similarity between *events* and *properties*. So although the ball has no particular similarity to the smashed window, the spatio-temporal coordinates of the impact of the ball at the window *are* similar (indeed, in this case, nearly identical to) the spatio-temporal coordinates of the epicenter of the shattering of the glass. Equally, although there is no similarity of many aspects of a three-dimensional human finger (e.g., its patterns of blood flow) to any two-dimensional pattern of ink on paper, there may be a similarity of the *surface pattern* upon the skin of the fingertip to the pattern of ink – so much so that a particular fingerprint may be viewed as causally connected, with almost complete certainty, to that specific finger of a specific person (either directly, when the fingerprint is taken, or indirectly, through the reproduction of this original fingerprint). Our claim is that many similarities between cause and effect can be understood as involving the sharing of large amounts of information between them. For example, consider the rich shared patterns in the dynamics of a moving hand and the shadow that it casts, or the patterns on tiny grooves on a fingertip and the inky lines of the fingerprint that it leaves, or the spatio-temporal information concerning the impact of a stone and the appearance and location of a fracture on the window it strikes. In such cases, we argue, the sharing of information provides strong prima facie evidence for a causal link between cause and effect. Naturally, any specific inference from informational similarity between two events to a causal connection between them may be incorrect. It may be a coincidence that the lights come on just when I throw the switch; if so, the temporal similarity between these events is likely to

---

[4] Although the converse clearly cannot hold – many similar things have no direct causal link between them, as follows from the earlier comment on third causes. We suggest that direct causal connections are those in which the modification of the putative "cause" transmits information about the putative effect."

mislead the cognitive system. Moreover, of course, the nature of the causal relationship may be misconstrued – the police might correctly infer that there is a causal link between a fingerprint on my window and a suspect, but they may not realize that this link is mediated by a third party who has planted the prints to incriminate the suspect.

We note also that the proposal that some, or perhaps even much, human learning is one-shot may resolve a paradox about human causal learning, rather analogous to the paradox of human reasoning that we discussed in the previous section. The paradox is that children appear to be able to learn extremely rapidly and effectively about the causal structure in an astoundingly complex physical and social world. Moreover, both children and adults frequently appear to find causal information perceptually evident in *single* exposures to events, even apparently highly unfamiliar events such as one geometric shape "chasing" or "leading" another (Michotte, 1963).

Yet people struggle with the most rudimentary contingency learning. For example, people have great difficulty assessing the degree of causal control they have over an event (e.g., Alloy & Abramson, 1979) and are typically equalled or outperformed by simple linear regression models (e.g., Cheng, 1997; Gluck & Bower, 1988). One explanation, of course, is that children do not really *learn* the causal structure of the physical and social world – rather, it may be innately specified (e.g., Sperber et al., 1995). However, there may be another resolution of the paradox. Perhaps the crucial difference, we suggest, may be that in laboratory tasks the information relating cause and effect is typically very strongly reduced – often to essentially the contingency information used in the associative models of learning. In this bid for careful control, it may be that experimenters are depriving people of the rich informational cues that support their highly successful ability to learn about the causal structure of the real world.

Moreover, it is also possible that the same issue explains some of the difference between the apparent ease with which people rapidly learn the causal properties of new objects (e.g., learning how to wield an unfamiliar object, even without visual input; Carello & Turvey, 2000) and the difficulty experts have with apparently vastly simpler inference problems, where conclusions must be drawn from experience of verbally specified data and hence, typically, informationally very reduced data (again, experts are typically outperformed by simple linear statistical models, or even by arguably simpler one-reason decision models; Gigerenzer & Goldstein, 1996; Gigerenzer, Todd, & the ABC Resesearch Group, 1999; see also Chater et al., 2003).

DISCUSSION

We have argued that causality is central to understanding human reasoning and learning. In particular, we have argued that human reasoning is reliable to the extent that the reasoner has recourse to a "mental mechanism" that models relevant aspects of the causal structure of the world. When reasoners are faced with, for example, verbal reasoning problems, in which the underlying causal structure is unclear to them, their reasoning performance collapses. How are causal relations learned? We suggest that they are typically learned from very rich informational relationships between causes and effects, which make possible one-shot learning; indeed, our abilities to learn over multiple trials, from relatively impoverished stimuli, as typically studied in the associative learning tradition (e.g., Cheng, 1997; Cheng & Novick, 1992; Gallistel & Gibbon, 2000; Kakade & Dayan, 2002; Shanks, 1995) may be quite modest.

Taken at face value, the viewpoint that we have developed in this chapter runs counter to connectionist models of cognition (see, e.g., Rumelhart & McClelland, 1986) and statistical models of perception, cognition, language acquisition, and social judgement (e.g., Bod, Hay, & Jannedy, 2003; Fiedler, 2000; Gigerenzer & Murray, 1987). These models typically involve aggregating over large amounts of data, rather than on focusing on a single learning episode. An emphasis on one-shot learning seems more compatible with views in which the acquisition of knowledge is not a matter of a continuous buildup of associative relationships, but, rather, involves the construction of all-or-none chunks of knowledge, as in the classical symbolic view of mind (e.g., Anderson, 1983; Fodor, 1975; Newell, 1990).

It is possible, of course, that a sharp distinction between these modes of learning is not appropriate. There are a range of hybrid cognitive architectures in which structural relationships are all or nothing; these structures are also associated with "weights" that encode their frequency of occurrence – and these weights may be aggregated over many learning episodes [e.g., as represented by links in a Bayesian network (Pearl, 1988) or production rules (e.g., Anderson, 1993)]. Thus, in terms of the framework developed in this chapter, it is possible that, although the structure of mental mechanisms representing causal relationships is discrete, such mechanisms are also associated with frequency or probabilistic information. This might be crucial, for example, in determining which of several possible causal structures is more plausible. It is also possible, of course, that different mental processes aggregate information in different ways. In particular, a natural suggestion is that mental processes that entail one-shot learning may involve a rich analysis of a single episode; this may be computationally intensive and hence might occur only serially, demanding significant attentional resources. Thus, one might suspect that, at one time, only one piece of information can be learned in one shot. By contrast, it might be that associative mechanisms are less demanding of attention and that many pieces of associative information can be continuously modified simultaneously (e.g., information about word frequency, bi-gram frequency, phoneme frequency, and so on, might all be modified, while processing a new sentence). Whether such a separation stands up to empirical research is not clear – although this divide might be viewed as consonant with two-process theories of human cognition that have been widely discussed (e.g., Evans & Over, 1996; Sloman, 1996).

Overall, we suggest that the framework of mental mechanisms, though undeveloped as a psychological theory, may be an important direction for future research on how people perceive, learn, represent, and reason the causal structure of the physical and social world in which they are engaged.

References

Ali, N., Chater, N., & Oaksford, M. (2004). *Causal interpretations affect conditional reasoning*. Manuscript. Department of Psychology, University of Warwick.

Alloy, L. B., & Abramson, L. Y. (1979). Judgment of contingency in depressed and nondepressed students: Sadder but wiser? *Journal of Experimental Psychology: General, 108*, 441–485.

Anderson, J. R. (1983). *The architecture of cognition*. Cambridge, MA: Harvard University Press.

Anderson, J. R. (1993). *Rules of the mind*. Hillsdale, NJ: Lawrence Erlbaum.

Bod, R., Hay, J., & Jannedy, S. (Eds.) (2003). *Probabilistic linguistics*. Cambridge, MA: MIT Press.

Braine, M. D. S. (1978). On the relation between the natural logic of reasoning and standard logic. *Psychological Review, 85*, 1–21.

Carello, C., & Turvey, M. T. (2000). Rotational dynamics and dynamic touch. In M. Heller (Ed.), *Touch, representation and blindness* (pp. 27–66). Oxford: Oxford University Press.

Chater, N., Oaksford, M., Nakisa, R., & Redington, M. (2003). Fast, frugal and rational: How rational norms explain behavior. *Organizational Behavior and Human Decision Processes, 90*, 63–86.

Chater, N., & Vitányi, P. (2003). The generalized universal law of generalization. *Journal of Mathematical Psychology, 47*, 346–369.

Cheng, P. W. (1997). From covariation to causation: A causal power theory. *Psychological Review, 104*, 367–405.

Cheng, P. W., & Holyoak, K. J. (1985). Pragmatic reasoning schemas. *Cognitive Psychology, 17*, 391–416.

Cheng, P. W., & Holyoak, K. J. (1989). On the natural selection of reasoning theories. *Cognition, 33*, 285–313.

Cheng, P. W., & Novick, L. R. (1992). Covariation in natural causal induction. *Psychological Review, 99*, 365–382.

Cosmides, L. (1989). The logic of social exchange: Has natural selection shaped how humans reason? Studies with the Wason selection task. *Cognition, 31*, 187–276.

de Houwer, J., & Beckers, T. (2002). A review of recent developments in research and theory on human contingency learning. *Quarterly Journal of Experimental Psychology, 55B*, 289–310.

Dickinson, A. (2001). Causal learning: An associative analysis. *Quarterly Journal of Experimental Psychology, 54B*, 3–25.

Edgington, D. (1995). On conditionals. *Mind, 104*, 235–329.

Evans, J. St. B. T., Newstead, S. E., & Byrne, R. M. J. (1993). *The psychology of deductive reasoning*. Hove, UK: Lawrence Erlbaum.

Evans, J. St. B. T., & Over, D. (1996). *Rationality and reasoning*. Hove, UK: Psychology Press.

Fiedler, K. (2000). Beware of samples! A cognitive-ecological sampling approach to judgment biases. *Psychological Review, 107*, 659–676.

Fiedler, K., Brinkmann, B., Betsch, T., & Wild, B. (2000). A sampling approach to biases in conditional probability judgments: Beyond base-rate neglect and statistical format. *Journal of Experimental Psychology: General, 129*, 399–418.

Fodor, J. A. (1968). *Psychological explanation*. New York: Random House.

Fodor, J. A. (1975). *The language of thought*. New York: Crowell.

Fodor, J. A. (1983). *Modularity of mind*. Cambridge, MA: MIT Press.

Gallistel, C. R., & Gibbon, J. (2000). Time, rate and conditioning. *Psychological Review, 107*, 289–344.

Gigerenzer, G., & Goldstein, D. G. (1996). Reasoning the fast and frugal way: Models of bounded rationality. *Psychological Review, 103*, 650–669.

Gigerenzer, G., & Murray, D. J. (1987). *Cognition as intuitive statistics*. Hillsdale, NJ: Lawrence Erlbaum.

Gigerenzer, G., Todd, P. M., & the ABC Research Group (1999). *Simple heuristics that make us smart*. New York: Oxford University Press.

Gilovich, T., Medvec, V. H., & Chen, S. (1995). Commission, omission, and dissonance reduction: Coping with regret in the 'Monty Hall' problem. *Personality and Social Psychology Bulletin, 21*, 182–190.

Gluck, M., & G. Bower (1988). From conditioning to category learning: an adaptive netword model. *Journal of Experimental Psychology: General, 117*, 227–247.

Glymour, C., & Cooper, G. (1999). *Causation, computation and discovery*. Cambridge, MA: MIT Press.

Hahn, U., Chater, N., & Richardson, L. B. C. (2003). Similarity as transformation. *Cognition, 87*, 1–32.

Henle, M. (1962). On the relation between logic and thinking. *Psychological Review, 69*, 366–378.

Hume, D. (2004). *An enquiry concerning human understanding*. T. L. Beauchamp (Ed.). Oxford: Oxford University Press (originally published 1748).

Johnson-Laird, P. N. (1983). *Mental models: Towards a cognitive science of language, inference, and consciousness*. Cambridge, UK: Cambridge University Press.

Johnson-Laird, P. N., & Byrne, R. M. J. (1991). *Deduction*. Hillsdale, NJ: Lawrence Erlbaum.

Johnson-Laird, P. N., & Byrne, R. M. J. (2002). Conditionals: A theory of meaning, pragmatics, and inference. *Psychological Review, 109*, 646–678.

Kakade, S., & Dayan, P. (2002). Acquisition and extinction in autoshaping. *Psychological Review, 109*, 533–544.

Kelley, H. H. (1972). Attribution in social interaction. In E. E. Jones, D. E. Kanouse, H. H. Kelley, R. E. Nisbett, S. Valins, & B. Weiner (Eds.), *Attribution: Perceiving the causes of behaviour* (pp. 1–26). Morristown, NJ: General Learning Press.

Krauss, S., & Wang, X. T. (2003). The psychology of the Monty Hall problem: Discovering psychological mechanisms for solving a tenacious brain teaser. *Journal of Experimental Psychology: General, 132*, 3–22.

Lauritzen, S. L., & Speigelhalter, D. J. (1988). Local computations with probabilities on graphical structures and their applications to expert systems. *Journal of the Royal Statistical Society: Series B, 50,* 157–224.

Leslie, A. M. (1987). Pretense and representation: The origins of "theory of mind". *Psychological Review, 94,* 412–426.

Leslie, A. M., & Keeble, S. (1987). Do six-month-old infants perceive causality? *Cognition, 25,* 265–288.

Li, M., & Vitányi, P. (1997). *An introduction to Kolmogorov complexity theory and its applications* (2nd edition). Berlin: Springer-Verlag.

McDermott, D. (1987). A critique of pure reason. *Computational Intelligence, 3,* 151–237.

Michotte, A. (1963). *The perception of causality.* New York: Basic Books.

Newell, A. (1990). *Unified theories of cognition.* Cambridge, MA: Harvard Universtiy Press.

Oaksford, M., & Chater, N. (1991). Against cognitive science. *Mind and Language, 6,* 1–38.

Oaksford, M., & Chater, N. (1998). *Rationality in an uncertain world: Essays on the cognitive science of human reasoning.* Hove, UK: Psychology Press.

Oaksford, M., & Chater, N. (2002). Commonsense reasoning, logic and human rationality. In R. Elio (Ed.), *Common sense, reasoning and rationality* (pp. 174–214). Oxford: Oxford University Press.

Pearl, J. (1988). *Probabilistic reasoning in intelligent system: Networks of plausible inference.* San Mateo, CA: Morgan Kaufmann.

Pearl, J. (2000). *Causality: Models, reasoning, and inference.* Cambridge: Cambridge University Press.

Poletiek, F. H. (2001). *Hypothesis testing behaviour.* Hove, UK: Psychology Press.

Putnam, H. (1974). The 'corroboration' of theories. In P. A. Schilpp (Ed.), *The philosophy of Karl Popper* (pp. 221–240). La Salle, IL: Open Court.

Quine, W. V. O. (1951). Two dogmas of empiricism. *The Philosophical Review, 60,* 20–43.

Rescorla, R. A., & Wagner, A. R. (1972). A theory of Pavlovian conditioning: Variations in the effectiveness of reinforcement and nonreinforcement. In A. H. Black & W. F. Prokasy (Eds.), *Classical conditioning II: Current research and theory* (pp. 64–99). New York: Appleton.

Rips, L. J. (1994). *The psychology of proof: Deductive reasoning in human thinking.* Cambridge, MA: MIT Press.

Rumelhart, D. E., & McClelland, J. L. (Eds.) (1986). *Parallel distributed processing* (Vols. 1 and 2). Cambridge, MA: MIT Press.

Shanks, D. R. (1995). *The psychology of associative learning.* Cambridge, UK: Cambridge University Press.

Shimojo, S., & Ichikawa, S. (1989). Intuitive reasoning about probability: Theoretical and experimental analyses of the problem of three prisoners. *Cognition, 32,* 1–24.

Simpson, E. H. (1951). The interpretation of interaction in contingency tables. *Journal of the Royal Statistical Society, Series B, 13,* 238–241.

Sloman, S. A. (1996). The empirical case for two systems of reasoning. *Psychological Bulletin, 119,* 3–22.

Spelke, E. S. (1998). Nativism, empiricism, and the origins of knowledge. *Infant Behavior and Development, 21,* 181–200.

Sperber, D., Premack, D., & Premack, A. J. (Eds.) (1995). *Causal cognition: A multidisciplinary debate.* New York: Oxford University Press.

Spirtes, P., Glymour, C., & Scheines, R. (1993). *Causation, prediction, and search,* New York: Springer-Verlag.

Wagner, A. R. (1981). SOP. A model of autonomatic memory processing in animal behavior. In N. E. Spear & R. R. Miller (Eds.), *Information processing in animals: Memory mechanisms* (pp. 5–47). Hillsdale, NJ: Lawrence Erlbaum.

Wellman, M. P., & Henrion, M. (1993). Explaining "explaining away." *IEEE Transactions on Pattern Analysis and Machine Intelligence, 15,* 287–292.

PART IV

WHAT INFORMATION CONTENTS ARE SAMPLED?

11

What's in a Sample? A Manual for Building Cognitive Theories

Gerd Gigerenzer

PREVIEW

How do you build a model of mind? I discuss this question from the point of view of sampling. The idea that the mind samples information – from memory or from the environment – became prominent only after researchers began to emphasize sampling methods. This chapter provides a toolbox of potential uses of sampling, each of which can form a building block in a cognitive theory. In it I ask four questions: *who* samples, *why*, *what*, and *how*.

Who: In the social sciences (in contrast to the natural sciences), not only researchers sample, but so do the minds they study. *Why*: I distinguish two goals of sampling, hypotheses testing and measurement. *What*: Researchers can sample participants, objects, and variables to get information about psychological hypotheses, and minds may sample objects and variables to get information about their world. *How*: I distinguish four ways to sample: (i) no sampling, (ii) convenience sampling, (iii) random sampling from a defined population, and (iv) sequential sampling. These uses of sampling have received unequal attention. The prime source of our thinking about sampling seems to be R. A. Fisher's small-sample statistics, as opposed to the use of random sampling in quality control, the use of sequential sampling, and the use of sampling for measurement and parameter estimation. I use this legacy to identify potentials of sampling in adaptive cognition that have received little attention.

In his *Opticks*, Isaac Newton (1952/1704) reported experiments with prisms to demonstrate that white light consists of spectral colors. Newton did not sample, nor was he interested in means or variances. In his view, good experimentation had nothing to do with sampling. Newton was not antagonistic to sampling, but he used it only when he thought it was appropriate, as in quality control. In his role as the master of the London Royal Mint, Newton conducted routine sampling inspections to determine

239

whether the amount of gold in the coins was too little or too large. Just as in Newton's physics, experimentation and statistics were hostile rivals in nineteenth-century physiology and medicine. The great experimenter Claude Bernard used to ridicule the use of samples; his favorite example was that it is silly to collect the urine of one person, or of even a group of persons, over a twenty-four-hour period, because it is not the same before and after digestion, and averages are reifications of unreal conditions (Gigerenzer et al., 1989, p. 129). When B. F. Skinner demonstrated the effects of reinforcement schedules, he used one pigeon at a time, not two dozen. Although Skinner did not sample pigeons, William Estes (1959) pointed out that his theory assumed that his pigeons sampled information about the consequences of their behavior.

These cases illustrate some of the perplexing faces of sampling. What is in a sample? Why coins but not prisms or urine? Why did we come to believe that sampling and experimentation are two sides of the same coin, whereas Newton, Bernard, and Skinner did not? Why did Skinner not sample pigeons but assume that pigeons sample information? In this chapter, I try to put some order into the puzzling uses and nonuses of sampling. In the introduction to this volume, Fiedler and Juslin distinguished various forms of cognitive sampling, such as internal versus external sampling (e.g., memory versus the Internet), and the unit size of the objects of sampling. In contrast, I will focus on the evolution of the ideas of sampling – from the statistical toolbox to theories of mind. The selection of tools that ended up in cognitive theories and those that did not is in part historical accident. For instance, the tools that researchers happened to be familiar with had the best chances of being selected. What I hope to achieve with this chapter is not a complete taxonomy of sampling tools, but rather a view into the toolbox that can help in rethinking the possibilities of sampling and in using the toolbox creatively when building theories of mind. I will proceed to answer the question "What's in a sample?" by asking *who* samples, *what*, *why*, and *how*?

WHO SAMPLES?

I begin with the observation that the answer to the question of *who* samples information is different in the cognitive sciences than in the fields from which statistical sampling theory actually emerged – astronomy, agriculture, demographics, genetics, and quality control. In these noncognitive sciences, the researcher alone may sample (Figure 11.1). For instance, an astronomer may repeatedly measure the position of a star, or an agricultural researcher may fertilize a sample of plots and measure the average number of potatoes grown. Sampling concerns objects that are measured on some variable. Why would that be different in the cognitive sciences?

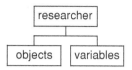

FIGURE 11.1. The structure of the potential uses of sampling in the noncognitive sciences. Researchers may sample objects (such as electrons) to measure these on variables (such as location and mass).

In the cognitive sciences (in contrast to the natural sciences), there are two "classes" of people who can engage in sampling: researchers and the participants of their studies (Figure 11.2). Whether and how researchers draw samples is generally seen as a methodological question. Whether and how researchers think that the minds of their participants engage in sampling of information is treated as a theoretical question. The labels "methodological" and "theoretical" suggest that both questions are unrelated and should be answered independently. After all, what do theories of cognitive processes have to do with the methods to test these theories?

I do not believe that these two issues are generally independent of each other. My hypothesis is that there is a significant correlation (not a one-to-one relation) in cognitive psychology between researchers' sampling practices and the role of sampling in their theories of mind. This hypothesis is an extension of my work on the tools-to-theories heuristic. The general thesis is twofold (Gigerenzer, 1991, 2000):

1. *Discovery*: New scientific tools, once entrenched in a scientist's daily practice, suggest new theoretical metaphors and concepts.
2. *Acceptance*: Once proposed by an individual scientist (or a group), the new theoretical concepts are more likely to be accepted by the scientific community if their members are also users of the new tool.

Note that Sigmund Freud, I. P. Pavlov, and the Gestalt psychologists, as well as the "father" of experimental psychology, Wilhelm Wundt, did not sample participants, and sampling played no role in their theories of mind.

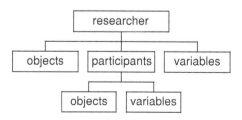

FIGURE 11.2. The structure of the potential uses of sampling in the cognitive sciences. Researchers may sample stimulus objects, participants, or variables, and their participants may themselves sample objects and variables.

All this changed after the unit of investigation ceased to be the individual person, and instead became the group mean – a process that started in the applied fields such as educational psychology (Danziger, 1990). Harold Kelley (1967), for instance, who used sampling and Fisher's analysis of variance to analyze his data, proposed that the mind attributes a cause to an effect in the same way, by sampling information and an intuitive version of analysis of variance. The community of social psychologists who also used analysis of variance as a routine tool accepted the theory quickly, and for a decade it virtually defined what social psychology was about. In contrast, R. Duncan Luce (1988) rejected routine use of analysis of variance as "mindless hypothesis testing in lieu of doing good research" (p. 582), and his theories of mind differed as a consequence. For instance, being familiar with the statistical tools of Jerzy Neyman and Egon S. Pearson and their doctrine of random sampling, Luce (1977) proposed that the mind might draw random samples and make decisions just as Neyman–Pearson theory does.

In summary, I propose that if researchers sample, they are likely to assume in their theories that the mind samples as well. If they do not, their view of cognitive processes typically also does not involve sampling. Moreover, the specific kind of sampling process that researchers use is likely to become part of their cognitive theories.

WHAT'S IN A SAMPLE?

In the cognitive sciences, the object of sampling can be threefold: participants, objects, and variables (Gigerenzer, 1981). Researchers can sample participants, stimulus objects, or variables. Today, participants are sampled habitually, objects rarely, and variables almost never. In addition, the minds under study can sample objects and variables. In cognitive theories, minds mostly sample objects, but rarely variables. This results in five possible uses of sampling in psychology (Figure 11.2).

My strict distinction between the cognitive and noncognitive sciences is an idealization; in reality there are bridges. The astronomers' concern with the "personal equation" of an observer illustrates such a link. Astronomers realized that researchers had systematically different response times when they determined the time a star travels through a certain point. This led to the study of astronomers' personal equations, that is, the time that needed to be subtracted to correct for their individual reaction times. In this situation, the object of sampling was both the researchers and their objects, such as stars (Gigerenzer et al., 1989).

WHY SAMPLING?

I distinguish two goals of sampling: hypotheses testing and measurement. Take significance tests as an example, where a sample statistic – such as *t* or

F – is calculated. In the early nineteenth century, significance tests were already being used by astronomers (Swijtink, 1987). Unlike present-day psychologists, astronomers used the tests to reject data (so-called outliers), not to reject hypotheses. At least provisionally, the astronomers assumed that a hypothesis (such as normal distribution of observational errors around the true position of a star) was correct and mistrusted the data. In astronomy, the goal was precise measurement, and this called for methods to identify bad data. In psychology, researchers trusted the data and mistrusted the hypotheses; that is, the goal became hypothesis testing, not measurement, following the influence of Fisher.

Hypothesis testing and measurement are concepts taken from statistical theory, and the obvious question is whether they are also good candidates for understanding how the mind works. Whatever the right answer may be, hypothesis testing has been widely assumed to be an adaptive goal of cognition, including in numerous studies that tried to show that people make systematic errors when testing hypotheses. Note that measurement has not been as extensively considered and studied as a goal of cognition (with some exceptions, such as the work of Brunswik, 1955), which is consistent with the fact that researchers tend to use their sampling tools for hypothesis testing rather than measurement.

HOW TO SAMPLE?

Sampling is not sampling. I distinguish four ways of how to sample, beginning with the nonuse of sampling.

Study Ideal Types, Not Samples

Newton thought that the science of optics was close to mathematics, where truth can be demonstrated in one single case, and he loathed researchers who replicated his experiments. Similarly, the most influential psychologists made their fame by studying one individual at a time. Freud's Anna O., Wundt's Wundt (the "father" of experimental psychology served as experimental subject), Pavlov's dog, Luria's mnemonist Shereshevski, and Simon's chess masters are illustrations. They represent ideal types, not averages. They may also represent distinct individual types, such as brain patients with specific lesions. Note that the ideal type approach does not mean that only one individual is studied. There may be several individuals, such as Freud's patients or Skinner's pigeons. The point is that the fundamental unit of analysis is $N = 1$, the singular case.

It is of a certain irony that Fisher's only psychological example in his influential *Design of Experiments* (1935) concerns the analysis of a Lady who claimed that she could tell whether the tea fusion or the milk was poured first into a cup of tea. This single-case study of extraordinary sensory abilities did not become the model for experimental research. Fisher sampled objects, not participants, as in Figure 11.1. Psychologists generally interpreted his methodology to be about sampling participants, not objects.

In his seminal book *Constructing the Subject*, Danziger (1990) argues that the reason why American psychologists turned away from studying individuals in the 1930s and 1940s and embraced means as their new "subject" had little to do with the scientific goals of our discipline. In contrast, this move was largely a reaction to university administrators' pressure on professors of psychology to show that their research was useful for applied fields, specifically educational research, which offered large sources of funding. The educational administrator was interested in questions such as whether a new curriculum would improve the *average* performance of pupils, and not in the study of the laws of the individual mind. Danziger provides detailed evidence that sampling of participants started in the applied fields but not in the core areas of psychology, and in the United States rather than in Germany, where professors of psychology were not, at that time, under pressure to legitimize their existence by proving their practical usefulness. Some of these differences still prevail today: Social psychologists tend to sample dozens or hundreds of undergraduates for five to ten minutes, whereas perceptual psychologists tend to study one or a few participants, each individually and for an extended time.

Convenience Samples

In the 1920s, Ronald A. Fisher (1890–1962) was chief statistician at the agricultural station in Rothamsted. Before Fisher, agricultural researchers had little sense for sampling. For instance, in the mid-nineteenth century, the British agriculturist James F. W. Johnston tried to determine which fertilizer was the best for the growth of turnips. He fertilized one plot, which yielded 24 bushels, and compared this result with those from three plots without fertilizer, which respectively yielded 18, 21, and 24 bushels of grain. Johnston understood that turnips naturally show up to 25% variation from plot to plot and that the average difference of about 10% that he observed was therefore not indicative of a real improvement. What Johnston did not understand was the importance of sample size – that this variability becomes less and less important as the number of plots on which the average is based increases (Gigerenzer et al., 1989, chapter 3).

Fisher's major contribution was to unite the rival practices of scientific experimentation and statistics. From Newton to Bernard to Skinner, this connection, as mentioned, had not existed. Fisher turned the two rival practices into two sides of the same coin, introducing randomized trials to agriculture, genetics, and medicine. By way of parapsychology and education, his ideas also conquered experimental psychology. The marriage between statistics and experimentation also changed statistics, from the general emphasis on large samples to Fisher's small-sample statistics. The idea of basing inferences on small samples – as in the typical experiment – was highly controversial. The statistician Richard von Mises (1957, p. 159) predicted that "the heyday of small sample theory... is already past." It was not past, however; Fisher prevailed.

Fisher's position emphasized some aspects of sampling – sample size, significance, and random assignment – and left out others. Most importantly, the concept of random sampling from a defined population had no place in Fisher's (1955) theory. Fisher's samples were not randomly drawn from a defined population. There was no such population in the first place. A sample whose population is not known is called a *convenience sample*. Fisher's liberal interpretation of how to sample became entrenched in psychology: The participants in psychological experiments are seldom randomly sampled, nor is a population defined.

Fisher did not think that convenience samples are a weakness. He held that in science there is no known population from which repeated sampling can be done. In a brilliant move, Fisher proposed to view any sample as a random sample from an *unknown hypothetical infinite population*. This solution has puzzled many statisticians: "This is, to me at all events, a most baffling conception"(Kendall, 1943, p. 17). However, Fisher's ideas about sampling were not the last word. Fisher had two powerful rivals, both of whom he detested.

Random Samples

The earliest quasi-random sampling procedure I know of is the trial of the Pyx (Stigler, 1999). The trial is a ceremony that goes back to the Middle Ages, the final stage of a sampling inspection scheme for the control of the quality of the London Royal Mint's production. The word Pyx refers to the box in which the sample of coins was collected, to determine whether the coins were too heavy or too light and contained too much or too little gold. As mentioned before, Newton served as master at the Royal Mint from 1699 until his death in 1727. The same Newton who did not use sampling for scientific experimentation supervised sampling for the purpose of quality control. The trial of the Pyx employed a form of sampling that is different from a convenience sample. It used a random sample drawn from a defined population, the total production of the Mint in one or a few years.

In the twentieth century, hypotheses testing that used random sampling from a defined population was formalized by the Polish statistician Jerzy Neyman and the British statistician Egon S. Pearson, the son of Karl Pearson. In their theory of hypotheses testing, one starts with two hypotheses (rather than one null hypothesis) and the probabilities of the two possible errors, Type I and Type II, from which the necessary sample size is calculated. A random sample is then drawn, after which one of the two hypotheses is accepted, and the other is rejected (in Fisher's scheme, the null can only be rejected, not accepted). Neyman and Pearson believed that they had improved the logic of Fisher's null hypothesis testing. Fisher (1955) did not think so. He thought that those who propose sampling randomly from a defined population and calculating sample size on the basis of cost–benefit trade-offs mistake science for quality control. He compared the Neyman–Pearsonians to Stalin's five-year plans, that is, to Russians who confuse technology with producing knowledge.

Sequential Sampling

A third line of sampling is sequential sampling, which had the status of a military secret during World War II and was later made public by Abraham Wald (1950). In comparison to Fisher's and Neyman and Pearsons's theories, sampling is sequential, not simultaneous. Whereas the sample size in Neyman–Pearson tests is fixed, calculated from a desired probability of Type I and Type II error, there is no fixed sample size in sequential sampling. Rather, a stopping criterion is calculated on the basis of the desired probabilities of Type I and Type II errors, and one continues to sample until it is reached. Sequential sampling has an advantage: It generally results in smaller sample sizes for the same alpha and power. Fisher was not fond of sequential sampling, for the same reasons that he despised Neyman–Pearson's theory. Although sequential sampling can save time and money, researchers in psychology rarely know and use it.

Which of these ideas of sampling have shaped psychological methods and theories of mind? I will now discuss each of the five possibilities for sampling in Figure 11.2.

DO RESEARCHERS SAMPLE PARTICIPANTS?

Do psychologists use individuals or samples as the unit of analysis? In the nineteenth and early twentieth centuries, the unit of analysis was clearly the individual. This changed in the United States during the 1920s, 1930s, and 1940s, when experimental studies of individuals were replaced by the treatment group experiment (Danziger, 1990). The use of samples of individuals began in the applied fields, such as education, and spread from there to the laboratories. The strongest resistance to this change in research practice came from the core of psychological science, perceptual research, where to the present day one can find reports of individual data rather than averages. Nonetheless, sampling participants has largely become the rule in psychology, and its purpose is almost exclusively hypothesis testing, or more precisely, null hypothesis testing. The use of samples for the measurement of parameters is comparatively rare.

How do researchers determine the size of the sample? Psychologists generally use rules of thumb ("25 in each group might be good enough") rather than the cost–benefit calculation prescribed by Neyman and Pearson. For instance, Cohen (1962) analyzed a volume of a major journal and found no calculation of sample size depending on the desired probabilities of Type I and Type II errors. When Sedlmeier and Gigerenzer (1989) analyzed the same journal twenty-four years later, nothing had changed: Sample size was still a matter of convenience, and as a consequence, the statistical power was embarrassingly low – a fact that went unnoticed.

Do researchers draw random samples of participants from a defined population? Experimental studies in which first a population is defined, then a random sample is drawn, and then the members of the sample are randomly assigned to the treatment conditions are extremely rare (e.g., Gigerenzer, 1984). When is sequential sampling of participants used? Virtually never. In summary, when researchers sample participants, they have perfectly internalized Fisher's ideas about sampling – except that, as already mentioned, Fisher sampled objects, not participants.

This almost exclusive reliance on convenience samples and Fisher's analysis of variance creates many of the problems that other uses of sampling tried to avoid. Researchers do not know the power of their tests; measuring constants and curves does not seem to be an issue; they waste time and money by never considering sequential sampling; and when they conclude that there is a true difference in the population means, nobody knows what this population is.

Why sample participants and analyze means if there is no population in the first place? Why not analyze a few individuals? In 1988, I spent a sabbatical at Harvard and had my office next to B. F. Skinner's. I asked him over tea why he continued to report one pigeon rather than averaging across pigeons. Skinner confessed that he once tried to run two dozen pigeons and feed the data into an analysis of variance, but he found that the results were less reliable than with one pigeon. You can keep one pigeon at a constant level of deprivation, he said, but you lose experimental control with twenty-four. Skinner had a point, which W. Gosset, the inventor of the t test, made before: "Obviously the important thing ... is to have a low real error, not to have a 'significant' result at a particular station. The latter seems to me to be nearly valueless in itself" (quoted in Pearson, 1939, p. 247). The real error can be measured by the standard deviation of the measurements, whereas a p value reflects sample size. One can get small real errors by increasing experimental control, rather than by increasing sample size. Experimental control can reveal individual differences in cognitive strategies that get lost in aggregate analyses of variance (e.g., Gigerenzer & Richter, 1990; Robinson, 1950).

In summary, psychologists' sampling of participants follows Fisher's convenience samples. Alternative sampling procedures are practically nonexistent. I believe that it is bad scientific practice to routinely use convenience samples and their averages as units of analysis. Rather, the default should be to analyze each individual on its own. This allows researchers to minimize the real error, to recognize systematic individual differences, and – last but not least – to know one's data.

DO RESEARCHERS SAMPLE OBJECTS?

Fisher made no distinction between the analysis of participants and objects. Do researchers sample stimulus objects in the same way they sample participants? The answer is "no": The classic use of random sampling for measurement in psychophysics has declined, and concern with sampling of objects is rare compared with sampling of participants.

In the astronomer's tradition, the use of random sampling for measurement is the first major use of sampling in psychophysics. In Fechner's work, samples were used to measure absolute and relative thresholds. In Thurstone's (1927) law of comparative judgment, an external stimulus corresponds to an internal normal distribution of subjective values, and a particular encounter with the stimulus corresponds to a randomly drawn subjective value from this distribution. The goal of repeated presentation of the same stimuli is to obtain psychological scales for subjective quantities. As Luce (1977) noted, there is a close similarity between the mathematics in Thurstone's law of comparative judgment and that in signal detection theory, but a striking difference in the interpretation. Thurstone used random variability for measurement, whereas in signal detection theory the mind is seen as an intuitive statistician who actively samples objects (Gigerenzer & Murray, 1987). The use of sampling for measurement has strongly declined since then, owing to the influence of Stevens and Likert, who promoted simple techniques such as magnitude estimation and rating scales that dispensed with the repeated presentation of the same stimulus. A tone, a stimulus person, or an attitude question is presented only once, and the participant is expected to rate it on a scale from, say, one to seven. Aside from research in perception and measurement theory, sampling of objects for the purpose of measuring subjective values and attitudes has been largely driven out of cognitive psychology (see, e.g., Wells & Windschitl, 1999).

As a consequence, Egon Brunswik (e.g., 1955) accused his colleagues of practicing a double standard by being concerned with the sampling of participants but not of stimulus objects. He argued that "representative" sampling of stimuli in natural environments is indispensable for studying vicarious functioning and the adaptation of cognition to its environment. For Brunswik, representative sampling meant random sampling from a defined population. In a classic experiment on size constancy, he walked with the participant through her natural environment and asked her at random intervals to estimate the size of objects she was looking at.

Like Fechner and Thurstone, Brunswik was concerned with measurement, but not with the construction of subjective scales. He understood cognition as an adaptive system and measured its performance in terms of "Brunswik ratios" (during his Vienna period, e.g., for measuring size constancy) and later (while at Berkeley) by means of correlations. He was not concerned with repeated presentations of the same object, nor with random sampling from any population, but with random sampling of objects from a natural population. Brunswik was influenced by the large-sample statistics of Karl Pearson. Pearson, who invented correlation statistics together with Galton, was involved in an intense intellectual and personal feud with Fisher. The clash between these two towering statisticians replicated itself in the division of psychology into two methodologically opposed camps: the large-sample correlational study of intelligence and personality, using the methods of Galton, Pearson, and Spearman, and the small-sample experimental study of cognition, using the methods of Fisher. The schism between these two scientific communities has been repeatedly discussed by the American Psychological Association (e.g., Cronbach, 1957) and still exists in full force today. Intelligence is studied with large samples; thinking is studied with small samples. The members of each community tend not to read and cite what the others write. Brunswik could not persuade his colleagues from the experimental community to take the correlational statistics of the rival discipline seriously. His concept of representative sampling died in the no-man's land between the hostile siblings. Even after the Brunswikian program was revived a decade after Brunswik died (Hammond, 1966; Hammond & Stuart, 2001), the one thing that is hard to find in neo-Brunswikian research is representative sampling.

But does it matter if researchers use random (representative) sampling or a convenience sample that is somehow selected? The answer depends on the goal of the study. If its goal is to measure the accuracy of perception or inaccuracy of judgment, then random sampling matters; if the goal is to test the predictions of competing models of cognitive mechanism, random sampling can be counterproductive, because tests will have higher power when critical items are selected (Rieskamp & Hoffrage, 1999). For claims about cognitive errors and illusions, the sampling of stimulus objects does matter (Gigerenzer & Fiedler, 2004). Research on the so-called overconfidence bias illustrates the point.

In a large number of experiments, participants were given a sample of general knowledge questions, such as "Which city has more inhabitants, Hyderabad or Islamabad?" Participants chose one alternative, such as "Islamabad," and then gave a confidence judgment, such as "70%," that their answer was correct. Average confidence was substantially higher than the proportion correct; this was termed "overconfidence bias" and attributed to a cognitive or motivational flaw. How and from what population the questions were sampled were not specified in these studies. As the story goes, one of the first researchers who conducted these studies went through almanacs and chose the questions with answers that surprised them. However, one can always demonstrate good or bad performance, depending on what items one selects. When we introduced random sampling from a defined population (cities in Germany), "overconfidence bias" largely or completely disappeared (Gigerenzer, Hoffrage, & Kleinbölting, 1991). The message that one of the most "stable" cognitive illusions could be largely due to researchers' sampling procedure was hard to accept, however, and was debated for years (e.g., by Griffin & Tversky, 1992). Finally, Juslin, Winman, & Olsson (2000) published a seminal review of more than 100 studies showing that "overconfidence bias" is practically zero with random sampling, but substantial with selected sampling (for a discussion of sampling in overconfidence research, see also Klayman et al., this volume; Hoffrage & Hertwig, this volume).

In summary, whereas sampling of participants has become institutionalized in experimental psychology, sampling of stimulus objects has not. Except for a few theories of measurement, which include psychophysics and Brunswik's representative design, it is not even an issue of general concern.

DO RESEARCHERS SAMPLE VARIABLES?

Now we enter no-man's land. Why would a researcher sample variables, and what would that entail? Few theories in psychology are concerned with how the experimenter samples the variables on which participants judge objects. One is personal construct theory (Kelly, 1955). The goal of the theory is to analyze the "personal constructs" people use to understand themselves and their world. George Kelly's emphasis on the subjective construction of the world precludes the use of a fixed set of variables, such as a semantic differential, and imposition of it on all participants. Instead, Kelly describes methods that elicit the constructs relevant for each person. One is to present triples of objects (such as mother, sister, and yourself), and to ask the participant first which of the two are most similar, then what it is that makes them so similar, and finally what makes the two different from the third one.

Unlike when sampling participants and objects, situations in which a population of variables can be defined are extremely rare. In Kelly's attempts to probe individual constructs, for instance, the distinction between convenience samples and random or representative samples appears blurred. If the goal of the research is to obtain statements about the categories or dimensions in which people see their world, then the researcher needs to think of how to sample the relevant individual variables.

I turn now to theories of how minds sample. According to our scheme, minds can sample along two dimensions: objects and cues (variables).

DO MINDS SAMPLE OBJECTS?

The idea that the mind samples objects to compute averages, variances, or test hypotheses only emerged after the institutionalization of inferential statistics in psychology (1940–1955), consistent with the tools-to-theories heuristic (Gigerenzer & Murray, 1987). From Fechner to Thurstone, probability was linked with the measurement of thresholds and the construction of scales of sensation, but not with the image of the mind as an intuitive statistician who draws samples for *cognitive inferences* or *hypothesis testing*. One of the first and most influential theories of intuitive statistics was signal detection theory (Tanner & Swets, 1954), which transformed Neyman–Pearson theory into a theory of mind.

There seem to be two main reasons for this late emergence of the view that the mind actively engages in sampling, The first is described by tools-to-theories: Only after a combination of Fisher's and Neyman–Pearson's statistical tools became entrenched in the methodological practices of psychologists around 1950 did researchers begin to propose and accept the idea that the mind might also be an intuitive statistician who uses similar tools (Gigerenzer, 1991). The second reason is the influence of Stanley S. Stevens, who rejected inferential statistics (Gigerenzer & Murray, 1987, chapter 2). For instance, in the first chapter of his *Handbook of Experimental Psychology*, Stevens (1951) included a section entitled "probability" (pp. 44–47), whose only purpose seems to be to warn the reader of the confusion that might result from applying probability theory to anything, including psychology. He was deeply suspicious of probabilistic models on the grounds that they can never be definitely disproved. Like David Krech and Edwin G. Boring, Stevens stands in a long tradition of psychologists who are determinists at heart.

Yet many current theories in cognitive and social psychology still do not incorporate any models of sampling. Consistent with this omission, most experimental tasks lay out all objects in front of the participants and thereby exclude information search in the first place. This tends to create cognitive theories with a blind spot for how people sample information and when they stop. This in turn creates a blind spot for the situations in which the mind does and does not sample, including when there might be evolutionary reasons to rely only on a single observation.

When Is It Adaptive Not to Sample?

Although Skinner did not sample pigeons, as mentioned before, his view about operant conditioning can be seen as a theory of information sampling. Specifically, this interpretation is invited by his variable reinforcement schedules, where an individual repeatedly exhibits a behavior (such as pecking in pigeons and begging in children) and samples information about consequences (such as food). Skinner's laws of operant behavior were designed to be general-purpose, that is, to hold true for all stimuli and responses. This assumption is known as the *equipotentiality* hypothesis. Similarly, when Thorndike found that cats were slow in learning to pull strings to escape from puzzle boxes, he concluded that learning occurs by trial and error, and he hoped that this would be a general law of learning. If all stimuli were equal, minds should always sample information to be able to learn from experience. The assumption that all stimuli are equal is also implicit in many recent versions of reinforcement learning (e.g., Erev & Roth, 2001). Consider William Estes, who was one of the first to formulate Skinner's ideas in the language of sampling:

All stimulus elements are equally likely to be sampled and the probability of a response at any time is equal to the proportion of elements ... that are connected to it. ... On any acquisition trial all stimulus elements sampled by the organism become connected to the response reinforced on that trial. (Estes, 1959, p. 399)

Is the assumption of the equivalence of stimulus objects in sampling correct? Are there stimulus objects that an organism does not and should not sample? John Garcia is best known for his challenge of the equipotentiality hypothesis. For instance, he showed that in a single trial, a rat can learn to avoid flavored water when it is followed by experimentally induced nausea, even when the nausea occurs two hours later. However, the same rat has great difficulty learning to avoid the flavored water when it is repeatedly paired with an electric shock immediately after the tasting:

From the evolutionary view, the rat is a biased learning machine designed by natural selection to form certain CS-US [conditional stimulus – unconditional stimulus] associations rapidly but not others. From a traditional learning viewpoint, the rat was an unbiased learner able to make any association in accordance with the general principles of contiguity, effect, and similarity. (Garcia y Robertson & Garcia, 1985, p. 25)

The evolutionary rationale for one-trial learning as opposed to sampling stimulus objects is transparent. Learning by sampling and proportionally increasing the probability of response can be dangerous or deadly when it comes to food, diet, and health. To avoid food poisoning, an organism can have a genetically inherited aversion against a food, or a genetically coded preparedness to learn a certain class of associations in one or a few instances.

Genetically coded preparedness shows that sampling cannot and should not be an element of all cognitive processes. Rather, whether an organism samples (a so-called bottom-up process) or does not (a top-down process) largely depends on the past and present environmental contingencies. A mind can afford to learn some contingencies, but not all – sampling can be overly dangerous. One-trial learning amply illustrates the adaptive nature of cognitive processes, which codes what will be sampled and what will not.

Convenience Sampling

One class of models developed after the inference revolution assumes that the mind samples information to test hypotheses, just as researchers came to do. Consider the question of how the mind attributes a cause to an event, which has been investigated in the work of Piaget and Michotte. In Michotte's (1963/1946) view, for instance, causal attribution was a consequence of certain spatio-temporal relationships; that is, it was determined "outside" the mind and did not involve inductive inference based on samples of information (see also Chater & Oaksford, this volume). After analysis of variance became institutionalized in experimental psychology, Harold Kelley (1967) proposed that the mind attributes a cause to an event just as researchers test causal hypotheses: by analyzing samples of covariation information and calculating F ratios (F for Fisher) in an intuitive analogy to analysis of variance. Note that the new ANOVA mind used the tests for rejecting hypotheses while trusting the data, parallel to the way researchers in psychology use ANOVA. If Kelley had lived a century and a half earlier, he might have instead looked to the astronomers' significance tests. As pointed out earlier, the astronomers assumed (at least, provisionally) that the hypothesis was correct but mistrusted the data. If this use of sampling had been taken as an analogy, the mind would have appeared to be expectation-driven rather than data-driven.

Kelley's causal attribution theory illustrates how Fisher's ANOVA was used to model the mind's causal thinking, assuming that the mind uses convenience samples for making inductive inferences about causal hypotheses.

As clear as the distinction between convenience and random sampling is in statistical theory, it is less so in theories that assume that the mind samples objects. Is the sample of people a tourist encounters on a trip to Beijing a random sample or a convenience sample? It may depend on whether the tour guide has planned all encounters ahead, or whether the tourist strolls through the city alone, or whether the tour guide has picked a random sample of Beijing tenth-graders to meet with.

Random Sampling

Psychophysics has been strongly influenced by Neyman–Pearson theory. Under the name of signal detection theory, it became a model of how the mind detects a stimulus against noise or a difference between two stimuli, and it replaced the concepts of absolute and relative thresholds.

Neyman's emphasis on random sampling from a defined population, as in quality control, became part of the cognitive mechanisms. For instance, Luce (1977; Luce & Green, 1972) assumed that a transducer (such as the human ear) transforms the intensity of a signal into neural pulse trains in parallel nerve fibers, and the central nervous system (CNS) draws a random sample of all activated fibers. The size of the sample is assumed to depend on whether or not the signal activates fibers to which the CNS is attending. From each fiber in the sample, the CNS estimates the pulse rate by either counting or timing, and these numbers are then aggregated into a single internal representation of the signal intensity. In Luce's theory, the mind was pictured as a statistician of the Neyman and Pearson school, and the processes of random sampling, inference, decision, and hypotheses testing were freed of their conscious connections and seen as unconscious mechanisms of the brain.

Sequential Sampling

The former First Lady, Barbara Bush, is reported to have said, "I married the first man I ever kissed. When I tell this to my children they just about throw up" (quoted in Todd & Miller, 1999). Is one enough, just as in Garcia's experiments, or should Barbara Bush have sampled more potential husbands? After Johannes Kepler's first wife died of cholera, he immediately began a methodological search for a replacement. Within two years, he investigated eleven candidates and finally married Number 5, a woman who was well educated but not endowed with the highest rank or dowry. Are eleven women a large enough sample? Perhaps too large, because the candidate Number 4, a woman of high social status and with a tempting dowry, whom friends urged Kepler to choose, rejected him for having toyed with her too long (Todd & Miller, 1999). Swiss economists Frey and Eichenberger (1996) asserted that people do not sample enough when seeking a mate, taking the high incidence of divorce and marital misery as evidence. In contrast, Todd and Miller (1999) argued that given the degree of uncertainty – one never can know how a prospective spouse will turn out – the goal of mate search can only be to find a fairly good partner, and they showed that under certain assumptions, Kepler's sample was large enough.

Mate search is essentially sequential for humans, although there are female birds that can inspect an entire sample of males lined up simultaneously. Since sequential sampling has never become part of the statistical tools used by researchers in psychology, one might expect from the tools-to-theories heuristic that minds are not pictured as performing sequential sampling either. This is mostly, but not entirely, true.

Cognitive processes that involve sequential sampling have been modeled in two different ways: optimizing models and heuristic models. Optimizing models are based on Abraham Wald's (1950) statistical theory, which has a stopping rule that is optimal relative to given probabilities of Type I and Type II errors (e.g., Anderson, 1990; for a discussion of optimal stopping rules see Busemeyer & Rapoport, 1988). Many of these models have been applied to psychophysical tasks, such as judging which of two lines is longer. In the case of a binary hypothesis (such as line *A* or *B*; marry or not marry), the basic idea of most sequential models is the following: A threshold is calculated for accepting one of the two hypotheses, based on the costs of the two possible errors, such as wrongly judging line *A* as larger, or wrongly deciding that to marry is the better option. Each reason or observation is then weighted and sampling of objects is continued until the threshold for one hypothesis is met, at which point the search is stopped and the hypothesis is accepted. These models are often presented as *as-if* models, whose task is to predict the outcome rather than the process of decision making, although it has been suggested that the calculations might be performed unconsciously.

Heuristic models of sequential sampling assume an aspiration level rather than optimization. Their goal is to model the process and the outcome of judgment or decision making. For instance, in Herbert Simon's (1955, 1956) models of satisficing, a person sequentially samples objects (such as houses or potential spouses) until encountering the first one that meets an aspiration level. In Reinhard Selten's (2001) theories of satisficing, the aspiration level can change with the duration of the sampling process.

Can sequential sampling ever be random? In statistical theory, the answer is yes. One draws sequentially from a population, until the stopping rule applies. In the case of mental sampling, it is much harder to decide whether a sequential search process should count as random. Consider, for instance, a satisficer who sequentially encounters potential spouses or houses until finding one that exceeds her aspiration level. In most cases, the sequential sample will be a convenience sample rather than a random sample from a defined population.

The relative rarity of sequential sampling in models of the mind goes hand in hand with experimenters' preference for tasks that do not provide an opportunity for the participants to sample objects: All objects are already displayed in front of the participant. Few experiments address the following questions: (i) When does the mind sample simultaneously versus sequentially? (ii) Is there an order in sequential search, that is, is the search random or systematic? (iii) How is sequential search stopped, that is, what determines when a sample is large enough?

DOES THE MIND SAMPLE VARIABLES?

Satisficing refers to a class of heuristics that apply to situations in which an aspiration level is given and the objects or alternatives are sampled sequentially. Alternatively, the objects can be given and the variables (cues, reasons, or features) need to be searched. Examples include choosing between two job offers (paired comparison) and classifying patients as high-risk or low-risk (categorization). As I mentioned, cognitive theories that model how minds sample objects are few, but those that model how minds sample variables are even rarer. For instance, models of similarity generally assume that the variables (features, cues, etc.) are already given and then postulate some way in which individual features are combined to form a similarity judgment – city block distance and feature matching are illustrations of this. However, in everyday situations, the features are not always laid out in front of a person but need to be searched for, and since there is typically a large or infinite number of features or cues, cognition may involve sampling features. Sampling cues or features can occur inside or outside of memory (e.g., on the Internet).

Unlike for sequential sampling of objects, there seem to be no optimizing models but only heuristic models for sampling of variables. There are two possible reasons. First, it is hard to think of a realistic population of variables, in contrast to a population of objects. Two job candidates, for instance, can vary on many different cues, and it is hard to define a population of cues. Second, the large number of cues makes optimizing models such as Bayes's rule or full classification trees computationally intractable, because the number of decision nodes increases exponentially with the number of cues in a full tree. Thus, even optimizing models need to use heuristic simplifications, as in Bayesian trees (Martignon & Laskey, 1999).

Heuristic models of sequential sampling include two major classes: one-reason decision making and tallying. Each heuristic consists of a search rule that specifies the direction of sampling, a stopping rule that specifies when sampling is terminated, and a decision rule. "Take the Best" (Gigerenzer & Goldstein, 1996) is an example of a heuristic that employs ordered search and one-reason decision making.

Take the Best

1. Search by validity: Search through cues in order of their validity. Look up the cue values of the cue with the highest validity first.
2. One-reason stopping rule: If one object has a positive cue value (1) and the other does not (0 or unknown), then stop the search and proceed to Step 3. Otherwise exclude this cue and return to Step 1. If no more cues are found, guess.
3. One-reason decision making: Predict that the object with the positive cue value (1) has the higher value on the criterion.

The validity of a cue i is defined as $v_i = R_i / P_i$, where R_i = number of correct predictions by cue i, and P_i = number of pairs where the values of cue i differ between objects. "Take The Best" typically samples a small number of cues. Now consider an example for a tallying heuristic, which relies on adding but not on weighing (or order).

Tallying

1. Random search: Search through cues in random order. Look up the cue values.
2. Stopping rule: After $m(1 < m \leq M)$ cues, stop the search and determine which object has more positive cue values (1), and proceed to Step 3. If the number is equal, return to Step 1 and search for another cue. If no more cues are found, guess.
3. Tallying rule: Predict that the object with the higher number of positive cue values (1) has the higher value on the criterion.

Here M refers to the total number of cues and m refers to the number of cues searched for.

The literature discusses various versions of tallying, such as unit-weight models in which all cues ($m = M$) or the m significant cues are looked up (Dawes, 1979; Einhorn & Hogarth, 1975). Unlike as-if models, which predict outcomes only, these models of heuristics predict process and outcome and can be subjected to a stronger test. A discussion of these and similar heuristics and the situations in which they are accurate can be found in Gigerenzer et al. (1999) and Gigerenzer & Selten (2001).

In summary, cognitive sampling of cues or features is a process that has been given little attention. Just as for sampling of objects, heuristic models exist that formulate stopping rules to determine when such a sample is large enough.

WHAT'S IN A SAMPLE?

Shakespeare has Juliet ask; "What's in a name?" What is in a name uncovers what the name means to us, and by analogy, what is in a sample reveals what sampling means to us. The taxonomy proposed in this chapter distinguishes two subjects of sampling (experimenter versus participant), two purposes of sampling (measurement versus hypotheses testing), three targets of sampling (participants, objects, and variables), and four ways of how to sample ($N = 1$, i.e., no sampling; convenience sampling; random sampling; and sequential sampling). As in Brunswik's representative design, these dimensions do not form a complete factorial design; for instance, participants do not sample participants. Among the logically possible uses of sampling, some are realized in practice, whereas others are not or are realized only by a minority. Is the resulting picture of the actual uses and nonuses of sampling one of chaos, orderly chaos, or reasonable choice? Is the overreliance on Fisher's convenience sampling in methodology a good or bad thing, and is the relative neglect of sequential sampling in both methodology and cognitive theories reasonable or unreasonable? Why is so little attention paid to the mind's sampling of features? Whatever the reader's evaluation is, a toolbox can open one's eyes to the missed opportunities or blind spots of sampling. There may be other paths to a toolbox of methods for sampling; the present one has a deliberate bias toward the evolution of the various ideas of sampling and the intellectual inheritance we owe to competing statistical schools. This historical window allows us to understand the current patchwork of sampling in both methodology and theory along with the possibilities of designing new theories of mind that overcome the historical biases we inherited.

References

Anderson, J. R. (1990). The adaptive character of thought. Hillsdale, NJ: Erlbaum.

Brunswik, E. (1955). Representative design and probabilistic theory in a functional psychology. *Psychological Review, 62*, 193–217.

Busemeyer, J. R., & Rapoport, A. (1988). Psychological models of deferred decision making. *Journal of Mathematical Psychology, 32*, 91–134.

Chater, N., & Oaksford, M. (2005). Mental mechanisms: Speculations on human causal learning and reasoning. In K. Fiedler & P. Juslin (Eds.), *Information sampling and adaptive cognition*. Cambridge, UK: Cambridge University Press.

Cohen, J. (1962). The statistical power of abnormal-social psychological research: A review. *Journal of Abnormal and Social Psychology, 65*, 145–153.

Cronbach, L. J. (1957). The two disciplines of scientific psychology. *American Psychologist, 12*, 671–684.

Danziger, K. (1990). *Constructing the subject: Historical origins of psychological research*. Cambridge, UK: Cambridge University Press.

Dawes, R. M. (1979). The robust beauty of improper linear models in decision making. *American Psychologist, 34*, 571–582.

Einhorn, H. J., & Hogarth, R. M. (1975). Unit weighting schemes for decision making. *Organizational Behavior and Human Performance, 13*, 171–192.

Erev, I., & Roth, A. E. (2001). Simple reinforcement learning models and reciprocation in the prisoner's dilemma game. In G. Gigerenzer & R. Selten (Eds.), *Bounded rationality: The adaptive toolbox* (pp. 215–231). Cambridge, MA: MIT Press.

Estes, W. K. (1959). The statistical approach to learning theory. In S. Koch (Ed.), *Psychology: A study of science* (Vol. 2, pp. 380–491). New York: McGraw-Hill.

Fisher, R. A. (1935). *The design of experiments*. Edinburgh: Oliver and Boyd.

Fisher, R. A. (1955). Statistical methods and scientific induction. *Journal of the Royal Statistical Society, Series B, 17*, 69–78.

Frey, B. S., & Eichenberger, R. (1996). Marriage paradoxes. *Rationality and Society, 8*, 187–206.

Garcia y Robertson, R., & Garcia, J. (1985). X-rays and learned taste aversions: Historical and psychological ramifications. In T. G. Burish, S. M. Levy, & B. E. Meyerowitz (Eds.), *Cancer, nutrition and eating behavior: A biobehavioral perspective* (pp. 11–41). Hillsdale, NJ: Lawrence Erlbaum.

Gigerenzer, G. (1981). *Messung und Modellbildung in der Psychologie* [Measurement and modeling in psychology]. Munich: Ernst Reinhardt (UTB).

Gigerenzer, G. (1984). External validity of laboratory experiments: The frequency-validity relationship. *American Journal of Psychology, 97*, 185–195.

Gigerenzer, G. (1991). From tools to theories: A heuristic of discovery in cognitive psychology. *Psychological Review, 98*, 254–267.

Gigerenzer, G. (2000). *Adaptive thinking: Rationality in the real world*. New York: Oxford University Press.

Gigerenzer, G., & Fiedler, K. (2004). *Minds in environments: The potential of an ecological approach to cognition*. Manuscript submitted for publication.

Gigerenzer, G., & Goldstein, D. G. (1996). Reasoning the fast and frugal way: Models of bounded rationality. *Psychological Review, 103*, 650–669.

Gigerenzer, G., Hoffrage, U., & Kleinbölting, H. (1991). Probabilistic mental models: A Brunswikian theory of confidence. *Psychological Review, 98*, 506–528.

Gigerenzer, G., & Murray, D. J. (1987). *Cognition as intuitive statistics*. Hillsdale, NJ: Lawrence Erlbaum.

Gigerenzer, G., & Richter, H. R. (1990). Context effects and their interaction with development: Area judgments. *Cognitive Development, 5*, 235–264.

Gigerenzer, G., & Selten, R. (Eds.) (2001). *Bounded rationality: The adaptive toolbox*. Cambridge, MA: MIT Press.

Gigerenzer, G., Swijtink, Z., Porter, T., Daston, L., Beatty, J., & Krüger, L. (1989). *The empire of chance. How probability changed science and everyday life*. Cambridge, UK: Cambridge University Press.

Gigerenzer, G., Todd, P. M., & the ABC Research Group (1999). *Simple heuristics that make us smart*. New York: Oxford University Press.

Griffin, D. & Tversky, A. (1992). The weighing of evidence and the determinants of confidence. *Cognitive Psychology, 24*, 411–435.

Hammond, K. R. (1966). *The psychology of Egon Brunswik*. New York: Holt, Rinehart & Winston.

Hammond, K. R., & Stewart, T. R. (Eds.) (2001). *The essential Brunswik: Beginnings, explications, applications*. New York: Oxford University Press.

Hoffrage, U., & Hertwig, R. (2005). Which world should be represented in representative design? In K. Fiedler & P. Juslin (Eds.), *Information sampling and adaptive cognition*. Cambridge, UK: Cambridge University Press.

Juslin, P., Winman, A., & Olssen, H. (2000). Naive empiricism and dogmatism in confidence research: A critical examination of the hard-easy effect. *Psychological Review, 107*, 384–396.

Kelley, H. H. (1967). Attribution theory in social psychology. In D. Levine (Ed.), *Nebraska symposium on motivation* (Vol. 15, pp. 192–238). Lincoln: University of Nebraska Press.

Kelly, G. A. (1955). *The Psychology of personal constructs*. New York: Norton.

Kendall, M. G. (1943). *The advanced theory of statistics* (Vol. 1). New York: Lippincott.

Klayman, J., Soll, J., Juslin, P., & Winman, A. (2005). Subjective confidence and the sampling of knowledge. In K. Fiedler & P. Juslin (Eds.), *Information sampling and adaptive cognition*. Cambridge, UK: Cambridge University Press.

Luce, R. D. (1977). Thurstone's discriminal processes fifty years later. *Psychometrika, 42*, 461–489.

Luce, R. D. (1988). The tools-to-theory hypothesis. Review of G. Gigerenzer and D. J. Murray, "Cognition as intuitive statistics". *Contemporary Psychology, 32*, 151–178.

Luce, R. D., & Green, D. M. (1972). A neural timing theory for response times and the psychophysics of intensity. *Psychological Review, 79*, 14–57.

Martignon, L., & Laskey, K. B. (1999). Bayesian benchmarks for fast and frugal heuristics. In G. Gigerenzer, P. M. Todd, & the ABC Research Group, *Simple heuristics that make us smart* (pp. 169–188). New York: Oxford University Press.

Michotte, A. (1963). *The perception of causality.* London: Methuen. (Original work published 1946.)

Newton, I. (1952). *Opticks: Or a treatise of the reflections, refractions, inflections and colours of light.* New York: Dover. (Original work published 1704.)

Pearson, E. S. (1939). "Student" as statistician. *Biometrika, 30*, 210–250.

Rieskamp, J., & Hoffrage, U. (1999). When do people use simple heuristics and how can we tell? In G. Gigerenzer, P. M. Todd, & the ABC Research Group, *Simple heuristics that make us smart* (pp. 141–167). New York: Oxford University Press.

Robinson, W. (1950). Ecological correlation and the behavior of individuals. *American Sociological Review, 15*, 351–357.

Sedlmeier, P., & Gigerenzer, G. (1989). Do studies of statistical power have an effect on the power of studies? *Psychological Bulletin, 105*, 309–316.

Selten, R. (2001). What is bounded rationality? In G. Gigerenzer & R. Selten (Eds.), *Bounded Rationality: The adaptive toolbox* (pp. 13–36). Cambridge, MA: MIT Press.

Simon, H. A. (1955). A behavioral model of rational choice. *Quarterly Journal of Economics, 69*, 99–118.

Simon, H. A. (1956). Rational choice and the structure of environments. *Psychological Review, 63*, 129–138.

Stevens, S. S. (Ed.) (1951). *Handbook of experimental psychology.* New York: Wiley.

Stigler, S. M. (1999). *Statistics on the table: The history of statistical concepts and methods.* Cambridge, MA: Harvard University Press.

Swijtink, Z. G. (1987). The objectification of observation: Measurement and statistical methods in the nineteenth century. In L. Krüger, L. Daston, & M. Heidelberger (Eds.), *The probabalistic revolution, Vol. I: Ideas in history* (pp. 261–285). Cambridge, MA: MIT Press.

Tanner, W. P., Jr., & Swets, J. A. (1954). A decision-making theory of visual detection. *Psychological Review, 61*, 401–409.

Thurstone, L. L. (1927). A law of comparative judgment. *Psychological Review, 34*, 273–286.

Todd, P. M., & Miller, G. F. (1999). From pride and prejudice to persuasion: Satisficing in mate search. In G. Gigerenzer, P. M. Todd, & the ABC Research Group, *Simple heuristics that make us smart* (pp. 287–308). New York: Oxford University Press.

von Mises, R. (1957). *Probability, statistics, and truth.* London: Allen and Unwin. (Original work published 1928.)

Wald, A. (1950). *Statistical decision functions.* New York: Wiley.

Wells, G. L., & Windschitl, P. D. (1999). Stimulus sampling and social psychological experimentation. *Personality and Social Psychology Bulletin, 25*, 1115–1125.

12

Assessing Evidential Support in Uncertain Environments

Chris M. White and Derek J. Koehler

In this chapter, we explore a classic problem in psychology: How do individuals draw on their previous experience in an uncertain environment to make a prediction or diagnosis on the basis of a set of informational cues? Intuitive predictions based on multiple cues are often highly sensitive to environmental contingencies yet at the same time often exhibit pronounced and consistent biases.

Our research has focused on judgments of the probability of an outcome based on binary cues (e.g., present/absent), in which the diagnostic value of those cues has been learned from direct experience in an uncertain environment. For example, in medical diagnosis, we might predict which disease a patient has based on a symptom that a patient does or does not exhibit. We have developed a model that captures both the strengths and weaknesses of intuitive judgments of this kind (the Evidential Support Accumulation Model, ESAM; Koehler, White, & Grondin, 2003). The model assumes that the frequency with which each cue is observed to have co-occurred with the outcome variable of interest is stored. The perceived diagnostic value of each cue, based on these frequencies, is calculated in a normatively appropriate fashion and integrated with the prior probability of the outcome to arrive at a final probability judgment for a given outcome.

ESAM is designed to account for people's behavior in uncertain environments created in the lab that are, effectively by definition, sampled without bias. In addition, we believe that these types of uncertain environments (those in which multiple cues imperfectly predict which of multiple outcomes will occur) can be viewed as a representative sample of the environments that people actually encounter in their lives. (In addition, we sample a number of different environments in the simulations that we report.) Therefore, factors of external sampling cannot explain the observed judgmental biases. The most important of these biases is that probabilities assigned to competing hypotheses given a particular pattern of symptoms consistently sum to more than 1 and that this sum increases as the number of present symptoms increases.

Since external sampling cannot explain these biases, we attribute them to the effects of internal sampling. In this chapter we show that when the amount of information that the model samples is reduced, relatively small decrements in performance occur. We also summarize previous research showing that sampling biases incorporated into the model allow it to reproduce a number of effects exhibited in people's judgments. Therefore, based on the current results, we conclude that the internal sampling biases posited by the model may not be overly harmful to the accuracy of the judgments it produces in at least some environments and judgment tasks.

The model is selective with regard to the sampling of informational cues in three respects. First, ESAM disregards conditional dependencies among cues. Second, the model is selectively attuned to present rather than absent cue statuses. And third, a more restricted set of cues is sampled when assessing the extent to which the evidence supports the alternative hypotheses than when assessing the support for the focal hypothesis of the judgment. As a result, judgments produced by the model are systematically biased even when they are based on experience in the form of an objectively representative information sample.

In summary, our work concentrates on the selective internal sampling of cue information. The unit of sampling is at the cue level, which is the lowest level at which it could realistically be in such a task, as is the case for the "Take the Best" heuristic (Gigerenzer and Goldstein, 1999). Toward the end of this chapter we discuss work by other researchers who assume that the unit of internal sampling for this type of task is that of entire exemplars (Dougherty, Gettys, & Ogden, 1999; Juslin & Persson, 2002).

This chapter is organized as follows: After describing ESAM and showing how certain cues are selected for sampling, we review recent experimental results that are consistent with the basic predictions of the model. Following this, we explore how ESAM's performance varies (relative to that of a Bayesian model) based on the number of cues that are sampled in different types of multiple-cue learning environments. We then briefly review the work of other researchers in this area and offer some general conclusions.

THE MODEL

Support Theory

ESAM was developed within the theoretical framework of support theory (Tversky & Koehler, 1994; Rottenstreich & Tversky, 1997). According to support theory, judged probability is assigned to descriptions of events, referred to as *hypotheses*, rather than directly to set-theoretic events as in probability theory. Support theory is thus nonextensional, allowing different probability judgments to be assigned to different descriptions of the same event, reflecting the observation that intuitive judgments of an event's probability are systematically influenced by the way in which that event is described (e.g., Fischhoff, Slovic, & Lichtenstein, 1978).

Support theory consists of two basic assumptions. The first is that judged probability reflects the relative support for the focal and alternative hypotheses:

$$P(A, B) = \frac{s(A)}{s(A) + s(B)}. \tag{12.1}$$

That is, the judged probability of focal hypothesis A rather than alternative hypothesis B is given by the evidential support for A, denoted $s(A)$, normalized relative to that available for B, denoted $s(B)$. If, for example, A and B represent two mutually exclusive diseases from which a patient might be suffering, the judged probability that the patient has disease A rather than disease B, denoted $P(A, B)$, is assumed to reflect the balance of evidential support for A versus that for B.

Support theory's second assumption is that if A is an implicit disjunction (e.g., the patient has a respiratory infection) that refers to the same event as an explicit disjunction of exclusive hypotheses A_1 and A_2 (e.g., the patient has a viral respiratory infection or a bacterial respiratory infection), denoted $(A_1 \vee A_2)$, then

$$s(A) \leq s(A_1 \vee A_2) \leq s(A_1) + s(A_2). \tag{12.2}$$

That is, the support of the implicit disjunction A is less than or equal to that of the explicit disjunction $A_1 \vee A_2$ (because how an event is described affects its assessed support), which in turn is less than or equal to the sum of support of its components when assessed individually (Rottenstreich & Tversky, 1997). In short, unpacking the implicit disjunction A into its components A_1 and A_2 can only increase its support, and hence its judged probability (cf. Fischhoff et al., 1978). The relationship between the support of A and its components A_1 and A_2 is said to be *subadditive*, in the sense that the whole receives less than the sum of its parts.

Support theory implies that, whenever an elementary hypothesis is evaluated relative to all of its alternatives taken as a group (referred to as the *residual*), the weight given to an alternative included implicitly in the residual is generally less than what it would have received had it been evaluated in isolation. Consider a case in which there are three elementary hypotheses: A, B, and C. For instance, suppose a patient is suffering from one (and only one) of three possible flu strains. According to support theory, when a person is asked to judge the probability that the patient is suffering from Flu Strain A, the resulting "elementary" probability judgment $P(A, \bar{A})$ is determined by the evidential support for Flu Strain A normalized relative to that for its complement (the residual not-A, represented $\overline{A}$). In this case, its complement is an implicit disjunction of Flu Strains B and C. Support theory implies that packing these alternatives together in an implicit disjunction (i.e., the residual) generally results in a loss of support, thereby increasing A's judged probability.

As a result, if separate elementary judgments are obtained of the probability of hypotheses A, B, and C, the total probability

$$T = P(A, \overline{A}) + P(B, \overline{B}) + P(C, \overline{C}) \tag{12.3}$$

assigned to the three elementary hypotheses will generally exceed 1, in violation of probability theory. The degree of subadditivity associated with the set of elementary judgments can be measured by the extent to which the total probability T assigned to them exceeds 1; the greater the value of T, the greater the degree of subadditivity.

A more precise measure of the degree of subadditivity associated with a single judgment is given by a discounting factor $w_{\overline{A}}$ that reflects the degree to which support is lost by packing individual hypotheses into the residual $\overline{A}$:

$$s(\overline{A}) = w_{\overline{A}}[s(B) + s(C)]. \tag{12.4}$$

Support theory's assumption of subadditivity (12.2) implies $w_{\overline{A}} \leq 1$. Lower values of $w_{\overline{A}}$ reflect greater subadditivity, that is, greater loss of support as a result of packing hypotheses B and C into the residual $\overline{A}$.

Koehler, Brenner, and Tversky (1997) offered a simple linear-discounting model according to which the support for the alternatives included in the residual is discounted more heavily as the support for the focal hypothesis of the elementary judgment increases; in other words, $w_{\overline{A}}$ decreases as $s(A)$ increases. This model captures the intuition that when the focal hypothesis is well supported by the available evidence, people feel less compelled to consider how the evidence might also support its alternatives than when the focal hypothesis is not well supported by the evidence. We refer to this phenomenon as *enhanced residual discounting*.

ESAM

Support theory describes the translation of support into probability but does not specify how support is assessed in the evaluation of the available evidence, which is assumed to vary across different judgment tasks and domains. Recently, we developed ESAM as a model of the support assessment process underlying judgments of probability based on patterns of binary (present/absent) cues, where the diagnostic value of those cues and the base rate of the relevant outcomes have been learned from previous experience in an uncertain environment (Koehler et al., 2003).

The model can be thought of as characterizing how a sample of information (i.e., previous experience) obtained in an uncertain environment is represented and subsequently used to assess the likelihood of a particular outcome. More specifically, it is assumed that previous experience in the environment is represented in the form of frequency counts of co-occurrences between informational cue values and outcomes of interest to the judge. Hence a sample of information, in these terms, consists of a set of counts of how frequently a particular outcome has been observed in conjunction with each available informational cue value. For simplicity, we assume that these frequency counts, obtained from experience in the probabilistic cue-based environment, are encoded and later retrieved without error (although this assumption could obviously be refined based on research on frequency estimation; see Sedlmeier & Betsch, 2002).

Because ESAM evaluates the available evidence on a cue-by-cue basis rather than in terms of the entire cue pattern as a whole, it takes as input the co-occurrence frequency with which each cue value (present or absent) has been observed in the presence of each possible hypothesis (or outcome) under evaluation. Although ESAM can readily accommodate any number of hypotheses and cues (and cues with more than two possible values), we will focus on the case of three possible hypotheses and six binary cues. This case corresponds to several experiments, reviewed in the "Experimental Tests" section, involving the diagnosis of simulated "patients" suffering from one of three possible flu strains on the basis of the presence or absence of six discrete symptoms. In this case, the model requires for each of the six cues a count of how frequently that cue was observed as being present and as being absent with each of the three possible hypotheses (i.e., how often a particular symptom was present or absent in conjunction with each of the three possible flu strains). Here, a symptom's absence refers to when a patient is known to not have the symptom; it does not refer to the situation in which the status of the symptom is unknown.

The frequency with which cue C is observed as present in cases where hypothesis H holds is denoted $f_1(C, H)$; the frequency with which cue C is observed to be absent in cases where hypothesis H holds is denoted $f_0(C, H)$. The number of competing hypotheses (or possible outcomes) is denoted N_H and the number of available cues is denoted N_C. Finally, let $f(H)$ represent the overall frequency, or "base rate," with which hypothesis H holds in the set of stored observations.

Diagnostic Value of Each Cue. ESAM assumes that the cue pattern serving as the basis of the probability judgment is assessed one cue at a time, with the diagnostic implication of each observed cue value being evaluated with respect to a target hypothesis. In accord with the Bayesian approach to subjective probability, a piece of evidence is said to be diagnostic with respect to a hypothesis to the extent that the introduction of the evidence justifies a change in the probability of that hypothesis relative to its prior or base-rate probability.

The diagnostic value $d_1(C, H)$ of the presence of cue C with respect to a particular hypothesis H is given as follows:

$$d_1(C, H) = \frac{f_1(C, H)/f(H)}{\sum_j [f_1(C, H_j)/f(H_j)]}. \tag{12.5}$$

The diagnostic value $d_0(C, H)$ of the absence of cue C with respect to a particular hypothesis H is given by a parallel expression (and if there were more than two cue statuses, more parallel expressions could be used for each of the statuses). The value of d varies between 0 and 1. This calculation can be thought of as distributing one "unit" of diagnostic value among the set of competing hypotheses, with hypotheses implicated by the cue's presence receiving a larger share than hypotheses upon which the cue's presence casts doubt.

In this manner, the model assumes that the judge is sensitive to the diagnostic value of individual cues without necessarily being sensitive to the diagnostic value of cue patterns. Specifically, this calculation of diagnostic value is insensitive to conditional dependence among cues. Consequently, the model will not capture configural cue processing effects (Edgell, 1978, 1980). The calculation of diagnostic value is also uninfluenced by the base rate or prior probability of the hypothesis in question, as is the case for the likelihood ratio in Bayes's rule.

According to ESAM, the diagnostic implication of each cue value constituting the cue pattern is individually assessed and then summed to arrive at an overall assessment of the diagnostic value of the cue pattern taken as a whole for a particular hypothesis. It is assumed that, because of their higher salience, present cue values are given greater weight than are absent cue values in the summation process. The diagnostic value of cue pattern $\mathbf{C}$ for hypothesis H, denoted $d_\mathbf{C}(H)$, is given by

$$d_\mathbf{C}(H) = (1 - \delta) \overbrace{\sum_i d_1(C_i, H)}^{\text{present cues}} + \delta \overbrace{\sum_i d_0(C_i, H)}^{\text{absent cues}}$$

$$\text{for cues } C_i, i = 1, \ldots, N_\mathbf{C} \text{ in } \mathbf{C}. \tag{12.6}$$

The free parameter δ (which can vary between 0 and 1) represents the weight placed on absent cues relative to that placed on present cues in the cue pattern. Relative underweighting of absent cues is indicated by $\delta < 1/2$ with $\delta = 0$ representing the special case of complete neglect of absent cues. As well as δ being able to represent the amount of weight placed on each set of cues, it could also represent the relative probabilities of cues of each status being sampled when assessing support for a hypothesis. We interpret this parameter in the latter sense in the Simulations section of this chapter. Note that the value of $d_C(H)$ will generally tend to increase with the number of cues constituting the cue pattern, with a maximum value of N_C.

In general, the weight placed on cues of each status when assessing support for a hypothesis depends on the salience of each status. Therefore, when cues have substitutive statuses (e.g., gender) they will likely have equal salience and therefore receive equal weighting when assessing support. In addition, when cues have more than two statuses, one could either estimate more weighting factors or simply assume equal weighting.

Base-Rate Sensitivity. ESAM's diagnostic value calculation is insensitive to the base rates of the competing hypotheses because it controls for the base-rate frequency $f(H)$ of the hypothesis under evaluation. ESAM accommodates potential base-rate sensitivity of the support assessment process in the translation from diagnostic value to support. The support for hypothesis H conveyed by cue pattern C, denoted $s_C(H) \geq 0$, is given by

$$s_C(H) = \left[\alpha \left(\frac{f(H)}{\sum_j f(H_j)} - \frac{1}{N_H} \right) + (1 - \alpha)d_C(H) \right]^\gamma . \qquad (12.7)$$

The free parameter α (which can vary between 0 and 1) provides a measure of the extent to which the support for hypothesis H, which is determined primarily by the diagnostic value $d_C(H)$ of the cue pattern for that hypothesis, is adjusted in light of its base rate (i.e., observed relative frequency in comparison with the alternative hypotheses). The adjustment is additive, as Novemsky and Kronzon (1999) found to be the case in human judgments, rather than multiplicative, as it is in the Bayesian calculations. The adjustment is positive in the case of high-base-rate hypotheses whose relative frequency exceeds $1/N_H$, the value expected under a uniform partition; the adjustment is negative for low-base-rate hypotheses. This convention was implemented for computational simplicity because in the special case of equal base rates this adjustment is zero and the parameter α drops out of the model. With unequal base rates, α reflects the judge's sensitivity to this consideration.

After combining the diagnostic value $d_C(H)$ of the cue pattern C in implicating hypothesis H with an adjustment in light of H's base rate, the resulting value is then exponentiated to arrive at the support for the hypothesis conveyed by the cue pattern (cf. Tversky & Koehler, 1994). The exponent γ is a free parameter (which can take any positive value) that influences the extremity of the resulting judgments. Its value can be interpreted as a measure of judgmental confidence, that is, the confidence with which the judge relies on his or her previous experience in evaluating the evidence.

Assessing Support of the Residual Hypothesis. Recall that, according to support theory, the residual hypothesis (i.e., the collection of alternatives to the focal hypothesis) receives less support than the sum of the support its component hypotheses would have received had they been evaluated individually. Koehler et al. (1997) suggested that the judge is less likely to fully evaluate the extent to which the evidence supports alternative hypotheses as support for the focal hypothesis increases. In ESAM, this assumption of enhanced residual discounting is implemented by restricting the number of cues that are sampled in accumulating support for (alternatives included in) the residual. Specifically, in contrast to the computation of support for the focal hypothesis, in which the diagnostic value of each cue in the cue pattern makes a contribution, it is assumed that only a subset of cues are sampled with regard to their contribution to the support for an alternative hypothesis included in the residual.

For simplicity, we assume that, given a particular level of support for the focal hypothesis, the probability q that a cue will be sampled and its diagnostic value added in the calculation of support for each alternative hypothesis included in the residual is the same for all of the cues. Since we fit the model to judgments aggregated across multiple trials and/or participants, this probability can be implemented in the form of a discounting weight that reflects the proportion of its full diagnostic value, on average, that a given cue will contribute to the support for a hypothesis included in the residual. This discounting weight, denoted $q_{\overline{H}}$ is assumed to be inversely proportional to the support for the focal hypothesis H:

$$q_{\overline{H}} = \frac{1}{\beta\, s(H) + 1}. \tag{12.8}$$

The free parameter β (which can take any nonnegative value) determines how quickly $q_{\overline{H}}$ decreases as $s(H)$ increases.

Support for the residual is then given by

$$s_C(\overline{H}) = \sum_{H_j \text{ in } \overline{H}} \left[\alpha \left(\frac{f(H_j)}{\sum_i f(H_i)} - \frac{1}{N_H} \right) + (1 - \alpha) q_{\overline{H}} d_C(H_j) \right]^{\gamma}. \tag{12.9}$$

In other words, the support for each alternative hypothesis included in the residual is determined by the sum of the diagnostic value contributed by each cue, which is discounted to reflect the restricted set of cues consulted in evaluating support for hypotheses included in the residual. The support for the residual as a whole is given by the sum of the support thus calculated of each alternative hypothesis that it includes.

The discounting can be implemented in this way because we are only interested in aggregated judgments. If we were to attempt to model individual judgments on a trial-by-trial basis, with probability $q_{\overline{H}}$ each cue would either be sampled or not when assessing the diagnostic implications of the cue pattern for the alternative hypotheses. The free parameter β therefore affects how many cues are sampled when assessing the support for the alternative hypotheses.

In summary, ESAM describes how the support for a hypothesis is assessed on the basis of probabilistic cues when information regarding the usefulness of the cues is available from previous observations represented in the form of stored frequency counts. The model has four free parameters: α (base-rate adjustment), β (enhanced residual discounting), γ (judgmental extremity), and δ (absent cue weighting). According to the model, the diagnostic value of each cue for a particular hypothesis is assessed independently and then summed over the cues constituting the cue pattern. The free parameter δ reflects the relative probability of sampling present versus absent cues. If a cue is sampled, its diagnostic value is included in the summation process. The free parameter δ therefore represents the differing psychological saliencies of each cue status. The free parameter α reflects the extent to which the diagnostic value assessment is adjusted in light of the base rate of the hypothesis under evaluation. The free parameter γ reflects the extremity of the resulting support estimates and reflects the confidence that a person has in his or her assessment of the evidence. The free parameter β reflects the degree to which, as the support for the focal hypothesis increases, the set of cues sampled in assessing the support for hypotheses included in the residual is restricted. We can therefore see that the free parameters δ and β control what proportion of the available information is sampled when computing the probability for the focal hypothesis given a certain pattern of cues. In the Simulations section we find that reasonably accurate responses are generated even when the proportion of information sampled is relatively low.

Normative Benchmark

In what ways would a Bayesian approach differ from that of ESAM in evaluating the implications of a pattern of cues for a particular hypothesis in light of previous experience with those cues? Although there are a number of more or less complicated approaches that could be developed from a Bayesian perspective (e.g., see Martignon & Laskey, 1999), we will consider only the simplest one here, which relies heavily on the assumption of conditional independence of cue values. If the cue values constituting the cue pattern are conditionally independent, then one can readily calculate the probability of observing any particular cue pattern given that a designated hypothesis holds (e.g., the probability of observing a particular pattern of symptoms given that the patient has a designated flu strain) as the product of the conditional probabilities of each individual cue given that hypothesis. This calculation serves as the basis for evaluating the likelihood ratio in the Bayesian approach, which is then combined with the prior probability of the hypothesis in question to arrive at an assessment of its posterior probability (i.e., its probability in light of the cue pattern). Assuming that both the conditional probabilities of each cue value and the overall prior probability of each hypothesis are estimated from a set of stored frequency counts summarizing previous experience with the cues, the probability of a hypothesis H given a pattern of cue values C is given by,

$$P(H \mid \text{cue pattern C}) = \frac{f(H) \prod\limits_{i=1}^{N_C} \frac{f(C_i,H)}{f(H)}}{\sum\limits_{j=1}^{N_H} \left[f(H_j) \prod\limits_{i=1}^{N_C} \frac{f(C_i,H_j)}{f(H_j)} \right]} \quad \text{for } C_i \text{ in cue pattern C,}$$

(12.10)

where $f(C_i, H)$ is the frequency with which cue value C_i (absent or present) was previously observed in conjunction with hypothesis H.

The numerator of (12.10) can be viewed as corresponding to the extent to which, in the Bayesian analysis, hypothesis H is supported in light of the available evidence. The product term in the numerator corresponds to the diagnostic value of the evidence, which is adjusted in light of the base rate or prior probability of the hypothesis in question as reflected by $f(H)$. The same calculation is used to assess the support conveyed by the cue pattern for each of the competing hypotheses. As in ESAM, the probability assigned to the hypothesis is given by its normalized support relative to its alternatives. Unlike in ESAM, of course, there is no accommodation in the Bayesian framework for discounting of support arising from packing together the alternatives to the focal hypothesis in the residual. That is, in contrast to ESAM in particular and support theory in general, the normative Bayesian framework produces judgments that are necessarily extensional (i.e., bound by rules of set inclusion) and additive (i.e., over decomposition of events into subsets).

Another key difference between ESAM and the Bayesian model (12.10) just described is that the Bayesian model integrates individual cue values (and considerations of hypothesis base rate or prior probability) in a multiplicative manner, whereas ESAM uses an additive integration form. As a consequence, in ESAM's additive framework, support tends to increase with the number of cues consulted, whereas in the Bayesian model it tends to decrease. Furthermore, in the integration process, ESAM accommodates differential weighting of cue absence and cue presence, whereas in the normative Bayesian approach cue absence and cue presence are logically interchangeable. Because ESAM is not a generalization of the Bayesian model (12.10) outlined here, there are no parameter values for which ESAM will exactly reproduce the corresponding judgments derived from the Bayesian approach. [A generalization of the Bayesian model is offered by Koehler et al. (2003).] ESAM does tend to produce judgments that correlate highly with the corresponding Bayesian values, however, when $\beta = 0$ and $\delta = \frac{1}{2}$. The former represents the special case of ESAM that produces additive judgments; and the latter places equal weight – as in the Bayesian model – on cue absence and cue presence. The performance of this version of ESAM, which samples all of the available information, is reported in the first simulation in the Simulations section later in the chapter.

An alternative – also arguably normative – approach adopts a frequentist perspective, in which the current cue pattern serving as evidence is assessed against previous observations that exactly match that pattern. The judged probability of a designated hypothesis is given by the proportion of previous cases matching the cue pattern in which the hypothesis held (e.g., the proportion of previous patients with an identical set of symptoms who suffered from the hypothesized flu strain). This approach represents the starting point in development of exemplar-based models of classification learning (e.g., Brooks, 1978; Medin & Schaffer, 1978) and has the advantage of being able to accommodate cue structures for which conditional independence does not hold. Both ESAM and the Bayesian model outlined here, by contrast, assume conditional independence of cues. The frequentist approach does, however, require a relatively large sample of previous observations to produce reliable probability estimates. It also requires stored frequency counts for every possible cue pattern, the number of which increases exponentially with the number of cues. As descriptive models, then, either ESAM or the Bayesian model outlined here might be more useful in producing reasonably accurate judgments in the face of small sample sizes and limited memory capacity.

EXPERIMENTAL TESTS

We have found ESAM to closely reproduce probability judgments made in several multiple-cue probability learning experiments conducted by Koehler (2000) and Koehler et al. (2003). All of these experiments employ a simulated medical diagnosis task in which participants attempt to diagnose which of three possible flu strains a patient is suffering from based on the set of symptoms exhibited by that patient.

General Method

These experiments generally proceed in two phases. In an initial learning phase, participants learn about cue–outcome (i.e., symptom–flu strain) associations from direct, trial-by-trial learning experience in the uncertain environment. On a typical learning trial, the participant is presented with a patient characterized in terms of a set of four to six binary (present/absent) symptoms, and on the basis of this information the participant makes a forced-choice diagnosis of the flu strain from which the patient is suffering. After making their diagnosis, participants are told which flu strain the patient actually has. The learning phase consists of a set of 240–300 such trials, which can be viewed as the information sample drawn from the uncertain environment on which subsequent inferential judgments must be based. Sampling from the uncertain environment is conducted such that participants encounter an unbiased sample presented in a passive, sequential manner. In the second, judgment phase of the experiment, participants are presented with additional patients and, for each, judge the probability that the patient is suffering from a designated flu strain. The judgment phase instructions emphasize that the flu strain designated as the target of judgment on a particular trial is selected arbitrarily and that its designation should not be taken as having any informational value regarding the patient's diagnosis.

Typical Results

Across the set of judgment trials, participants eventually assign a probability to each of the three competing hypotheses (flu strains) contingent on each possible combination of cue values (symptoms). There are two aspects of the results of these studies that are of particular interest. First, as predicted by support theory, the probability judgments tend to be systematically subadditive; that is, the sum of the probabilities assigned to the three competing hypotheses given a particular pattern of cue values consistently exceeds one, in contrast to the requirement of probability theory that the total probability assigned to a mutually exclusive and collectively exhaustive set of events should add to one. Second, the degree of subadditivity for a set of judgments increases with the number of present cue values (symptoms) in the cue pattern on which the judgments are based. That is, the total probability assigned to the three flu strains is higher for patients exhibiting many symptoms than for patients exhibiting fewer symptoms.

Koehler (2000) found these systematic biases to hold even when (a) the training phase required participants to make probability judgments rather than forced-choice diagnoses and (b) the judgment phase required participants to make judgments of absolute frequency rather than of probability or relative frequency. In light of many criticisms of work in the heuristics and biases tradition that judgmental biases arise from nonrepresentative item sampling or use of problems that otherwise contradict the previous experience of research participants, it is notable that we have observed such pronounced and systematic biases in judged probability even when the judgments are based on a directly experienced representative sample of information from the judgment environment.

Fitting ESAM to the Data

ESAM was developed as a model of how probability judgments are generated in this type of task. It could be extended to generate choices or diagnoses, as participants give in the initial learning phase of our studies, and this is done in the Simulations section. However, our previous work only investigated subjective probability judgments.

ESAM readily reproduces the key experimental results from our studies, namely that the probability judgments are generally subadditive and that the degree of observed subadditivity increases with the number of present cue values in the cue pattern on which the judgments are based. That latter result follows from ESAM's assumptions of absent-cue underweighting (i.e., $\delta < 1/2$; absent cue values are given less weight than present cue values) and of enhanced residual discounting (i.e., $\beta > 0$; support for alternative hypotheses included in the residual is discounted more heavily as support for the focal hypothesis increases). The result of these assumptions is that, as the number of present cue values increases, support for the focal hypothesis also tends to increase, and consequently the degree of subadditivity increases (i.e., support for specific alternative hypotheses included in the residual is more heavily discounted).

Koehler et al. (2003) fit ESAM to the judgment data from four multiple-cue probability learning experiments. The best-fitting values of three of ESAM's four free parameters (α was only applicable in one experiment) varied moderately across the four experiments (β varied from 0.10 to 0.66, γ from 1.14 to 2.96, and δ varied least: from 0.22 to 0.27), with some of the variability being accounted for by some of the differences between the learning environments used in each experiment. ESAM was able to account for a large amount of the variability in the observed judgments for which the Bayesian model was unable to account. For example, Figure 12.1 shows the observed versus predicted probability judgments in Experiment 1 (Koehler et al., 2003).

Testing ESAM's Assumptions

Koehler et al. (2003) also compared ESAM's fit to that of a number of alternative models, each of which differed from ESAM by a single key assumption. The alternative models tended to fit the data less well, providing some corroboration for each of ESAM's key assumptions. Specific models exhibiting poorer data fits were those that (a) consider entire cue patterns rather than single cues, (b) fail to place greater weight on cue presence than on cue absence, (c) use different methods for calculating the diagnostic value of a cue, (d) do not adjust for the base rate of a hypothesis, and (e) assume no relationship between the support for the focal hypothesis and discounting of support for its alternatives. Further analyses showed

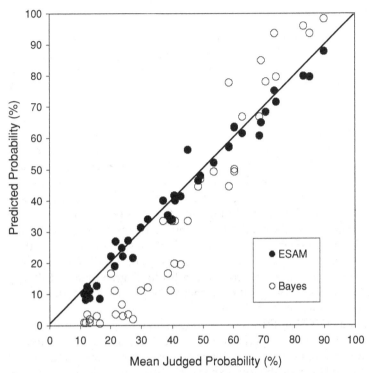

FIGURE 12.1. Mean probability judgments versus ESAM and Bayesian predictions in Experiment 1 of Koehler et al. (2003).

that the model's simplifying assumption of perfectly accurate frequency counts of previous cue–hypothesis co-occurrence proved to cost relatively little in terms of fit to the data.

ESAM's Accuracy

Despite the systematic biases introduced by some of its underlying assumptions, ESAM was found to produce reasonably accurate judgments in the experiments we reported as evaluated by a comparison to the corresponding Bayesian values. The accuracy achieved by ESAM is impressive because the model ignores or underweights some information (e.g., absent cues) used in the Bayesian approach and also uses simpler (e.g., additive instead of multiplicative) rules for combining the implications of different pieces of evidence. The model nonetheless can produce reasonably accurate judgments because it is capable of evaluating and aggregating the diagnostic implications of each cue in the cue pattern. Indeed, the model incorporates the basic definition of diagnosticity used in the Bayesian approach and, as with Bayes's rule, integrates considerations of the diagnosticity of the evidence and of the base rate or prior probability of the hypothesis. As a result, ESAM and the Bayesian approach tend to produce highly correlated output. This observation is consistent with previous research showing that as long as they correctly identify and integrate the direction of each cue's diagnostic implication, information-integration models will often produce output that corresponds quite closely to the output of the corresponding normative model even if they differ in the weights attached to the cues and in the specific form of the combination rule they employ (e.g., Anderson, 1981; Dawes, 1979).

SIMULATIONS

Following a long line of research in the heuristics and biases tradition, we have developed a descriptive model of subjective probability that uses only a selected subset of the evidence available from the environment as a basis for judgment. As a consequence, its judgments exhibit consistent and predictable biases relative to the standard normative benchmarks. This cost is balanced by the potential benefit of circumventing some of the informational and computational complexities associated with the normative calculations. A natural question one might ask, then, concerns the relative costs and benefits of ESAM's simplifying assumptions. Specifically, we wondered whether it might be possible to characterize the types of judgment environments in which the model renders reasonably accurate judgments (at considerable informational and computational savings) versus the types of environments in which the model performs relatively poorly.

In this section, we first describe the different judgment environments used to assess the relative costs and benefits of ESAM's simplifying assumptions and then define how the performance of ESAM is assessed by comparing it to the output of the Bayesian analysis (12.10). Having established this, we evaluate for each environment how close ESAM's judgments come to the Bayesian benchmark when all of the available information is sampled, and subsequently evaluate the drop in ESAM's performance when only sampling a subset of the information. In general, it is found that ESAM incurs only small costs in performance when not all of the information is sampled, which may be outweighed by the reduction in computational requirements that results.

Environments

We investigated ESAM's performance across a wide range of environments by varying the number of cues, the number of hypotheses, and whether or not all of the cues were equally predictive. We used either

three or six hypotheses and three, six, or twelve conditionally independent cues. In some environments each cue was as predictive as all of
the others (we refer to these as the flat cue diagnosticity environments),
and in others the predictive value of the cues varied between the cues
(we refer to these as the J-shaped cue diagnosticity environments, following Gigerenzer & Goldstein, 1999). The base rates of all of the hypotheses
were equal in all environments, and we used binary (present/absent) cues.
Experiment 3 by Koehler et al. (2003) used an environment extremely similar to the six-cue, three-hypothesis, J-shaped cue diagnosticity environment
used here; the results of that experiment are discussed in the following in
light of the current findings.

The diagnostic value of a cue is determined by the proportion of times
that the cue is present in conjunction with each of the competing hypotheses, as shown in Tables 12.1 and 12.2 for each of the flat and J-shaped cue
diagnosticity environments, respectively. In all environments, a given cue
was present (a) most often when one hypothesis occurred (for example, the
first cue might be present 85% of the time that hypothesis 1 occurred), (b)
the complementary proportion of times when another hypothesis occurred
(for example, 15% of the time that hypothesis 3 occurred), and (c) in graded
proportions between these two extreme values (for example, between 85%
and 15%) when the other hypotheses occurred (for example, if there were
only three hypotheses then the cue was present 50% of the time that hypothesis 2 occurred). The frequency with which each of the other cues was
present shared this structure, but each cue was present most often in conjunction with a different hypothesis (for example, the second cue might be
present 85% of the time that hypothesis 2 occurred, etc.). In this way, a "set"
of cues can be constructed with the presence of each cue in the set most
strongly implicating a different hypothesis. Therefore, the number of cues
in a set equaled the number of hypotheses. Across all of the hypotheses,
all cues within a set were equally diagnostic. Only complete sets of cues
were used, and so environments in which there would be fewer cues than
hypotheses were not possible.

We measured the diagnosticity of each cue by computing the average
value of "delta-p" across the hypotheses for that cue (see Wasserman,
Dorner, & Kao, 1990). Delta-p is equal to the difference in the conditional
probability of a given hypothesis between when the cue is present and
when it is absent. We took an average value of this statistic across all of
the hypotheses in the environment to obtain a measure of the cue's overall diagnosticity, which we refer to as a "mean delta-p," as defined by the
equation,

$$\overline{\Delta P_C} = \left[\sum_i |P(H_i \mid C+) - P(H_i \mid C-)| \right] \Big/ N_H, \qquad (12.11)$$

where $C+$ denotes the presence of cue C, and $C-$ its absence.

TABLE 12.1. *Probability of Each Cue Being Present Given the Occurrence of Each Hypothesis in Flat Cue Diagnosticity Environments*

Number of Hypotheses	Number of Cues	Cue	Hypothesis #1	#2	#3	#4	#5	#6	Mean ΔP
3	3	A	0.85	0.50	0.15				0.311
		B	0.15	0.85	0.50				
		C	0.50	0.15	0.85				
3	6	A	0.85	0.50	0.15				0.311
		B	0.15	0.85	0.50				
		C	0.50	0.15	0.85				
		D	0.85	0.50	0.15				0.311
		E	0.15	0.85	0.50				
		F	0.50	0.15	0.85				
3	12	A	0.85	0.50	0.15				0.311
		B	0.15	0.85	0.50				
		C	0.50	0.15	0.85				
		D	0.85	0.50	0.15				0.311
		E	0.15	0.85	0.50				
		F	0.50	0.15	0.85				
		G	0.85	0.50	0.15				0.311
		H	0.15	0.85	0.50				
		I	0.50	0.15	0.85				
		J	0.85	0.50	0.15				0.311
		K	0.15	0.85	0.50				
		L	0.50	0.15	0.85				
6	6	A	0.85	0.71	0.57	0.43	0.29	0.15	0.140
		B	0.15	0.85	0.71	0.57	0.43	0.29	
		C	0.29	0.15	0.85	0.71	0.57	0.43	
		D	0.43	0.29	0.15	0.85	0.71	0.57	
		E	0.57	0.43	0.29	0.15	0.85	0.71	
		F	0.71	0.57	0.43	0.29	0.15	0.85	
6	12	A	0.85	0.71	0.57	0.43	0.29	0.15	0.140
		B	0.15	0.85	0.71	0.57	0.43	0.29	
		C	0.29	0.15	0.85	0.71	0.57	0.43	
		D	0.43	0.29	0.15	0.85	0.71	0.57	
		E	0.57	0.43	0.29	0.15	0.85	0.71	
		F	0.71	0.57	0.43	0.29	0.15	0.85	
		G	0.85	0.71	0.57	0.43	0.29	0.15	0.140
		H	0.15	0.85	0.71	0.57	0.43	0.29	
		I	0.29	0.15	0.85	0.71	0.57	0.43	
		J	0.43	0.29	0.15	0.85	0.71	0.57	
		K	0.57	0.43	0.29	0.15	0.85	0.71	
		L	0.71	0.57	0.43	0.29	0.15	0.85	

TABLE 12.2. *Probability of Each Cue Being Present Given the Occurrence of Each Hypothesis in J-Shaped Cue Diagnosticity Environments*

Number of Hypotheses	Number of Cues	Cue	Hypothesis #1	#2	#3	#4	#5	#6	Mean ΔP
3	6	A	0.85	0.50	0.15				0.311
		B	0.15	0.85	0.50				
		C	0.50	0.15	0.85				
		D	0.675	0.50	0.325				0.155
		E	0.325	0.675	0.50				
		F	0.50	0.325	0.675				
3	12	A	0.85	0.50	0.15				0.311
		B	0.15	0.85	0.50				
		C	0.50	0.15	0.85				
		D	0.675	0.50	0.325				0.155
		E	0.325	0.675	0.50				
		F	0.50	0.325	0.675				
		G	0.5875	0.50	0.4125				0.078
		H	0.4125	0.5875	0.50				
		I	0.50	0.4125	0.5875				
		J	0.544	0.50	0.456				0.039
		K	0.456	0.544	0.50				
		L	0.50	0.456	0.544				
6	12	A	0.85	0.71	0.57	0.43	0.29	0.15	0.140
		B	0.15	0.85	0.71	0.57	0.43	0.29	
		C	0.29	0.15	0.85	0.71	0.57	0.43	
		D	0.43	0.29	0.15	0.85	0.71	0.57	
		E	0.57	0.43	0.29	0.15	0.85	0.71	
		F	0.71	0.57	0.43	0.29	0.15	0.85	
		G	0.675	0.605	0.535	0.465	0.395	0.325	0.070
		H	0.325	0.675	0.605	0.535	0.465	0.395	
		I	0.395	0.325	0.675	0.605	0.535	0.465	
		J	0.465	0.395	0.325	0.675	0.605	0.535	
		K	0.535	0.465	0.395	0.325	0.675	0.605	
		L	0.605	0.535	0.465	0.395	0.325	0.675	

This metric shows how the flat and J-shaped cue diagnosticity environments differed. In flat cue diagnosticity environments, the cues in one set were as diagnostic as the cues in the other sets. In the J-shaped cue diagnosticity environments, the cues in the first set were the most diagnostic and the diagnosticity of the cues in the other sets fell off with a J-shaped distribution. Obviously, J-shaped environments were only possible when there was at least two sets of cues. It has been argued that most natural environments exhibit this J-shaped pattern of cue diagnosticities (Gigerenzer & Goldstein, 1999), and it is in this type of environment that ignoring many of the cues may not harm the accuracy of the probability judgments to a very large degree if the least diagnostic cues are ignored (as shown by the "Take the Best" heuristic and other frugal algorithms; see Gigerenzer & Goldstein, 1999).

In some of what we call J-shaped cue diagnosticity environments there are only two values of cue diagnosticity. Technically, of course, with only two diagnosticity levels it is not really appropriate to refer to the distribution as J-shaped, as at least three levels would be needed to produce this or any other kind of nonlinear pattern. Therefore, the reader may wish to replace the term J-shaped distribution in these cases with the less specific but more appropriate term "descending."

Accuracy

There are a number of ways to measure the accuracy of a model that generates probability judgments for multiple hypotheses; here we concentrate on four that we found to be particularly informative. The first two involve choosing which hypothesis is the most likely to be correct given a particular cue pattern; these choices are then compared to which hypothesis is the most likely based on the Bayesian calculations. The other two methods measure how close ESAM's probability judgments are to the probability judgments generated using the Bayesian calculations.

The simplest way to choose which hypothesis is correct for a certain cue pattern is to choose the hypothesis with the highest probability assigned to it. The expected accuracy of classifying a cue pattern will then equal the Bayesian probability of that hypothesis being the correct answer. Therefore, if the Bayesian probability of a certain hypothesis being correct given a certain set of cue values is 0.60, and ESAM assigns this hypothesis a probability of 0.65, then this method chooses this hypothesis every time and is correct on 60% of the trials involving that cue pattern. This is a *maximizing* method because choosing the hypothesis with the highest judged probability every time the cue pattern is encountered maximizes the expected proportion of correct choices.

The second method of choosing a hypothesis is *probability matching*. With this method, one uses the judged probabilities of each hypothesis as the probability of choosing that hypothesis (Luce, 1959). Therefore, if ESAM estimates the probability of hypotheses 1, 2, and 3 to be 0.65, 0.25, and 0.10 respectively, then the three hypotheses are chosen with those probabilities. If the Bayesian probabilities of the hypotheses are 0.60, 0.35, and 0.05, then the proportion of times that ESAM chooses the correct hypothesis will equal the probability of choosing each hypothesis multiplied by the probability of the chosen hypothesis being correct, summed over all of the hypotheses. Therefore, in this example ESAM would choose the correct hypothesis with probability $(0.65 \times 0.60) + (0.25 \times 0.35) + (0.10 \times 0.05) = 0.4825$. As explained previously, if β is greater than zero then ESAM produces probability judgments that sum to more than one. In these situations, we assume that the probability of choosing each hypothesis is the probability assigned to that hypothesis divided by the sum of all of the probabilities given to that cue pattern (which normalizes the predicted probabilities to sum to one).

For both choice methods, the mean accuracy is computed by taking the expected accuracy on trials involving each cue pattern and weighting each by the proportion of times that the cue pattern would be encountered in the judgment environment (i.e., as if the cue patterns were sampled in a representative manner). The difference between these two methods is that the accuracy of the maximizing strategy depends only on the rank order of the probability judgments, whereas the accuracy of the probability matching strategy depends on the actual value of the probability judgments. Therefore, out of the two strategies the probability matching strategy is arguably more informative in the sense that it is more sensitive to the judgments produced by the model.

To identify a benchmark of ideal accuracy, the accuracy of combining all of the information in a Bayesian manner must be assessed. With the maximizing strategy, the Bayesian would always choose the hypothesis with the highest probability and would be correct with a probability equal to the Bayesian probability of that hypothesis (0.60 in our example given here). With the probability matching strategy, each category would be chosen with a probability equal to the Bayesian probability and would have that same probability of being correct (0.485 in our example). Another informative benchmark is chance-level accuracy, which can be found by computing the expected accuracy of choosing randomly among the competing hypotheses. This yields a mean accuracy of $1/N_H$ (0.333 in our example).

As can be seen in Table 12.3, these benchmarks for the maximum accuracy (Bayesian responding) and chance-level accuracy (random responding) are different in each environment, making comparisons of choice accuracy across environments difficult unless we control for this factor. We do this by calculating, for each environment, where ESAM's performance falls along the range between the chance-level and maximum choice accuracy. This is done by subtracting the chance-level accuracy from ESAM's accuracy, and dividing this by the difference between the maximum and chance-level accuracies in that environment. This yields an accuracy measure ranging between 0 and 1 that can be compared across different environments. To illustrate, in the example already discussed the Bayesian probability matching strategy resulted in an accuracy of 0.485, ESAM achieved a probability matching accuracy of 0.4825, and giving a random response would yield an

TABLE 12.3. *Accuracy Achievable by a Bayesian Responder and a Random Responder*

Cue Diagnosticity	Number of Hypotheses	Number of Cues	Bayesian Accuracy		Random Responder	
			Maximize	Probability Match	Accuracy	RMSE
Flat	3	3	0.765	0.660	0.333	0.306
		6	0.858	0.807	0.333	0.338
		12	0.956	0.937	0.333	0.373
	6	6	0.606	0.455	0.167	0.207
		12	0.748	0.656	0.167	0.239
J-shaped	3	6	0.790	0.694	0.333	0.309
		12	0.800	0.703	0.333	0.310
	6	12	0.643	0.503	0.167	0.213
Means						
Flat			0.854	0.800		
J-shaped			0.744	0.633		
	3		0.804	0.720	0.333	0.327
	6		0.666	0.538	0.167	0.220

accuracy of 0.333 on average. Therefore, the rescaled probability matching choice accuracy of ESAM would be (0.4825 − 0.333)/(0.485 − 0.333) = 0.984. A random responder would achieve an accuracy of 0 by this measure, whereas a Bayesian responder would achieve an accuracy of 1.

The other two methods of computing the accuracy of the probability judgments measure how close the probability judgments generated by ESAM are to the Bayesian set of probability judgments. The first of these takes all of the probability judgments and computes a correlation between the probability judgments generated by ESAM and the Bayesian set of probabilities. The second method takes the square root of the mean squared difference between each Bayesian and ESAM-generated probability, also known as a root-mean-squared error (RMSE). The RMSE is the more sensitive measure of these two, but only when ESAM's judgments for each cue pattern sum to one (i.e., when there is no subadditivity). When ESAM's judgments for each cue pattern sum to more than one then the RMSE increases accordingly. In this case, the correlation may be a more informative measure as it is insensitive to scale differences.

The bottom of Table 12.3 shows the accuracy achievable by a random and Bayesian responder for each type of environment. To compare across the two levels of one variable (e.g., flat versus J-shaped cue diagnosticity environments), the environments included in calculating a mean for one level of the variable must be identical to those included in calculating the mean for the other level of that variable. Therefore, since the J-shaped cue diagnosticity environments all have more cues than hypotheses, only the flat cue diagnosticity environments in which there are more cues than hypotheses are used to calculate the corresponding mean. This allows meaningful comparisons to be made across the variable of interest. The same logic is used in calculating the means reported in all of the tables and figures in this chapter.

ESAM's Parameters

The purpose of the simulations reported in this chapter is to assess how ESAM's performance varies based on the number of cues that are sampled in computing the probability judgments. To do this, we vary δ (the absent versus present cue weighting parameter) and β (the enhanced residual discounting parameter). All of the hypotheses have equal base rates in the environments investigated here so α (the base-rate information versus diagnostic information weighting parameter) drops out of the model.

The final parameter in ESAM is γ. Varying this affects the extremity of the probability judgments, and we have previously interpreted the value of this parameter as reflecting the amount of confidence in the judgments given (Koehler et al., 2003). This parameter has no effect on the number of cues sampled to make each judgment, and so we kept it constant across the different simulations. Since there is no obvious default value for this parameter, we determined the value to use for all of the simulations by finding the value that allowed ESAM to fit the Bayesian probabilities as closely as possible when there was no residual discounting and absent and present cues were weighted equally ($\beta = 0$, $\delta = 0.5$).

Because the diagnostic value of each cue is summed to obtain the total diagnostic value of the cue pattern for each hypothesis (see Eq. 12.7), the absolute size of the total diagnostic value of the cue pattern is affected by the number of cues in the environment; specifically, $d_C(H)$ varies from 0 to N_C. As a result, the value of γ that allows ESAM to best fit the Bayesian values in each environment depends on the number of cues in the environment. This complication can be avoided by changing the support calculation slightly, by exponentiating the total diagnostic value of the cue pattern $d_C(H)$ by the number of cues in the environment N_C:

$$d_C(H) = \left[(1 - \delta) \overset{\text{present cues}}{\sum_{i}} d_1(C_i, H) + \delta \overset{\text{absent cues}}{\sum_{i}} d_0(C_i, H) \right]^{N_C}$$

$$\text{for cues } C_i, i = 1, \ldots, N_C, \text{ in } \mathbf{C}. \quad (12.12)$$

Using this approach, the value of γ that allows ESAM to most closely fit the Bayesian values is very close to 1.35 in each of the environments used in our simulations. We therefore used this modified form of the total cue diagnosticity equation and a value of $\gamma = 1.35$ in all of the simulations reported here.

ESAM's Performance

We begin by assessing ESAM's performance when using all of the cues, and then examine how ESAM's simplifying assumptions (which cause certain cues to be removed from the calculations) impact the model's accuracy. We first examine the impact of ESAM's simplifying assumptions by only sampling present cues, and then proceed to sample only certain subsets of the cues when evaluating support for the alternative hypotheses. Initially, we choose which cues to sample at random as in the implementation of ESAM described previously, and then adopt a more systematic method of sampling only the most diagnostic cues. We then combine these two simplifying assumptions to assess ESAM's performance when only cues of one status are sampled, and only a subset of cues are sampled when assessing support for hypotheses in the residual. It is shown that even when a large proportion of the cues are ignored, the accuracy of the choices and the probability judgments produced by ESAM remain relatively high in most of the environments examined.

We investigated the validity of ESAM's third main simplifying assumption, that all cues are assumed to be conditionally independent, in

TABLE 12.4. *Accuracy of ESAM when All Cues Are Sampled*

Cue Diagnosticity	Number of Hypotheses	Number of Cues	Probability Judgments		Choice Probability Matching (0–1 Scale)
			Correlation	RMSE	
Flat	3	3	0.999	0.016	0.979
		6	1.000	0.005	0.996
		12	1.000	0.006	1.000
	6	6	0.997	0.015	0.972
		12	0.998	0.015	0.990
J-shaped	3	6	0.996	0.027	1.008
		12	0.996	0.031	1.024
	6	12	0.996	0.020	1.018
Means					
Flat			0.999	0.009	0.995
J-shaped			0.996	0.026	1.017
	3		0.999	0.013	0.994
	6		0.997	0.017	0.993

other work (White, 2002; White & Koehler, 2003). Participants' judgments showed minor sensitivity to violations of conditional independence in environments similar to those used here, and so we extended ESAM to allow for this observed minor sensitivity. However, we leave the investigation of the impact of ignoring conditional dependencies between cues on the model's performance to future work. In the current environments, all cues were conditionally independent.

Complete Cue Sampling. ESAM's performance when all of the available information is used is remarkably good considering that the information is combined additively rather than in the mathematically correct multiplicative manner. In Table 12.4, we see that all probability matching choice accuracies exceed 0.97 and all RMSEs are 0.031 or less (using parameter values of $\beta = 0$, $\gamma = 1.35$, and $\delta = 0.5$). This performance is again consistent with previous research showing that as long as the appropriate information is used then the exact method by which it is combined is relatively unimportant (e.g., Anderson, 1981; Dawes, 1979).

We report only the probability matching choice accuracy in this and all of the subsequent analyses because the maximizing technique almost always yields performance identical to that of a Bayesian responder. This observation is interesting in itself as it shows that a vast amount of the information can be ignored and yet a good decision regarding which of the hypotheses

TABLE 12.5. *Performance of ESAM when Only Cues of One Status Are Sampled*

Cue Diagnosticity	Number of Hypotheses	Number of Cues	Probability Judgments		Choice Probability Matching (0–1 Scale)
			Correlation	RMSE	
Flat	3	3	0.987	0.050	0.985
		6	0.984	0.061	0.996
		12	0.995	0.037	0.997
	6	6	0.968	0.054	0.993
		12	0.984	0.044	0.990
J-shaped	3	6	0.972	0.076	1.011
		12	0.984	0.059	1.024
	6	12	0.977	0.049	1.024
Means					
Flat			0.988	0.047	0.994
J-shaped			0.978	0.061	1.020
	3		0.981	0.062	0.997
	6		0.976	0.049	1.002

is the *most* likely can still be made. However, as the amount of information sampled decreases, accuracy deteriorates much more quickly when evaluating the *relative* likelihood of the hypotheses compared to when only identifying which hypothesis is the most likely.

Sampling Cues of Only One Status. As already discussed, people's judgments tend to focus on the cues that are present rather than on those that are absent. Therefore, we investigated how performance decreased when only cues that are present are sampled from the available information (in the environments studied here, this is logically equivalent to only sampling cues that are absent). ESAM's performance when only cues that are present are sampled is surprisingly close to that of a Bayesian responder. See Table 12.5; here parameter values of $\beta = 0$, $\gamma = 1.35$, and $\delta = 0$ or 1 used but identical results are obtained when either extreme value of δ is used.[1] Although the probability judgments are markedly less accurate (RMSEs increased from approximately 0.01–0.02 when all cues were sampled to approximately 0.05 when only present cues were sampled), the accuracy of the choices based on those judgments remained almost unchanged.

[1] The boundary values of δ (0.0 and 1.0) cannot actually be used owing to the support for all hypotheses equaling zero when all cues have the status that is being ignored, causing the probability of each hypothesis to be undefined. Therefore, we used values of $\delta = 0.0001$ and $\delta = 0.9999$ in the simulations to avoid this.

This pattern of results is due to some of the probabilities generated being more extreme than the Bayesian values, and thereby yielding a higher accuracy measure than the Bayesian probability matching strategy, and the other probabilities being less extreme than the Bayesian values, and thereby yielding a lower accuracy measure. Specifically, when only present cues are used to make the judgments then, when less than half of the cues are present, the judgments produced are overly extreme; when more than half of the cues are present, by contrast, the judgments are less extreme than the corresponding Bayesian values. This pattern of results can best be explained with an example: If, for a given cue pattern, the Bayesian probabilities are 0.60, 0.30, and 0.10 for hypotheses #1, #2, and #3, respectively, then the Bayesian probability matching choice accuracy would be 0.46. If more extreme values are produced by ESAM of 0.75, 0.20, and 0.05, then the probability matching choice accuracy would be .515, but if less extreme values are produced of 0.45, 0.35, and 0.20 for another cue pattern that has the same Bayesian probabilities, then the accuracy would be 0.395. Therefore, on average the choice accuracy of ESAM is about the same as the Bayesian responder (means of 0.455 and 0.46, respectively, in our example), but the RMSE of ESAM's probability judgments is quite high (0.11 in our example).

This explains why the mean accuracy of ESAM when only sampling cues of one status is close to that obtained when present and absent cues receive equal weighting whereas the RMSE is greater. Therefore, we can conclude that although attending to cues of only one status may compromise the accuracy of the probability judgments generated, it may (on average) not compromise the accuracy of choices made on the basis of those judgments (i.e., identification of the most likely hypothesis).

The median best-fitting parameter value of δ for the participants making probability judgments in Experiment 3 of Koehler et al. (2003) was 0.25. When δ is set to 0.25 in ESAM in our simulations using the almost equivalent environment, the results ($\delta = 0.25$, RMSE $= 0.044$, accuracy $= 1.01$) are understandably halfway between those achieved when absent cues are ignored ($\delta = 0.0$, RMSE $= 0.076$, accuracy $= 1.01$) and when absent and present cues are weighted equally ($\delta = 0.5$, RMSE $= 0.027$, accuracy $= 1.01$). The intermediate value of δ can be viewed as representing a trade-off between the number of cues sampled and the overall accuracy afforded by the judgment strategy.

Cue Sampling When Assessing Support for the Alternatives. Descriptively, the total of the judged probabilities assigned to a set of competing hypotheses typically exceeds 1. Support theory accounts for this by positing that the support for the alternative hypotheses that are included in the residual is discounted (Tversky & Koehler, 1994; Rottenstreich & Tversky, 1997). This principle is implemented in ESAM by way of fewer cues being sampled when computing the support for the hypotheses in the residual. We now investigate the effect that this assumption has on ESAM's performance.

The absolute support values for the hypotheses in each of the environments differ, and so the results obtained by using a fixed value of β across all of the environments are uninterpretable. Since a different proportion of cues is sampled for each judgment when assessing the support for the alternative hypotheses, the value of β was set so that the *mean* proportion of cues sampled when assessing the support for the alternative hypotheses was equal in each environment. Once Eq. 12.9 has been used to determine the proportion of cues to sample when assessing support for the alternative hypotheses, the actual cues to sample can be selected either at random or systematically. Koehler et al. (2003) assumed that the cues chosen to sample were selected randomly. Table 12.6 shows the results using this method.

There are a few interesting aspects of the results shown in Table 12.6. First, the mean total probability judgments are greater when there are six hypotheses than when there are three hypotheses. This effect has also been documented with human participants (see Tversky & Koehler, 1994, for a review), corroborating our account of the observed subadditivity. In addition, the mean total probability judgments increase as the proportion of cues sampled decreases, which is to be expected. Finally, when each judgment is based on less information (i.e., when the mean number of cues sampled to calculate the support for the alternative hypotheses decreases), the accuracy of those judgments decreases accordingly. This decrease in accuracy is seen in the correlation between ESAM's and the Bayesian probability judgments and also in the accuracy of the probability matching choices.

Reducing the number of cues sampled when assessing the support for a hypothesis will necessarily cause the judgments produced to be less accurate. However, this reduction in accuracy can be partially offset if, instead of randomly selecting which cues to sample, one chooses to sample the most diagnostic cues. To do this, an extra piece of computation is required to order the cues according to their diagnostic value. Specifically, we first order the sets of cues according to their overall diagnosticity and then within each set we order the individual cues according to how diagnostic each is for the hypothesis whose support is currently being calculated. In this way, a hierarchy of cues can be created. The appropriate proportion of cues to sample is then determined by starting with the cue highest on this ordering and including all cues down to a level that achieves the desired sampling proportion. This general process is similar to that used by the "Take the Best" algorithm (Gigerenzer & Goldstein, 1999), which orders the cues in terms of diagnosticity and then makes a decision based exclusively on the most diagnostic cue for which the two objects currently being assessed have different values.

TABLE 12.6. *Performance of ESAM when Only a Certain Proportion (a Random Subset) of the Cues Is Sampled to Compute the Support for the Alternative Hypotheses*

Cue Diagnosticity	Number of Hypotheses	Number of Cues	Proportion Sampled	Probability Judgments						Choice		
				Correlation			Total Probability Judgment			Probability Matching (0–1 scale)		
				0.9	0.8	0.6	0.9	0.8	0.6	0.9	0.8	0.6
Flat	3	3		1.00	0.99	0.83	1.18	1.35	1.82	0.96	0.88	0.61
		6		0.98	0.95	0.72	1.23	1.41	1.96	0.98	0.92	0.66
		12		0.97	0.91	0.67	1.22	1.41	1.91	0.98	0.94	0.75
	6	6		0.96	0.85	0.65	1.76	2.40	3.56	0.87	0.64	0.36
		12		0.89	0.76	0.56	1.92	2.60	3.71	0.75	0.52	0.28
J-shaped	3	6		0.97	0.92	0.75	1.29	1.53	1.98	0.97	0.88	0.60
		12		0.95	0.85	0.64	1.42	1.74	2.23	0.95	0.78	0.45
	6	12		0.89	0.76	0.55	2.08	2.85	4.04	0.76	0.52	0.27
Means												
Flat				0.95	0.87	0.65				0.90	0.79	0.56
J-shaped				0.94	0.84	0.65				0.89	0.73	0.47
	3			0.98	0.95	0.77	1.23	1.43	1.92	0.97	0.89	0.62
	6			0.92	0.79	0.59	1.92	2.62	3.77	0.79	0.56	0.30
Mean				0.95	0.87	0.57				0.90	0.76	0.42

The measure of cue diagnosticity used to establish this ordering relies on computations already carried out by ESAM. Specifically, given a particular hypothesis the diagnosticity of a cue being present and being absent is calculated using the appropriate version of Eq. 12.5. The cue's diagnosticity for that particular hypothesis is defined as the absolute difference between these two values. The cue's overall diagnosticity for the full set of hypotheses is then calculated as the mean of these absolute differences across the set of hypotheses.

Table 12.7 shows ESAM's performance using this systematic cue sampling method. Qualitatively, the results are similar to those produced when randomly selecting which cues to sample. The main difference is that the decrease in accuracy resulting from a reduction in the number of cues sampled is markedly less dramatic using the systematic cue sampling method than it is using the random cue sampling method. This difference is shown in Figure 12.2, in which the mean correlation and choice accuracy across all of the environments is plotted for each method of cue selection. This finding is to be expected since the most diagnostic information is sure to be retained when the cues are selected based on their diagnosticity whereas the most diagnostic information will often be discarded when the cues are selected randomly.[2]

As long as at least 90% of the cues (approximately) are sampled on average when assessing the support for the alternative hypotheses (selecting the subset of cues using the systematic method), the probability matching choice accuracy is maintained at around the maximum level of 1; see Figure 12.3. Below 90%, the accuracy starts to fall off reasonably quickly in most of the environments. The main reason for this decrease is that if the average proportion of cues sampled falls to lower levels, then if the focal hypothesis has at least a moderate amount of support then very few cues are sampled when assessing the support for the alternatives, thereby causing the focal hypothesis to receive a high probability judgment. This reduces accuracy because if two or more hypotheses have a moderate amount of support they will all be given a very high probability judgment, and so the model loses the power to differentiate between these hypotheses. Therefore, ESAM

---

[2] Comparing Tables 12.6 and 12.7 shows that selecting the cues to sample systematically yields lower mean total probability judgments relative to when the selection is done randomly. This is due to the alternative hypotheses not losing as much support when the sampling is done systematically because the cues that may contribute most to the support of the alternatives are certain to be included in the calculation of the support for those hypotheses. This contrasts with the random selection method in which the same cues are no more likely to be selected than are cues that contribute less to the support. Therefore, on average the support for the alternative hypotheses is greater when the cues are selected systematically than when selected randomly, despite the fact that the same proportion of cues is sampled with both methods.

TABLE 12.7. *Performance of ESAM when Only a Certain Proportion (a Systematic Subset) of the Cues Is Sampled to Compute the Support for the Alternative Hypotheses*

Cue Diagnosticity	Number of Hypotheses	Number of Cues	Proportion Sampled	Probability Judgments						Choice		
				Correlation			Total Probability Judgment			Probability Matching (0–1 scale)		
				0.9	0.8	0.6	0.9	0.8	0.6	0.9	0.8	0.6
Flat	3	3		1.00	0.92	0.93	1.11	1.55	1.58	1.02	0.79	0.81
		6		0.97	0.97	0.80	1.18	1.18	1.80	1.01	1.01	0.79
		12		1.00	0.93	0.69	1.02	1.24	1.87	1.00	0.97	0.80
	6	6		0.94	0.84	0.73	1.56	2.30	3.07	1.02	0.70	0.48
		12		0.87	0.77	0.58	1.64	2.16	3.47	0.82	0.63	0.34
J-shaped	3	6		0.97	0.94	0.73	1.22	1.44	1.97	1.02	0.92	0.62
		12		0.94	0.88	0.64	1.42	1.65	2.21	0.95	0.84	0.48
	6	12		0.87	0.75	0.57	1.92	2.75	3.85	0.84	0.55	0.30
Means												
Flat				0.95	0.89	0.69				0.94	0.89	0.64
J-shaped				0.93	0.86	0.65				0.94	0.77	0.47
	3			0.98	0.94	0.82	1.17	1.39	1.78	1.02	0.90	0.74
	6			0.89	0.79	0.63	1.71	2.40	3.46	0.89	0.63	0.37
Mean				0.94	0.88	0.71				0.96	0.80	0.58

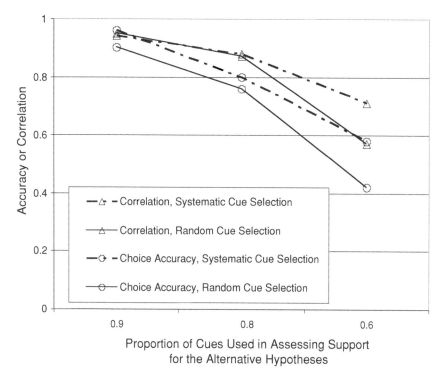

FIGURE 12.2. Random versus systematic selection of cues used in assessing support of the alternative hypotheses.

chooses each hypothesis that has a reasonable amount of support with almost equal frequencies.[3]

The median value of the observed mean total probability judgment in Experiment 3 of Koehler et al. (2003) was 1.31. When the value of β is set such that ESAM produces a mean total probability judgment of this size (given a similar environment) then the cue sampling proportion is 0.87, and the probability matching choice accuracy is 0.96. This implies that the number of cues sampled to compute the support for the alternative hypotheses, at least in this study, is set about as low as possible without sustaining any significant decrement in accuracy.

Cue Sampling When Assessing Support for the Alternatives and Sampling Cues of Only One Status. Because of its greater accuracy, we use the systematic method of cue sampling to assess ESAM's performance when both simplifying assumptions are implemented concurrently. Surprisingly,

---

[3] This is basically a problem in the extremity of the support values, and so may be affected by manipulating the γ parameter.

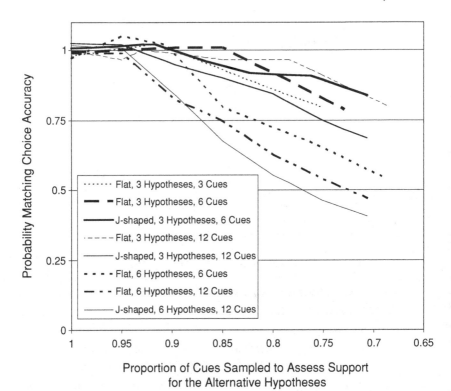

FIGURE 12.3. Probability matching choice accuracy versus proportion of cues sampled when assessing support for the alternative hypotheses.

although the correlation between ESAM's probability judgments and the Bayesian values decreases in comparison to the results of the simulations reported in the foregoing, the choice accuracy increases (see Table 12.8; the random method of cue selection yields a qualitatively similar pattern of results). The reason is again that by only sampling cues of a certain status the judgments produced become more extreme, causing the probability matching choice strategy to more closely resemble the maximizing choice strategy whereas the judgments themselves diverge from the Bayesian values more.

We can again investigate how the model performs when parameter values are used that allow the model to mimic the human data from Experiment 3 of Koehler et al. (2003). If values of $\delta = 0.25$ and a systematic cue sampling proportion of 0.88 are used (mean total probability judgment = 1.31), then the probability judgments generated correlate quite highly with the Bayesian probabilities ($r = 0.882$) and choice accuracy is reasonably high (0.935). If we assume that a value of $\delta = 0.25$ means that all of the present cues are sampled, and one-third of the absent cues are sampled

TABLE 12.8. *Performance of ESAM when Only a Certain Proportion (a Systematic Subset) of the Cues Is Sampled to Compute the Support for the Alternative Hypotheses, and Only Cues of a Certain Status within Those Sampled Are Used*

Cue Diagnosticity	Number of Hypotheses	Number of Cues	Proportion Sampled	Probability Judgments						Choice		
				Correlation			Total Probability Judgment			Probability Matching (0–1 scale)		
				0.9	0.8	0.6	0.9	0.8	0.6	0.9	0.8	0.6
Flat	3	3		0.92	0.75	0.81	1.16	1.46	1.57	0.99	0.80	0.83
		6		0.89	0.81	0.66	1.21	1.39	1.78	0.98	0.85	0.80
		12		0.85	0.79	0.63	1.23	1.35	1.79	0.94	0.90	0.77
	6	6		0.75	0.64	0.50	1.59	2.21	3.05	0.99	0.90	0.62
		12		0.72	0.60	0.47	1.55	2.06	3.05	0.92	0.79	0.60
J-shaped	3	6		0.83	0.74	0.60	1.29	1.49	1.83	0.94	0.86	0.74
		12		0.80	0.70	0.55	1.32	1.54	1.93	0.93	0.84	0.62
	6	12		0.66	0.55	0.44	1.74	2.35	3.23	0.94	0.80	0.59
Means												
	Flat			0.82	0.73	0.59				0.95	0.85	0.72
	J-shaped			0.76	0.66	0.53				0.94	0.83	0.65
	3			0.88	0.77	0.69	1.22	1.45	1.73	0.97	0.84	0.79
	6			0.71	0.60	0.47	1.63	2.21	3.11	0.95	0.83	0.60
Mean				0.80	0.70	0.58				0.95	0.84	0.70

293

$(\delta/[1 - \delta] = 0.25/0.75 = 1/3)$ then the model uses only 62% of the cue frequency counts used by the full Bayesian method, and yet it loses very little in terms of accuracy.

Conclusions

We have shown that when all of the available information characterizing previous observations in an uncertain environment is used, ESAM can produce choices and probability judgments that are remarkably close to those obtained by using the normatively correct Bayesian calculations. This is despite ESAM combining the information from the individual cues for each hypothesis additively rather than multiplicatively.

Interestingly, the accuracy of the choices made does not decrease when only cues of a certain status are sampled. At the same time, the probability judgments themselves are slightly less accurate, but this can be explained by the probabilities for certain cue patterns becoming more extreme whereas the probabilities for others become more conservative when this method of sampling is used. We also investigated the accuracy of the responses generated by ESAM when only some of the cues are sampled to assess the support for the alternative hypotheses. The decreases in accuracy were less pronounced when the cues not sampled were the least diagnostic cues rather than a random subset of the cues. It was shown that a small proportion of the cues can be ignored when assessing the support for the alternative hypotheses without any decrease in choice accuracy, whereas the total probability judgments for each cue pattern are significantly greater than one. When ESAM generates probability judgments that show biases similar to those exhibited by the participants in Experiment 3 of Koehler et al. (2003), only 62% of the cues are sampled to generate probability judgments that are highly accurate.

This research demonstrates how, when only a subset of the information available is sampled, reasonably accurate responses can still be generated. Furthermore, we have shown that when ESAM displays biases of a similar magnitude to those exhibited by humans (with the consequence that far less information is sampled than is available), the accuracy of the probability judgments and choices generated by ESAM is still close to that obtainable when all of the information is sampled.

OTHER MODELS

Other research investigating sampling subsets of cues includes Gigerenzer and Goldstein's (1999) "Take the Best" heuristic. They found that accurate choices could be made when only one cue or a few of the cues were considered when making each choice. Here, we have extended this approach of selectively sampling certain cues but used a quite different task with very different methods of choosing which cues to sample. However, we reached the same conclusion: Not all of the information needs to be sampled to make a reasonably accurate choice or judgment. In evaluating the diagnostic value of a cue pattern, ESAM integrates separate assessments of the implications of each individual cue value constituting the cue pattern, and hence it implicitly assumes conditional independence of cues. Effectively, then, ESAM can be viewed as employing a prototype representation of cue information, in which the interpretation of each cue value is uninfluenced by other cue values in the cue pattern. Exemplar-based models, by contrast, have the potential advantage of detecting and exploiting configural information in cases where conditional dependencies exist among cue values, but at the cost of substantially higher memory storage requirements. For example, in the case of six binary cues (i.e., symptoms) and three competing hypotheses as investigated in many of our experiments, counts must be maintained of the frequency with which each of the sixty-four possible cue patterns co-occur with each hypothesis in an exemplar representation. The prototype representation requires only twelve counts (i.e., an absence and a presence count for each cue) per hypothesis, rather than sixty-four, because it maintains counts for each separate cue value rather than for entire cue patterns. In essence, the prototype representation as employed by ESAM discards information that the exemplar representation retains.

Other researchers have studied ways in which entire exemplars could be selectively sampled. One instance of this is Dougherty et al.'s (1999) extension of Hintzman's (1988) MINERVA-2 memory model to decision-making tasks (MINERVA-DM), and in particular to judgments of conditional likelihood. The output of MINERVA-DM is based on imperfect exemplar representations (or memory traces). In one simulation, they provided three sets of simulated participants different amounts of experience with a domain, by storing 80, 200, or 600 memory traces, each representing a patient displaying ten present/absent symptoms that probabilistically predicted each of two diseases. This manipulation is functionally equivalent to storing all 600 traces and subsequently sampling 13.3%, 33.3%, or 100% of them. The resulting judgments exhibited less overconfidence when they were based on a larger amount of experience. The same pattern of results has been reported with human participants in multiple-cue judgment tasks (see Dougherty et al., 1999, for a discussion of the human data). However, again in line with the results presented in this chapter, the large differences in the amount of information sampled resulted in relatively small differences in the measure they used (overconfidence). However, Dougherty et al. (1999) did not report any direct measures of the accuracy of the judgments produced by their model.

Juslin and Persson (2002) suggested a modification of Medin and Schaffer's (1978) context model in which all exemplars are activated in parallel but are retrieved, or sampled, serially (PROBEX – PROBabilities from EXemplars). The order of retrieval of the exemplars is determined by the most similar exemplar to the current probe being the most likely to be sampled first. Once "a clear-enough conception of the estimated quantity has been attained" (p. 569) retrieval terminates. In addition, although the status of all the cues is known when each exemplar is encountered, the status of each cue in each exemplar has a certain probability of not being available upon retrieval due to an encoding failure or forgetting. This is functionally equivalent to sampling only a subset of the cues for each exemplar.

PROBEX was able to mimic human performance by only sampling a few exemplars before making a response. The researchers also claimed that the model was robust even when only a few exemplars are stored (Juslin, Nilsson, & Olsson, 2001; Juslin & Persson, 2002). However, they did not directly measure changes in performance over varying amounts of sampled information. Instead, it was suggested that

an important research program [would be] to explore exemplar models that make demands that are more modest on storage and retrieval. For example, can exemplar models that presume storage of only a subset of objects, or retrieval of only a few exemplars, provide as good fit to classification data as more demanding versions? (p. 598, 2002)

Given the interesting findings presented here regarding the effects of limited sampling in a cue-frequency-based model, ESAM, we agree with Juslin & Persson's (2002) suggestion that similar research should be conducted with an exemplar-based model.

SUMMARY

To make a reasonably accurate prediction or diagnosis in an uncertain environment on the basis of a set of informational cues and previous experience, not all of the available information needs to be sampled nor does it need to be combined in the mathematically correct way. We assessed the accuracy of a mathematical model of human probability judgment, the Evidential Support Accumulation Model (ESAM), in a variety of environments. ESAM is a model of how the evidence from multiple cues is combined to assess the support for conflicting hypotheses, with the evidence from each cue being combined additively. Derived from the principles of support theory, the support for the focal hypothesis is assessed by sampling more cues than when assessing the support for the alternative hypotheses, thereby mimicking the subadditivity observed in human probability judgments. In addition, ESAM assumes that cues of different statuses (e.g., present versus absent) that reflect differences in psychological salience are sampled with differing probabilities. In the eight simulated environments studied, when only a subset of the available information was sampled only a small decrement in the judgmental accuracy achieved by the model was observed.

References

Anderson, N. H. (1981). *Foundations of information integration theory.* New York: Academic Press.

Brooks, L. R. (1978). Non-analytic concept formation and memory for instances. In E. Rosch & B. B. Lloyd (Eds.), *Cognition and categorization* (pp. 169–211). Hillsdale, NJ: Lawrence Erlbaum.

Dawes, R. M. (1979). The robust beauty of improper linear models in decision making. *American Psychologist, 34,* 571–582.

Dougherty, M. R. P., Gettys, C. E., & Ogden, E. E. (1999). MINERVA-DM: A memory processes model for judgments of likelihood. *Psychological Review, 106* (1), 108–209.

Edgell, S. E. (1978). Configural information processing in two-cue probability learning. *Organizational Behavior & Human Decision Processes, 22,* 404–416.

Edgell, S. E. (1980). Higher order configural information processing in nonmetric multiple-cue probability learning. *Organizational Behavior & Human Decision Processes, 25,* 1–14.

Fischhoff, B., Slovic, P., & Lichtenstein, S. (1978). Fault trees: Sensitivity of estimated failure probabilities to problem representation. *Journal of Experimental Psychology: Human Perception & Performance, 4,* 330–334.

Gigerenzer, G., & Goldstein, D. G. (1999). Betting on one good reason: The take the best heuristic. In G. Gigerenzer, P. M. Todd, & the ABC Research Group (Eds.), *Simple heuristics that make us smart: Evolution and cognition.* London: Oxford University Press.

Hintzman, D. L. (1988). Judgments of frequency and recognition memory in a multiple-trace memory model. *Psychological Review, 95,* 528–551.

Juslin, P., Nilsson, H., & Olsson, H. (2001). Where do probability judgments come from? Evidence for similarity-graded probability. In J. Moore & K. Stenning (Eds.), *Proceedings of the twenty-third annual conference of the cognitive science society.* Hillsdale, NJ: Lawrence Erlbaum.

Juslin, P., & Persson, M. (2002). PROBabilities from Exemplars (PROBEX): A "lazy" algorithm for probabilistic inference from generic knowledge. *Cognitive Science, 26,* 563–607.

Koehler, D. J. (2000). Probability judgment in three-category classification learning. *Journal of Experimental Psychology: Learning, Memory, and Cognition, 26,* 28–52.

Koehler, D. J., Brenner, L. A., & Tversky, A. (1997). The enhancement effect in probability judgment. *Journal of Behavioral Decision Making, 10,* 293–313.

Koehler, D. J., White, C. M., & Grondin, R. (2003). An evidential support accumulation model of subjective probability. *Cognitive Psychology, 46,* 152–197.

Luce, R. D. (1959). *Individual choice behavior.* New York: Wiley.

Martignon, L., & Laskey, K. B. (1999). Bayesian benchmarks for fast and frugal heuristics. In G. Gigerenzer, P. M. Todd, & the ABC Research Group (Eds.), *Simple heuristics that make us smart: Evolution and cognition.* London: Oxford University Press.

Medin, D. L., & Schaffer, M. M. (1978). Context theory of classification learning. *Psychological Review, 85,* 207–238.

Novemsky, N., & Kronzon, S. (1999) How are base-rates used, when they are used: A comparison of additive and Bayesian models of base-rate use. *Journal of Behavioral Decision Making, 12,* 55–69.

Rottenstreich, Y., & Tversky, A. (1997). Unpacking, repacking, and anchoring: Advances in support theory. *Psychological Review, 104,* 406–415.

Sedlmeier, P., & Betsch, T. (2002). *ETC. Frequency processing and cognition.* Oxford: Oxford University Press.

Tversky, A., & Koehler, D. J. (1994). Support theory: A nonextensional representation of subjective probability. *Psychological Review, 101,* 547–567.

Wasserman, E. A., Dorner, W. W., & Kao, S. F. (1990). Contributions of specific cell information to judgments of inter-event contingency. *Journal of Experimental Psychology: Learning, Memory and Cognition, 16,* 509–521.

White, C. M. (2002). *Modelling the influence of cue dependencies and missing information in multiple-cue probability learning.* Master's thesis, University of Waterloo, Waterloo, Ontario, Canada.

White, C. M., & Koehler, D. J. (2003). *Modeling the influence of cue dependencies in multiple-cue probability learning.* Unpublished manuscript, University of Waterloo, Waterloo, Ontario, Canada.

13

Information Sampling in Group Decision Making

Sampling Biases and Their Consequences

Andreas Mojzisch and Stefan Schulz-Hardt

Groups often have to choose among several decision alternatives. For example, selection committees have to decide which of numerous applicants to hire, jury members have to decide whether the accused is guilty or not, medical decision-making teams have to recommend one of several possible treatments, and executive boards have to choose among different policy options. Such group decisions are normally preceded by a discussion of the merits of each decision alternative. This discussion can be understood as a process of information sampling: To prepare the final judgment or decision, the group samples information from their members' diverse information pools.

At this point, two of the core ideas underlying the whole book come into play. One of these is that information sampling is, in general, not a random process; rather, it is subject to systematic asymmetries originating from environmental or psychological constraints. The same is true for group discussions. In this chapter, we will deal with those two asymmetries that – in our view – are most characteristic and most specific for group information sampling, namely, the tendency to focus on so-called shared information (i.e., information known by all members prior to the discussion) and the tendency to predominantly discuss so-called preference-consistent information (i.e., information supporting the group members' individual decision preferences).

The second core idea of the book coming into play is that biased judgments need not necessarily reflect distorted cognitive processes but might rather be a consequence of biased sampling. In the corresponding chapters of this book, this idea has been outlined for subjective validity judgments (Freytag & Fiedler), availability biases (Dawes), biased group impressions (Meiser), and overconfidence (Klayman, Soll, Juslin, & Winman). Interestingly, whereas this idea seems to be rather new and innovative in individual judgment research, it has attracted much more attention in group judgment and group decision-making studies. In fact, when it was first observed that groups systematically fail to solve those tasks for which they have the unique potential to outperform individual decision makers as well as all kinds of voting schemes, the initial and most obvious explanation for this failure was biased sampling (Stasser & Titus, 1985). As a consequence, whereas in individual judgment and decision-making research the pendulum seems to swing from biased cognitive processes to biased sampling, in group information sampling research the opposite trend is visible. That is, group information sampling researchers, particularly in recent years, have had to learn that biased sampling does not explain biased group decisions completely and that biased cognitive processes, for example, also have to be taken into account to explain biased group decisions. Thus, the relationship between biased group sampling and biased (or suboptimal) group decisions will be the second focus of our chapter.

This chapter, in sum, is organized around four basic questions. The first two deal with the basic sampling biases in group information sampling: (1) Why are group members likely to introduce and repeat more shared than unshared information during the discussion? (2) Why are members likely to introduce and repeat more preference-consistent than preference-inconsistent information during the discussion?

The third and fourth questions deal with the relationship between these sampling biases and group decisions: (3) What are the consequences of biased information sampling for decision making in groups? (4) Why do groups sometimes fail to outperform individual decision makers even when information sampling during the discussion is unbiased?

BIASED SAMPLING IN FAVOR OF SHARED INFORMATION

In groups, each member typically knows some items of information before the discussion that other members do not know and, thus, members can potentially convey novel information to one another. Such uniquely held information is referred to as *unshared information*. In contrast, information available to all group members at the onset of the discussion has been termed *shared information* (Stasser & Titus, 1985).[1] By pooling their unshared information, groups have the potential to make a decision that is better informed than the prediscussion decisions (or preferences) of their individual members. Moreover, the pooling of unshared information can serve a corrective function when the information distribution prior to the discussion leads group members to favor a decision alternative that is suboptimal given the full set of information. In this case, group members have

---

[1] Of course, there is also partially shared information, that is, information known by some but not all group members prior to the discussion.

the potential to collectively piece together a complete and unbiased picture of the relative merits of each decision alternative.

However, one of the best replicated findings in small-group research is that group members are more likely to discuss shared information than unshared information. This asymmetry (or bias) consists of two components: The first is that group members introduce more of their shared than unshared information during the discussion; and the second is that, once introduced into the discussion, shared information is also repeated more often than unshared information (see Stasser & Birchmeier, 2003, for review). In simple terms, groups tend to talk about what their members already know rather than to pool unique information that others do not have. The most basic and most well known explanation for this bias is a probabilistic rather than psychological one, namely, the sampling advantage of shared information, as laid out in the so-called collective information sampling model.

Collective Information Sampling

Garold Stasser and his associates (Stasser & Titus, 1987; Stasser, Taylor, & Hanna, 1989) proposed a *collective information sampling model* that addresses how shared versus unshared information enters the group discussion. According to this model, the content of a group discussion is obtained by sampling from the pool of information that group members collectively hold. This information sampling occurs by group members recalling and mentioning items of information during the discussion. A key assumption of Stasser's (1992) collective information sampling model is that sampling a given item of information is a *disjunctive task*. According to Steiner's (1972) classic typology of group tasks, a disjunctive task can be defined as a task for which the whole group succeeds if any one group member succeeds. In this sense, it is assumed that only one member has to recall and mention an item of information to bring it to the attention of the group. Thus, a group will fail to discuss an item if, and only if, all group members who have access to this item fail to recall and mention it. Stated more formally, Stasser & Titus (1987) proposed that the probability $p(D)$ that a given item of information will be discussed during the group discussion can be expressed as a function of $p(M)$, the probability that any one member will mention the item, and n, the number of members who can potentially mention it:

$$p(D) = 1 - [1 - p(M)]^n. \tag{13.1}$$

The most important implication of Eq. 13.1 is that shared information has a *sampling advantage* over unshared information. This can easily be demonstrated if one keeps in mind that for unshared information n equals 1 and, thus, the probability that an unshared item will be discussed, $p(D_{us})$, equals $p(M)$. In contrast, for shared information, n equals the total number of group members and therefore the probability that a shared item will be discussed, $p(D_s)$, is higher than $p(M)$, except when $p(M)$ equals 0 or 1. To illustrate this, suppose that each member of a three-person group can recall and mention 50% of the items known prior to the discussion; that is, let $p(M) = .50$. In this case, Eq. 13.1 predicts that the probability that a shared item will be discussed is relatively high: $p(D_s) = 1 - [1 - .5)]^3 = .88$. In contrast, the probability of the group discussing an unshared item is substantially lower: $p(D_{us}) = p(M) = .50$. Equation 13.1 also allows for the possibility that a given item is partially shared. In the example here, if two group members had known an item before the discussion, then $n = 2$ and $p(D_{ps}) = 1 - [1 - .5)]^2 = .75$. Overall, Stasser (1992) has proposed a model that teaches a simple lesson: The larger the number of group members that have access to an item of information prior to the discussion, the more likely it is that the item will actually be discussed.

An important prediction that can be derived from this model is that the sampling advantage favoring shared information should depend on group size. In the aforementioned example, if group size were five rather than three, the probability of the group discussing a shared item would be .97, whereas the probability of discussing an unshared item would remain at .50. Thus, the disparity between the probabilities of discussing shared and unshared items is predicted to increase as group size increases. Of course, this line of reasoning makes the simplistic assumption that $p(M)$ remains the same as group size is increased. In fact, in many situations increasing group size will decrease $p(M)$, for example, because there will be less air time for members to mention their items if group size is increased.

The picture gets somewhat more complicated when taking into consideration that the sampling advantage favoring shared information is influenced by the value of $p(M)$. The sampling advantage becomes increasingly large as the value of $p(M)$ departs from either 0 or 1 and is maximized at a value of $p(M)$ that depends on group size. As Stasser and Titus (1987) noted, the maximum sampling advantage occurs when

$$p(M) = 1 - (1/n)^{[1/(n-1)]}, \tag{13.2}$$

where n is group size. Thus, in a three-person group, the sampling advantage favoring shared information is maximized when $p(M) = 1 - (1/3)^{[1/(3-1)]} = .42$. For example, if each member of a three-person group mentions about 42% of the information that he or she knows prior to the discussion, the sampling advantage of shared information is higher compared to a situation in which $p(M)$ is greater or less than .42. The implication is that interventions to facilitate information exchange will not necessarily lessen the sampling advantage favoring shared information but may sometimes even increase it. For example, if in a three-person group $p(M)$ is initially .30 (i.e., below the discussion advantage maximum value), an intervention that moves $p(M)$ closer to .42 will ultimately increase the sampling advantage of shared information. However, when $p(M)$ is initially at or above .42, increasing $p(M)$ by any amount should lessen the sampling advantage of shared information.

Stasser et al. (1989) examined the predictions derived from the collective information model by noting the frequencies with which shared and unshared information was discussed in three- or six-person groups. Prior to the discussion, participants received descriptions of three hypothetical candidates for the job of student body president. Information about the candidates was distributed such that some items were read by only one group member (unshared information) whereas other items were read by all members (shared information). Groups discussed the three candidates and were asked to decide which candidate was best suited to be student body president. To manipulate the level of $p(M)$, Stasser et al. (1989) instructed some of their groups to focus on recalling information and avoid expressing their candidate preferences until they felt that they had discussed all the relevant items (structured discussion). In line with the predictions of the collective information sampling model, it was observed that groups discussed proportionally more shared than unshared items. More importantly, the sampling advantage of shared information was more pronounced in six-person groups. Finally, at least for six-person groups, structuring the discussion resulted in more items being discussed, but this increase was predominantly a consequence of more shared items being discussed. Again, this result is what the sampling model predicts if $p(M)$ is relatively low. (Note that there were fifty-four items across the three candidates and that members were not allowed to keep their information sheets during the discussion.) Thus, structuring the discussion moved $p(M)$ closer to the discussion advantage maximum value.

Sequential Entry of Shared and Unshared Information into the Discussion

Larson, Foster-Fishman, & Keys (1994) proposed a dynamic extension of the collective information sampling model that predicts the sequential entry of shared and unshared information into the discussion. They argued that the collective information sampling model fails to take account of how the *sampling opportunities* for unshared and shared information change as the discussion unfolds. To illustrate this, suppose that a selection committee has to decide which of two candidates to hire for a job. The selection committee has three members, all of whom possess nine items of shared information and each of whom has an additional six items of unshared items about the candidates. At the start of the discussion, there are eighteen opportunities for unshared information to be mentioned (i.e., each of the three members can mention one of his or her six unshared items) and twenty-seven opportunities for shared information to be mentioned (i.e., nine items times the three members who can mention each item). All other things being equal, at the start of the discussion the probability of sampling a shared item is $27/(27 + 18) = .60$, whereas the probability of sampling an unshared item is $18/(27 + 18) = .40$. The crucial point is that these probabilities are not static but change each time a new item of information is contributed to the discussion. When a shared item is brought into the discussion, the number of opportunities to sample further, not-yet-discussed shared items is reduced by three (because all three committee members originally had access to the shared item that has just been mentioned). In contrast, each time an unshared item is brought into the discussion, the number of opportunities to sample further, as yet unshared items is reduced by only one (because only one member originally had access to the unshared item just mentioned). Consequently, the sampling opportunities for shared items decrease at a faster rate than the sampling opportunities for unshared items. This means that as the discussion proceeds, the probability of sampling additional, not-yet-discussed shared items should systematically decrease, whereas the probability of sampling additional, not-yet-discussed unshared items should increase. An important implication of this temporal pattern is that if the discussion continues for a sufficiently long time, the balance of sampling opportunities is predicted to shift to favor unshared items.

Larson et al. (1996) tested this prediction by tracking the discussion of medical decision-making teams who were given two hypothetical medical cases to diagnose. Each team consisted of one resident, one intern, and one medical student. The information for each case was presented on short videotapes showing the patient being interviewed by a physician. About half of the information for each case was given to all three team members, whereas the remaining half was unshared. It was found that, in general, shared items were more likely to be sampled during the discussion and were sampled earlier than unshared items. What is more, over the course of the discussion the probability of sampling new, not-yet-discussed shared items systematically decreased, whereas the probability of sampling unshared items systematically increased. For example, the first item introduced into discussion was almost always shared and over 70% of the items mentioned second were shared items too. However, the tenth item introduced into the discussion was, on average, equally likely to be a shared or unshared one. Henceforth unshared items were sampled more frequently than shared items. Thus, the results lend support to the dynamic version of the information sampling model by Larson (Larson et al., 1994). Moreover, the results demonstrate that even teams of trained professionals faced with decisions similar to those they make as part of their daily work are subject to biased information sampling.

Repeating Shared and Unshared Information

Up to now, we have focused our analysis on the question of whether a shared or unshared item is mentioned. However, in comparison to unshared information, not only is shared information, in general, more likely to be mentioned during the group discussion, but members are also more likely to repeat shared information than unshared information after it has been mentioned. In other words, shared items not only have a discussion advantage but also a *repetition advantage* over unshared items (e.g., Larson et al., 1994; Schulz-Hardt et al., 2004; Stasser et al., 1989).

It should be noted that Stasser's collective information sampling model (Stasser, 1992) does not address the repetition of previously discussed shared and unshared information. Once an unshared item has been mentioned during the discussion, any group member can, at least in principle, repeat it later on. Thus, if repeating information is viewed solely as a process of randomly sampling from members' memories, it would be predicted that previously discussed items of formerly unshared information are as likely to be repeated as previously discussed shared items.

What then accounts for the repetition advantage of shared information? One simple explanation is based on the assumption that the probability of a group member mentioning an item of information increases with each exposure to that item. Of course, if a shared item is mentioned for the first time during the discussion, this mention is by definition the second exposure of members to that item (because the first exposure of a shared item is prior to the discussion). In contrast, the first mention of an unshared item is the first exposure for all but one member (the member who initially owned the unshared item). Consequently, after the first mention, a shared item may still be easier to recall than a previously unshared one.

An alternative perspective is that shared information is more likely to be repeated than unshared information because it is valued more highly than unshared information. In fact, two recent studies (Greitemeyer, Schulz-Hardt, & Frey, 2003; Kerschreiter et al., 2004) provide conclusive evidence that shared information is judged to be more credible and more important than unshared information.

The evaluation bias in favor of shared information is founded on at least two inherent features of shared information: First, shared items are by definition held by all group members before the discussion and, thus, all group members view them as their own information. In contrast, all group members experience a large portion of the unshared items (namely, all but their own) as being presented by other group members during the discussion. Note that several lines of research suggest that decision makers treat the contributions of others differently than their own information. For example, recent studies on the utilization of advice suggest a "self/other effect" whereby people discount the weight of advice and favor their own opinion (e.g., Harvey & Fischer, 1997; Yaniv & Kleinberger, 2000; Yaniv, 2004). As previously shown, this asymmetric weighting of one's own and other information seems to arise from an information asymmetry: Decision makers are privy to their own information and thoughts, but not to those of the advisor (Yaniv, 2004). In the context of group decision making, Chernyshenko et al. (2003) observed that group members perceived their own discussed unshared items as more important than the discussed unshared items of others. With regard to perceived validity, van Swol, Savadori, & Sniezek (2003) found that information that group members owned (both shared and unshared) before the discussion was judged as more valid than other information. They dubbed this phenomenon *ownership bias.*

The second important inherent feature of shared information is that it can be socially validated. Thus, when a group member contributes a shared item to the discussion, all members can confirm its veracity. In contrast, unshared items cannot be corroborated by other group members and, hence, may be treated with skepticism. In keeping with this notion, the finding that shared information is perceived as more credible and more important than unshared information has been typically explained by the fact that it can be socially validated (e.g., Greitemeyer et al., 2003; Parks & Cowlin, 1996; Postmes, Spears, & Cihangir, 2001). However, caution is called for, since the evaluation bias in favor of shared information observed in previous studies may not necessarily be due to social validation but can also be explained in terms of ownership bias. A methodologically rigorous test of the social validation explanation has recently been conducted by Mojzisch et al. (2004). In this study, Mojzisch et al. (2004) showed that items that could be corroborated by other group members were evaluated more favorably, independent of whether those items were owned before the discussion or presented by another group member during the discussion. In essence, both enhanced memorability and greater perceived value of shared information may explain its repetition advantage over unshared information.

Mutual Enhancement

By building on and extending the social validation hypothesis, Wittenbaum, Hubbell, & Zuckerman (1999) proposed an additional process that promotes the dominance of shared information: the tendency for group members to positively evaluate one another's task capabilities when mentioning shared information. They called this process *mutual enhancement.* The mutual enhancement model is based on the notion that being exposed to shared information validates members' knowledge. The process by which mutual enhancement is proposed to promote the sampling advantage favoring shared information is as follows (Wittenbaum et al., 1999):

Step 1: Suppose that one group member communicates a shared item of information.

Step 2: The recipient recognizes the item and confirms its accuracy and task relevance. This ability to verify the accuracy and relevance of the item leads to higher ratings of the task capabilities of the communicator. At the same time, after hearing a shared item, the recipients feel better about their own task capability because another member found it important enough to mention an item they posses. Thus, the recipients of shared items think highly of both their own task capability and the communicator's task capability.

Step 3: As a result, members who communicate shared items receive encouragement from the other members for doing so. For example, members may lean forward and smile when a shared item is mentioned or they may comment on its importance.

Step 4: According to Wittenbaum et al. (1999), members who are reinforced for mentioning shared information continue to do so because they enjoy the encouragement and validation from other members.

Step 5: This interactive process of mutual encouragement and validation is proposed to boost the group's tendency to discuss shared information.

Wittenbaum et al. (1999) conducted three experiments to examine the proposed relationship between step 1 and step 2 of their model. In experiments 1 and 2, participants were led to believe that they were working on a personnel selection task with an assigned partner in another room. Participants first received the applications of two hypothetical academic job candidates. Before reading the curricula vitae of the candidates, participants were told what kind of information they had that their partner did not have. To exchange information about the candidates, participants were instructed to list ten items of information that they wanted to share with their partner. These lists were collected by the experimenter, who subsequently left the room, apparently to deliver the lists to the dyad partner. A little later the experimenter returned with handwritten lists ostensibly written by the participants' partner. These bogus lists contained either mostly shared or mostly unshared information. As predicted, participants judged shared information to be more accurate and influential than unshared information. Moreover, dyad members who received mostly shared information evaluated their own task competence and the task competence of their partners more highly than members who received mostly unshared information did. In experiment 3 this effect was replicated in face-to-face dyads.

Notwithstanding the notion that focusing discussion on shared information leads to mutual enhancement, an important question left unanswered by the study of Wittenbaum et al. (1999) is whether mutual enhancement actually promotes the sampling advantage of shared information (as outlined in step 3 and 4 of the model). Furthermore, Wittenbaum et al. (1999) used an information distribution in which sharedness and preference-consistency of information were partially confounded. Consequently, it is not possible to determine whether the finding that shared information was judged to be more accurate and influential than unshared information is due to social validation or due to preference-consistent information evaluation. To address these questions, Kerschreiter et al. (2004) conducted an extended replication of the Wittenbaum et al. (1999) study. In contrast to Wittenbaum et al. (1999), sharedness and preference-consistency of information were orthogonally manipulated. Furthermore, Kerschreiter et al. (2004) extended the experimental paradigm developed by Wittenbaum et al. (1999). Specifically, after receiving the bogus lists participants were asked to provide handwritten feedback to their dyad partners. For example, they were asked to write down how helpful the information that they had received from their dyad partner was. In return, participants received either positive or negative bogus feedback from their dyad partners and were led to expect that they would finally discuss the decision case face-to-face with their partners. Beforehand, they were asked to list ten items they regarded as particularly important for the final discussion. As expected, Kerschreiter et al. (2004) observed that shared information – independent of preference-consistency – was evaluated as being more credible and relevant than unshared information. Moreover, Kerschreiter et al. (2004) found that participants encouraged their partners more for communicating preference-consistent information than for communicating preference-inconsistent information. Note that the latter finding is not relevant to the mutual enhancement model since it makes no specific predictions concerning the consequences of communicating preference-consistent or preference-inconsistent information. However, in stark contrast to the predictions derived from the mutual enhancement model, Kerschreiter et al. (2004) found that participants encouraged their partners more for communicating *unshared* information than for shared information. Additionally, when asked which items they intended to exchange during the anticipated face-to-face discussion, participants primarily mentioned unshared and preference-consistent information (independent of the feedback participants had received from their partners). In other words, although group members may positively evaluate one another's task capabilities when mentioning shared information, there is no evidence that this mutual enhancement effect can account for the sampling advantage of shared information. Rather, the results by Kerschreiter et al. (2004) suggest that in groups where members know which information is shared and which is unshared the sampling advantage of shared information may be weakened because members intentionally share their unique information with each other. Interestingly, this is exactly what conversational norms imply, namely, to avoid telling others information that they already know (Grice, 1975).

BIASED SAMPLING IN FAVOR OF PREFERENCE-CONSISTENT INFORMATION

As outlined in the introduction of this chapter, information sampling in groups is typically not only biased in favor of shared information but also in favor of information that supports rather than contradicts group members' preferences. In other words, information sampling is biased in favor of preference-consistent information (i.e., advantages of the preferred alternative and disadvantages of the nonpreferred alternatives). Unfortunately, the distinction between preference-consistent and preference-inconsistent information has received relatively little attention in small-group research, probably because most of the research that followed in the footsteps of Stasser's classic studies (Stasser & Titus, 1985, 1987; Stasser et al., 1989) focused solely on the sampling advantage favoring shared information. Ironically, it was Stasser and Titus (1985, p. 1,470) who proposed that information supporting the group's preferred decision alternative may be more likely to enter the discussion than information opposing this alternative. Moreover, they suggested that the tendency to preferentially discuss preference-consistent information may explain the fact that group members often fail to correct their suboptimal initial preferences on the basis of information sampling during the discussion (Stasser & Titus, 1987).

It is important to note that in numerous studies on biased information sampling a prediscussion distribution of information was used in which sharedness and preference-consistency of information were at least partially confounded; that is, most of the shared information supported the group members' preferred decision alternative (e.g., Brodbeck et al., 2002; Dennis, 1996; Galinsky & Kray, 2004; Hollingshead, 1996; Kelly & Karau, 1999; Lam & Schaubroeck, 2000; Lavery et al., 1999; Stasser & Stewart, 1992; Stasser, Stewart, & Wittenbaum, 1995; Stasser & Titus, 1985; van Swol et al., 2003; Winquist & Larson, 1998). In the next section of this chapter, we will outline why this biased distribution of information has been used in so many studies. For now, it is sufficient to note that the confounding of sharedness and preference-consistency of information suggests that in the aforementioned studies the sampling advantage favoring shared information may, at least partially, reflect a tendency for group members to favor preference-consistent over preference-inconsistent information.

In what follows, we will distinguish three components of preference-consistent sampling of information. First, preference-consistent information is introduced and repeated more frequently than preference-inconsistent information. Second, group members frequently frame information in a preference-consistent way when it is mentioned. Finally, after having made a preliminary decision, groups have a tendency to selectively search for preference-consistent information at the expense of preference-inconsistent information.

The Bias toward Discussing Preference-Consistent Information

A clear-cut demonstration of preference-consistent information sampling has been reported by Dennis (1996). In this study, participants met in six-member groups to select one out of three students for university admission. Groups either interacted face-to-face or with a group support system (GSS). As expected, group members were more likely to exchange preference-consistent information than neutral information or preference-inconsistent information. This result was true for both GSS and non-GSS groups. Importantly, this sampling advantage for preference-consistent information cannot be attributed to the fact that preference-consistent items were predominantly shared among group members because tests were performed only on unshared items. Further evidence for the sampling advantage of preference-consistent information has been obtained by Schulz-Hardt et al. (2004). In this study, three-person groups had to decide which of four candidates was best suited for a job as airplane captain. Prior to the discussion, participants individually indicated their preference for one of the candidates. During the discussion, group members introduced more of their shared and preference-consistent information than of their unshared and preference-inconsistent information. Additionally, groups were more likely to repeat both shared and preference-consistent information than unshared and preference-inconsistent information.

What accounts for the sampling advantage of preference-consistent information? In our view, there are three relatively simple explanations. First, as suggested by Stasser & Titus (1985, p. 1,470), information sampling in groups is governed by a "norm of advocacy." That is, members act on the assumption that they should argue for their initial preference during the discussion. This tendency should be especially pronounced at the beginning of a group discussion. To illustrate this, consider how a typical group discussion starts. As group members are not normally aware of one another's initial preferences, group discussions frequently start with group members exchanging their preferences. Thereafter, group members typically explain why they favor that particular decision alternative. In other words, imagine how odd it would be if a group member first stated a preference for a particular decision alternative and subsequently enumerated the disadvantages of that alternative and the advantages of the other alternatives. As stated by Stasser & Titus (1987, p. 83), "playing devils' advocate, in the absence of explicit instructions to do so, is viewed as unusual, if not insincere, behavior."

A second possible explanation for the sampling advantage of preference-consistent information is that group members may be motivated to defend their own preferred alternative and to have this alternative adopted by the group. Consequently, they intentionally mention more information supporting than contradicting their initial preferences. This idea has been elaborated by Wittenbaum, Hollingshead, & Botero (in press). According to Wittenbaum et al. (in press), information sampling in groups should be conceived as a motivated process whereby members deliberately choose what information to mention and how to communicate it to other members to satisfy their individual goals. For example, a group member might sometimes remember a particular item but strategically choose not to communicate it to other members because that item contradicts his or her preferred decision alternative.

The third explanation teaches us that even if group members are purely accuracy motivated and do not follow any implicit or explicit advocacy norms, they will be more likely to exchange preference-consistent than preference-inconsistent information. To illustrate this explanation, consider an accuracy-motivated group member. As a matter of course, this person will be more likely to discuss those pieces of information that he or she feels are important and credible than items that seem to be irrelevant or unreliable. Now this is the point where bias comes in: Arguments that are consistent with a person's prior opinion are perceived to be stronger and more credible than those that are inconsistent with this opinion. This preference-consistent evaluation of information has consistently been demonstrated in research on the "prior-belief effect" (e.g., Edwards & Smith, 1996) and on "motivated reasoning" (e.g., Ditto et al., 1998).[2] Thus, what may happen in a group discussion is that members will mention more preference-consistent than preference-inconsistent arguments simply because preference-consistent arguments are perceived to be of higher quality than preference-inconsistent ones. Empirical evidence for this prediction has been provided in the aforementioned study by Greitemeyer et al. (2003). They showed that preference-consistent information was perceived as more credible and relevant and was intended to be

---

[2] Note that biased evaluation of information in favor of preference-consistent information may also be understood in terms of biased sampling, namely, biased sampling from memory. According to the disconfirmation model (Edwards & Smith, 1996), when people are confronted with an argument that is incompatible with their prior belief, they will engage in a deliberative memory search for information that will undermine the argument. In contrast, information that is consistent with one's beliefs is more likely to be accepted at face value. This more thorough testing process increases the likelihood of weaknesses in inconsistent arguments being detected. In addition, because people inevitably possess more arguments for than against their own opinions and thus have a skewed argument base, in the case of an inconsistent argument most of the information retrieved will be refutational in nature.

repeated more often than preference-inconsistent information. In addition, participants preferred to discuss and request more preference-consistent than preference-inconsistent information. Finally, the intention to mention and to repeat more of the preference-consistent information was at least partially mediated by a preference-consistent evaluation of information. Similar results have been obtained in the aforementioned study by Kerschreiter et al. (2004). In sum, the findings of Greitemeyer et al. (2003) and Kerschreiter et al. (2004) strongly suggest that even if group members are not motivated to strategically defend their own preferred alternative or to have this alternative adopted by the group, they are more likely to mention more preference-consistent than preference-inconsistent arguments – simply because preference-consistent arguments are judged to be of higher value than preference-inconsistent arguments.

Preference-Consistent Framing of Information

In experiments on information sampling, the content of group discussions is typically coded to identify which items of information were mentioned correctly. To be counted as a correct recall, items mentioned during the discussion have to be unambiguously understood to refer to one of the original items contained in the profile. This "all-or-nothing" approach to coding information sampling has recently been criticized by Wittenbaum et al. (in press). They argued that, by using this method of coding, the subtle frame that the communicator places on the piece of information is lost. According to Wittenbaum et al. (in press), members are frequently biased not only in their choice of which information to mention but also in the frame that they place on a particular item when it is mentioned. Specifically, they hypothesized that members spin information about their preferred alternative upward (making it seem more positive) and spin information about the nonpreferred alternatives downward (making it seem more negative). For example, a member of a selection committee may mention negative information about his or her preferred job candidate but may do so in a way that frames the information in a more positive light. For example, if a job candidate is said to be not very interested in receiving further training, a supporter of this candidate may communicate this piece of information to the other committee members but add that this information, of course, does not imply that the candidate would not participate in vocational training courses. Moreover, the supporter might note that the job candidate may not be interested in receiving further training because he or she already possesses excellent job competence. In contrast, a supporter of another candidate may argue that someone who is not interested in receiving further training must be a lazy person.

The conjecture that group members frame information in a preference-consistent way when they mention it was recently tested by Wittenbaum,

Bowman, & Hollingshead (2003). In this study, three-person groups had to select which of four hypothetical cholesterol-reducing drugs would be the best drug to market. When coding information, coders noted not only the proportion of preference-consistent and preference-inconsistent information mentioned but also the spin that group members placed on the information when they mentioned it. Two main findings were obtained: First, it was observed that members were more likely to mention preference-consistent than preference-inconsistent information. Second, members were more likely to spin upward items about their own preferred drug and to spin downward items about the drugs preferred by others than they were to spin downward items about their own preferred drug and to spin upward items about the preferred drugs of others.

Findings from this study have been taken to support the notion that information sharing is a strategic process (Wittenbaum et al., in press). We should, however, be cautious with this interpretation. As noted previously, it is also conceivable that group members act on the implicit assumption that they should argue for their initial preference during the discussion (Stasser & Titus, 1985). Moreover, preference-consistent framing of information can also be due to preference-consistent evaluation of information. In fact, it has been shown in many experiments that whenever people have developed a preference for a particular decision alternative they restructure the available information in favor of the preferred alternative (e.g., Greitemeyer & Schulz-Hardt, 2003; Russo, Medvec, & Meloy, 1996). In other words, preference-consistent framing of information can easily be explained without resorting to any strategic processes.

Biased Group Search for External Information

As outlined in the foregoing, groups typically fail to disclose uniquely held information and to enlarge the task-relevant knowledge of the individual members. However, there is another way in which groups can broaden their information base for decision making: by seeking information from sources outside the group. For example, after having made a preliminary decision, groups can call in an expert for a further opinion. Unfortunately, there is a growing body of evidence suggesting that such information seeking processes in groups are frequently unbalanced: Groups prefer information supporting the decision alternative currently favored by the group compared to information opposing it.

At the individual level, searching for preference-consistent information has been investigated extensively in dissonance studies on selective exposure of information. [For an overview see Frey (1986); for new evidence see Jonas et al. (2001).] Recently, Schulz-Hardt et al. (2000, 2002) demonstrated in two studies that this type of bias also occurs in groups. In the first study (Schulz-Hardt et al., 2000), participants first made an individual decision concerning an investment case or a decision problem from the area of economic policy. Based on this decision, groups were formed that either contained a minority and a majority faction (heterogeneous groups) or consisted of members favoring the same decision alternative (homogeneous groups). After having discussed the decision case and having reached a preliminary decision, group members were informed that additional pieces of information about the decision problem, namely, articles written by experts, could be requested to prepare the final group decision. Each article was summarized by a main thesis (one sentence), from which it was apparent whether the corresponding article supported or conflicted with the preliminary decision of the group. The main theses were written on a separate sheet of paper that was handed out to the group members. Groups were asked to mark those articles they wanted to read later on.

In all three experiments, Schulz-Hardt et al. (2000) observed a preference for information supporting the preliminary group decision. This bias was most pronounced in homogeneous groups and was still significant in groups in which the minority faction was relatively small compared to the majority faction (one group member versus four group members). In contrast, when faction sizes were more equal (three members versus two members or two members versus one) the bias was no longer significant. Of course, these findings need not necessarily reflect a group-level phenomenon; they could simply be due to an aggregation of individual preference-consistency biases. To test whether the observed differences between homogeneous and heterogeneous groups indeed reflect a group-level phenomenon, group members were asked to indicate their individual information requests prior to the discussion (Schulz-Hardt et al., 2000, experiment 3). Thus, a statisticized group baseline could be calculated by simply summing up the individual information requests. This baseline reflected how the group would seek information if nothing happened at the group level. Not surprisingly, homogeneous groups showed a somewhat higher bias in this statisticized group information search than heterogeneous groups. This can easily be explained by the fact that individual members were biased toward information supporting their preferences, and in heterogeneous groups the bias of the minority faction runs counter to the bias of the majority faction. More importantly, it could be demonstrated that homogeneous groups had an even higher information search bias than their statisticized baselines. In contrast, heterogeneous groups showed a smaller information search bias than their statisticized baselines. In other words, the differences between homogeneous and heterogeneous groups were not simply the result of aggregated individual biases but instead were due to social influence processes within groups.

Additional analyses revealed mediating mechanisms for this effect. First, initial preference heterogeneity increased the likelihood for disagreement immediately prior to the start of the information search. It was only in this case that preference heterogeneity debiased information search. Second, consistent disagreement led to lower confidence about the correctness of the group's decision and to lower commitment to the group's preliminary decision, and these two processes, in turn, mediated the effect of preference heterogeneity on biased information search.

THE CONSEQUENCES OF BIASED INFORMATION SAMPLING IN GROUPS

So far, everything we have said about biased information sampling in groups may have given the impression that predominantly sampling shared and preference-consistent information is inevitably dysfunctional. However, both sampling shared and sampling preference-consistent information may also be of particular value for the group. For example, discussing shared information might help group members to develop trust and to form a common ground (Clark & Brennan, 1991). Furthermore, a preference for information supporting a preliminary decision stabilizes people's preferences and fosters the group's ability to rapidly implement the decision (Beckmann & Kuhl, 1984).

With regard to decision quality, the consequences of biased information sampling crucially depend on the prediscussion distribution of information. To illustrate this dependency, consider an extremely simplified situation in which a three-person medical team is asked to discuss a patient case and develop a differential diagnosis. Suppose that there are two possible diagnoses, A and B, based on the symptoms of the patient. The correct diagnosis, A, is supported by ten symptoms, whereas diagnosis B is supported by only five symptoms.

Now contrast two different ways in which this entire set of symptoms can be distributed among the members of the clinical team. In an unbiased information distribution, there are four shared symptoms supporting diagnosis A and two shared symptoms supporting diagnosis B. Additionally, each team member receives two unshared symptoms supporting diagnosis A and one unshared symptom supporting diagnosis B (see Table 13.1). Assuming that each symptom is equally diagnostic, the information distributed to each member is in favor of diagnosis A (i.e., the correct diagnosis) because each member has access to six symptoms supporting diagnosis A and only three symptoms that support B. Thus, each member has a representative subset of the total pool of symptoms. In this case, a *manifest profile* (Lavery et al., 1999) discussing unshared items can broaden group members' knowledge, but it should not change their decision. Even if group members primarily discuss their shared information, they will (still) make a correct decision because shared and unshared information have the same decision implication. However, note that in these cases group decision making cannot "pay off" with regard to decision quality because each member

TABLE 13.1. *A Simplified Example of a Manifest Profile*

	Symptoms Supporting Diagnosis A	Symptoms Supporting Diagnosis B	Item Distribution	Decision Implication
Individual level				
X	A_1-A_4, A_5, A_6	B_1, B_2, B_3	6 A > 3 B	A
Y	A_1-A_4, A_7, A_8	B_1, B_2, B_4	6 A > 3 B	A
Z	A_1-A_4, A_9, A_{10}	B_1, B_2, B_5	6 A > 3 B	A
Group level				
X ∪ Y ∪ Z	A_1-A_{10}	B_1-B_5	10 A > 5 B	A

would have reached the correct decision based solely on what he or she knew prior to the discussion.

The situation changes dramatically if one considers an information distribution that has been termed a *hidden profile* (Stasser, 1988). In a hidden profile, shared and unshared information have a different decision implication, and the decision alternative implied by the unshared information is the correct one (relative to the entire set of information that is given to the group as a whole). Consequently, no group member can detect the best solution on the basis of his or her information set prior to the discussion. Stated differently, group members' individual information is not representative of the entire information set. Thus, groups can only arrive at a correct decision by pooling the unshared information during discussion.

To illustrate this, consider a simplified version of a hidden profile using the previous example (see Table 13.2). Again, there are ten symptoms supporting diagnosis A and only five symptoms supporting diagnosis B. Let us now suppose that only one of the items supporting diagnosis A is shared, whereas the remaining nine items are unshared. Let us further suppose that all of the information supporting diagnosis B is shared. In this case, each member should favor diagnosis B prior to the discussion

TABLE 13.2. *A Simplified Example of a Hidden Profile*

	Symptoms Supporting Diagnosis A	Symptoms Supporting Diagnosis B	Item Distribution	Decision Implication
Individual level				
X	A_1, A_2, A_3, A_4	$B_1, B_2, B_3 B_4, B_5$	4 A < 5 B	B
Y	A_1, A_5, A_6, A_7	$B_1, B_2, B_3 B_4, B_5$	4 A < 5 B	B
Z	A_1, A_8, A_9, A_{10}	$B_1, B_2, B_3 B_4, B_5$	4 A < 5 B	B
Group level				
X ∪ Y ∪ Z	A_1-A_{10}	B_1-B_5	10 A > 5 B	A

because each has access to only four items that support diagnosis A but five that support B. To recognize that there are more symptoms supporting diagnosis A than B, group members would have to exchange their unshared items.

Thus, in a hidden profile, a superior decision alternative exists, but its superiority is *hidden* from the individual group members because of the way in which information is distributed prior to the discussion. Note that hidden profiles are especially important in group decision making because they represent the prototype of situations in which discussion-based group decisions can potentially lead to higher decision quality compared to individual decisions or social combinations of individual decisions.[3]

Ironically, a great deal of research has shown that, in hidden profile situations, groups rarely discover the best alternative (e.g., Brodbeck et al., 2002; Dennis, 1996; Hollingshead, 1996; Kelly & Karau, 1999; Lam & Schaubroeck, 2000; Lavery et al., 1999; Schulz-Hardt et al., 2004; Stasser & Titus, 1985; van Swol et al., 2003; Winquist & Larson, 1998). Instead, groups largely opt for the decision alternative that is implied by their members' shared information – which is suboptimal on the basis of the entire set of information. For example, in the groundbreaking study by Stasser & Titus (1985), only 18% of the groups in the hidden profile condition arrived at the correct solution. In contrast, in the control condition where all decision-relevant information was shared among members, 83% of the groups selected the best alternative. In other words, groups typically fail to detect the correct solution in the very situation in which they have the potential to outperform individual decision makers.

Group-Level Explanations for the Failure of Groups to Solve Hidden Profiles

The group information sampling biases described in the previous sections of this chapter constitute what is probably the most well known explanation that has been offered to account for the failure of groups to discover hidden profiles (Stasser, 1992; Stasser & Titus, 1985). According to the biased information sampling model, the correct decision alternative is not detected because the group fails to exchange sufficient unshared information and hence the superiority of this alternative does not become evident. Note that both the tendency to primarily discuss shared information and the tendency to primarily discuss preference-consistent information

---

[3] Unfortunately, there have been no studies to date showing how hidden profiles are generated in natural field settings. However, of course, if one assumes that the distribution of information among group members has a random component in the real world, a certain amount of decision problems (which obviously has to be lower than 50%) can be expected to be hidden profiles.

interfere with the solution of hidden profiles, since under conditions of a hidden profile unshared information supports a decision alternative that is preferred by almost nobody upon entering the discussion. Hence, to solve a hidden profile, the group has to exchange information that is both unshared and inconsistent with most or all group members' prediscussion preferences. In other words, it is assumed that groups fail to solve hidden profiles because the group discussion is biased toward both shared and preference-consistent information (Stasser & Titus, 1985).

An entirely different explanation for the failure of groups to solve hidden profiles has been proposed by Gigone & Hastie (1993, 1997). This explanation focuses on *preference negotiation* and *premature consensus*. According to Gigone & Hastie, information exchange plays little or no role with regard to the group's decision. Instead, group members are presumed to use the discussion primarily to exchange their prediscussion preferences and to negotiate the group decision on the basis of these preferences. The process by which preference negotiation is proposed to hinder groups from solving a hidden profile is as follows: Because hidden profiles predispose members to prefer suboptimal decision alternatives upon entering the group discussion, the correct alternative typically has no proponents and is therefore ignored. Consequently, it is unlikely that preference exchange and negotiation will lead to a consensus in favor of the correct alternative. Rather, a suboptimal alternative that is preferred by all or most members prior to the discussion will prematurely be accepted.

Biased Information Evaluation as an Explanation for the Failure to Solve Hidden Profiles

From all we have said, a group that works "perfectly" – that is, one in which no premature preference negotiation takes place and all information about the decision problem is exchanged – should be quite successful at solving hidden profiles. However, recent evidence suggests that this assumption is overly simplistic. Even in the absence of any dysfunctional group processes, group members evaluate the information discussed in the light of their individual prediscussion preferences. Because preference-consistent information is systematically judged to be more important and credible than preference-inconsistent information, Greitemeyer & Schulz-Hardt (2003) predicted that group members may stick to their suboptimal prediscussion preferences even if all relevant information has been exchanged. This prediction was supported in two experiments. In both experiments participants first received individual information about three decision alternatives and were asked to indicate their preference. The individual information was either representative of the entire information set (manifest profile) or not representative (hidden profile). Subsequently, instead of discussing the decision case with other participants in a group, each participant received a protocol of a fictitious group discussion, with him- or herself and two other fictitious members as protagonists. In this discussion protocol, all available information about the decision alternatives was exchanged. The participants' task was to reach a final decision after having studied the discussion protocol. Additionally, for each piece of information discussed in the protocol, participants were asked to rate its importance, credibility, and valence (i.e., how positive or negative it is). As predicted, only a few participants in the hidden profile condition detected the best alternative. Moreover, Greitemeyer & Schulz-Hardt (2003) found evidence for biased evaluation of the information exchanged in the discussion protocol: Preference-consistent information was judged to be more important than preference-inconsistent information. In addition, information about the preferred decision alternative was judged as being more favorable than information about the nonpreferred decision alternatives. Finally, it was shown that biased information evaluation at least partially mediated the maintenance of an initial suboptimal preference. In conclusion, preference-consistent evaluation of information fosters the maintenance of an initial preference and hence impedes the solution of a hidden profile.

Again, as in the case of information sampling, information evaluation is not only biased in favor of preference-consistent information but also in favor of shared information (Greitemeyer et al., 2003; Kerschreiter et al., 2004). As outlined previously, the evaluation bias in favor of shared information may be due to social validation but can also be explained in terms of the "ownership bias". However, note that Mojzisch et al. (2004) found that items that could be corroborated by other group members were perceived to be of greater value and had a greater influence on participants' final decisions, independent of whether those items were owned prior to the discussion or presented by another group member during the discussion.

In summary, three processes are responsible for the failure of groups to solve hidden profiles, two operating at the group level and one operating at the individual level. First, groups frequently focus on negotiating the final decision on the basis of their members' initial preferences rather than exchanging their unique information. Since hidden profiles predispose group members to prefer suboptimal decision alternatives prior to the discussion, a premature consensus in favor of a suboptimal alternative is likely to emerge. Second, even if groups do exchange the relevant information, information sampling is biased against the best solution because groups predominantly discuss shared and preference-consistent information. Finally, even if the group exchanges all relevant information, biased evaluation of information in favor of both shared and preference-consistent information fosters the maintenance of an initial suboptimal decision preference. Figure 13.1 depicts the three processes proposed to account for the failure of groups to solve hidden profiles.

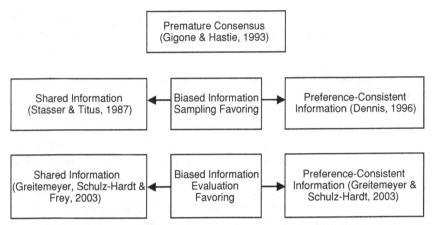

FIGURE 13.1. Explanations for the failure of groups to discover hidden profiles.

It is important to note that each of the three processes is sufficient but not necessary to hamper the group members' ability to solve the hidden profile. Moreover, the three processes are supplementary rather than contradictory. For example, although premature consensus on a suboptimal alternative could be counteracted by sufficient exchange of the critical unshared information, this corrective potential of group discussion frequently is not realized because shared information is more likely to be discussed than unshared information (Winquist & Larson, 1998). Similarly, the pooling of unshared information has the potential to correct suboptimal initial preferences, but biased information evaluation favoring both shared and preference-consistent information makes the likelihood of this corrective potential being exploited slight.

Hence groups have to overcome many hurdles to solve hidden profiles, since intrinsic features of group discussions and individual information evaluation work against the solution of hidden profiles. Thus, it is not very surprising that the search for interventions aimed at facilitating the solution of hidden profiles has not been very successful to date (see Stasser, 1999, and Stasser & Birchmeier, 2003, for reviews). Consequently, one important lesson that can be learned from recent research is that such interventions will only be successful if they are able to counteract all three processes outlined here as obstacles to solving hidden profiles.

Finally, we should point out that the same processes that impede the solution of hidden profiles may be beneficial in the case of manifest profiles, which obviously occur more frequently in natural field settings than hidden profiles. For example, in the case of a manifest profile, the discussion bias favoring shared information will have no detrimental effects on decision making but will help group members to better understand and relate to each other. Furthermore, other things being equal, the fact that other people have arrived at the same opinion can be taken as an indicator of the correctness of one's opinion (Chaiken & Stangor, 1987). Thus, in the case of a manifest profile, reliance on the degree of consensus as an indicator of correctness provides a fast and frugal heuristic in group decision making. Accordingly, the processes that impede the solution of hidden profiles are by no means pathological but rather functional under most conditions. However, because groups only have the potential to outperform individual decision makers (as well as any combinations of individual votes) if a hidden profile exists, these processes prevent groups from realizing their potential with regard to decision quality.

CONCLUSION

One significant appeal of group decision making stems from the idea that groups can benefit from the diversity of expertise, knowledge, and perspectives that their members bring to bear on a decision problem. Thus, information sampling is viewed as a key feature of the group decision-making process, and this sampling during the group discussion can serve an important corrective function if members enter the discussion with a biased subset of information supporting a suboptimal decision alternative.

However, as we have pointed out throughout this chapter, there is a growing body of evidence that raises doubts about group members' ability to efficiently pool their information. Specifically, the pooling of information is generally hampered by two discussion biases: First, groups discuss more of their shared than unshared information. Moreover, once introduced into the discussion, shared information is also repeated more often than unshared information. Second, information is introduced and repeated more frequently if it supports, rather than contradicts, members' prediscussion preferences.

Note that both discussion biases can easily be explained without assuming that group members are loafing, confirming or pursuing selfish goals. That is, shared information may be more likely to be discussed than unshared information simply because the former can be mentioned by more persons than the latter. Similarly, preference-consistent information may be more likely to be discussed than preference-inconsistent information because the former is judged to be of higher value than the latter or because group members act on the assumption that they should argue for their preferred alternative. Nonetheless, of course, in natural decision-making groups members might also strategically select what information to share with others, and such tendencies may increase these two discussion biases.

With regard to decision quality, we have argued that hidden profiles are particularly important because they represent the prototype of situations in which discussion-based group decisions potentially lead to higher decision quality compared to individual decisions or social combinations of individual decisions. Unfortunately, detecting the best decision alternative in a hidden profile requires information that is both unshared and preference-inconsistent to be exchanged. Consequently, the bias in favor of both shared and preference-consistent information impedes the solution of hidden profiles. Indeed, there is compelling evidence showing that biased group decisions can be explained by biased sampling processes. However, these biased sampling processes do not tell the whole story with regard to how suboptimal decisions in hidden profile situations emerge. They are sufficient to explain the phenomenon (i.e., groups that fail to discuss unshared information cannot solve the problem) but they are not necessary for the occurrence of these suboptimal decisions: Even if the group succeeds in pooling the critical information, biased evaluation of information in favor of both shared and preference-consistent information impedes the solution of hidden profiles.

Looking ahead, we envision several aspects that merit further research. For example, better understanding is needed of why shared information is repeated more frequently than unshared information. As noted in the foregoing, Stasser's collective information sampling model does not account for the repetition bias of shared information. In contrast, both enhanced memorability and greater perceived value of shared information may explain its repetition advantage over unshared information. Furthermore, an important direction for future research is to test the impact of the repetition bias of shared information on the group's decision. Thus, it is quite conceivable that the greater decision impact of shared information is not only because that shared information is more likely to be mentioned overall but also because shared information is repeated more frequently than unshared information. Indeed, there is preliminary evidence supporting this notion (van Swol et al., 2003, experiment 2).

Another avenue for future research is to take a closer look at biased information sampling favoring preference-consistent information. For example, Schulz-Hardt et al. (2004) found that the positive effect of disagreement among group members' prediscussion preferences on decision quality was mediated by the repetition bias in favor of preference-consistent information. Thus, the less likely group members were to repeat preference-consistent items the more likely they were to solve the hidden profile. Similarly, future research would do well to scrutinize preference-consistent framing of information.

Another line of research worth pursuing involves examining which factors influence biased evaluation of information in a group context. Note that to date research has focused primarily on investigating factors that influence the relative amounts of shared and unshared information that are discussed by groups. One of the few exceptions is the work by Postmes et al. (2001) showing that when groups have a norm of critical thought they are more likely to positively value unshared information and to solve a hidden profile than groups with a consensus norm. To reiterate, it is not discussion content per se that predicts the group's decision but the subjective evaluation and interpretation of discussion content. Consequently, we need to move beyond studying the process of information exchange and take a closer look at the process of information evaluation and integration in group decision making. As outlined in the introduction, this constitutes a shift that is somehow diametrically opposed to the one that this book documents for individual judgment and decision-making research, namely, to pay more attention to biased sampling processes when explaining judgment and decision asymmetries, biases, and errors. Although the lesson that each coin has two sides appears to be trivial, it nevertheless seems to be one that had, and still has, to be learned.

References

Beckmann, J., & Kuhl, J. (1984). Altering information to gain action control: Functional aspects of human information processing in decision making. *Journal of Research in Personality, 18*, 224–237.

Brodbeck, F. C., Kerschreiter, R., Mojzisch, A., Frey, D., & Schulz-Hardt, S. (2002). The dissemination of critical unshared information in decision-making groups: The effect of prediscussion dissent. *European Journal of Social Psychology, 32*, 35–56.

Chaiken, S., & Stangor, C. (1987). Attitudes and attitude change. *Annual Review of Psychology, 38*, 575–630.

Chernyshenko, O. S., Miner, A. G., Baumann, M. R., & Sniezek, J. A. (2003). The impact of information distribution, ownership, and discussion on group member judgment: The differential cue weighting model. *Organizational Behavior and Human Decision Processes, 91*, 12–25.

Clark, H. H., & Brennan, S. E. (1991). Grounding in communication. In L. B. Resnick, J. M. Levine, & S. D. Teasley (Eds.), *Perspectives on socially shared cognition* (pp. 127–149). Washington, DC: American Psychological Association.

Dennis, A. R. (1996). Information exchange and use in small group decision making. *Small Group Research, 27*, 532–550.

Ditto, P. H., Scepansky, J. A., Munro, G. D., Apanovitch, A. M., & Lockhart, L. K. (1998). Motivated sensitivity to preference-inconsistent information. *Journal of Personality and Social Psychology, 75*, 53–69.

Edwards, K., & Smith, E. E. (1996). A disconfirmation bias in the evaluation of arguments. *Journal of Personality and Social Psychology, 71*, 5–24.

Frey, D. (1986). Recent research on selective exposure to information. In L. Berkowitz (Ed.), *Advances in experimental social psychology* (Vol. 19, pp. 41–80). New York: Academic Press.

Galinsky, A. D., & Kray, L. J. (2004). From thinking about what might have been to sharing what we know: The role of counterfactual mind-sets on information sharing in groups. *Journal of Experimental Social Psychology, 40*, 606–618.

Gigone, D., & Hastie, R. (1993). The common knowledge effect: Information sharing and group judgment. *Journal of Personality and Social Psychology, 65*, 959–974.

Gigone, D., & Hastie, R. (1997). The impact of information on small group choice. *Journal of Personality and Social Psychology, 72*, 132–140.

Greitemeyer, T., & Schulz-Hardt, S. (2003). Preference-consistent evaluation of information in the hidden profile paradigm: Beyond group-level explanations for the dominance of shared information in group decisions. *Journal of Personality and Social Psychology, 84*, 322–339.

Greitemeyer, T., Schulz-Hardt, S., & Frey, D. (2003). Präferenzkonsistenz und Geteiltheit von Informationen als Einflussfaktoren auf Informationsbewertung und intendiertes Diskussionsverhalten bei Gruppenentscheidungen [Preference-consistency and sharedness of information as predictors of information evaluation and intended behavior in group discussions]. *Zeitschrift für Sozialpsychologie, 34*, 9–23.

Grice, H. P. (1975). Logic and conversation. In P. Cole & J. L. Morgan (Eds.), *Syntax and semantics, 3: Speech acts* (pp. 41–58). New York: Academic Press.

Harvey, N., & Fischer, I. (1997). Taking advice: Accepting help, improving judgment and sharing responsibility. *Organizational Behavior and Human Decision Processes, 70*, 117–133.

Hollingshead, A. B. (1996). The rank-order effect in group decision making. *Organizational Behavior and Human Decision Processes, 68*, 181–193.

Jonas, E., Schulz-Hardt, S., Frey, D., & Thelen, N. (2001). Confirmation bias in sequential information search after preliminary decisions: An expansion of dissonance theoretical research on "selective exposure to information." *Journal of Personality and Social Psychology, 80*, 557–571.

Kelly, J. R., & Karau, S. J. (1999). Group decision making: The effects of initial preferences and time pressure. *Personality and Social Psychology Bulletin, 25*, 1342–1354.

Kerschreiter, R., Schulz-Hardt, S., Faulmüller, N., Mojzisch, A., & Frey, D. (2004). *Psychological explanations for the dominance of shared and preference-consistent information in group discussions: Mutual enhancement or rational decision-making?* Working paper, Ludwig-Maximilians-University Munich.

Lam, S. S. K., & Schaubroeck, J. (2000). Improving group decisions by better pooling information: A comparative advantage of group decision support systems. *Journal of Applied Psychology, 85*, 565–573.

Larson, J. R., Christensen, C., Abbott, A. S., & Franz, T. M. (1996). Diagnosing groups: Charting the flow of information in medical decision-making teams. *Journal of Personality and Social Psychology, 71*, 315–330.

Larson, J. R., Jr., Foster-Fishman, P. G., & Keys, C. B. (1994). Discussion of shared and unshared information in decision-making groups. *Journal of Personality and Social Psychology, 67*, 446–461.

Lavery, T. A., Franz, T. M., Winquist, J. R., & Larson, J. R., Jr. (1999). The role of information exchange in predicting group accuracy on a multiple judgment task. *Basic and Applied Social Psychology, 21*, 281–289.

Mojzisch, A., Schulz-Hardt, S., Kerschreiter, R., Brodbeck, F. C., & Frey, D. (2004). *Social validation as an explanation for the dominance of shared information in group decisions: A critical test and extension.* Working paper, Dresden University of Technology.

Parks, C. D., & Cowlin, R. A. (1996). Acceptance of uncommon information into group discussion when that information is or is not demonstrable. *Organizational Behavior and Human Decision Processes, 66*, 307–315.

Postmes, T., Spears, R., & Cihangir, S. (2001). Quality of decision making and group norms. *Journal of Personality and Social Psychology, 80*, 918–930.

Russo, J. E., Medvec, V. H., & Meloy, M. G. (1996). The distortion of information during decisions. *Organizational Behavior and Human Decision Processes, 66*, 102–110.

Schulz-Hardt, S., Brodbeck, F. C., Mojzisch, A., Kerschreiter, R., & Frey, D. (2004). *Group decision making in hidden profile situations: Dissent as a facilitator for decision quality*. Working paper, Dresden University of Technology.

Schulz-Hardt, S., Frey, D., Lüthgens, C., & Moscovici, S. (2000). Biased information search in group decision making. *Journal of Personality and Social Psychology, 78*, 655–669.

Schulz-Hardt, S., Jochims, M., & Frey, D. (2002). Productive conflict in group decision making: Genuine and contrived dissent as strategies to counteract biased information seeking. *Organizational Behavior and Human Decision Processes, 88*, 563–586.

Stasser, G. (1988). Computer simulation as a research tool: The DISCUSS model of group decision making. *Journal of Experimental Social Psychology, 24*, 393–422.

Stasser, G. (1992). Pooling of unshared information during group discussion. In S. Worchel, W. Wood, & A. Simpson (Eds.), *Group process and productivity* (pp. 48–67). Newbury Park, CA: Sage.

Stasser, G. (1999). The uncertain role of unshared information in collective choice. In L. L. Thompson, J. M. Levine, & D. M. Messick (Eds.), *Shared ocgnition in organizations: The management of knowledge* (pp. 49–69). Hillsdale, NJ: Lawrence Erlbaum.

Stasser, G., & Birchmeier, Z. (2003). Group creativity and collective choice. In P. B. Paulus & B. A. Nijstad (Eds.), *Group creativity* (pp. 85–109). New York: Oxford University Press.

Stasser, G., & Stewart, D. (1992). Discovery of hidden profiles by decision-making groups: Solving a problem versus making a judgment. *Journal of Personality and Social Psychology, 63*, 426–434.

Stasser, G., Stewart, D. D., & Wittenbaum, G. M. (1995). Expert roles and information exchange during discussion: The importance of knowing who knows what. *Journal of Experimental Social Psychology, 31*, 244–265.

Stasser, G., Taylor, L. A., & Hanna, C. (1989). Information sampling in structured and unstructured discussions of three- and six-person groups. *Journal of Personality and Social Psychology, 57*, 67–78.

Stasser, G., & Titus, W. (1985). Pooling of unshared information in group decision making: Biased information sampling during discussion. *Journal of Personality and Social Psychology, 48*, 1467–1478.

Stasser, G., & Titus, W. (1987). Effects of information load and percentage of common information on the dissemination of unique information during group decision making. *Journal of Personality and Social Psychology, 53*, 81–93.

Steiner, I. D. (1972). *Group process and productivity*. New York: Academic Press.

van Swol, L. M., Savadori, L., & Sniezek, J. A. (2003). Factors that may affect the difficulty of uncovering hidden profiles. *Group Processes & Intergroup Relations, 6*, 285–304.

Winquist, J. R., & Larson, J. R. (1998). Information pooling: When it impacts group decision making. *Journal of Personality and Social Psychology, 74*, 371–377.

Wittenbaum, G. M., Bowman, J. H., & Hollingshead, A. B. (2003). *Strategic informa-tion sharing in mixed-motive decision-making groups.* Paper presented to the Small Group Division of the National Communication Association, November, 2003. Miami, FL.

Wittenbaum, G. M., Hollingshead, A. B., & Botero, I. C. (in press). *From cooperative to motivated information sharing in groups: Moving beyond the hidden profile paradigm.* Communication Monographs.

Wittenbaum, G. M., Hubbell, A. P., & Zuckerman, C. (1999). Mutual enhancement: Toward an understanding of the collective preference for shared information. *Journal of Personality and Social Psychology, 77,* 967–978.

Yaniv, I. (2004). Receiving other people's advice: Influence and benefit. *Organiza-tional Behavior and Human Decision Processes, 93,* 1–13.

Yaniv, I., & Kleinberger, E. (2000). Advice taking in decision making: Egocentric dis-counting and reputation formation. *Organizational Behavior and Human Decision Processes, 83,* 260–281.

14

Confidence in Aggregation of Opinions from Multiple Sources

David V. Budescu

To the memory of my friend and colleague, Janet A. Sniezek (1951–2003)

This chapter summarizes research related to several interrelated questions regarding the process by which single decision makers (DMs) aggregate probabilistic information regarding a certain event from several, possibly asymmetric, advisors who rely on multiple and, possibly overlapping and correlated, sources of information. In particular I seek to understand and characterize (a) the nature of the aggregation rules used by DMs and (b) the factors that affect the DMs' confidence in the final aggregate.

The chapter starts with a short literature overview whose main goal is to set the stage by placing this work within the broad area of information aggregation. Next I present a descriptive model of confidence in information integration that is based on two principles: (a) People combine multiple sources of information by applying simple averaging rules; and (b) the DM's level of confidence in the aggregate is a monotonically decreasing function of its perceived variance. Some of the model's predictions regarding the structural and natural factors that affect the DM's confidence are discussed and tested with empirical data from four experiments (Budescu & Rantilla, 2000; Budescu et al., 2003). The results document the relation between the DM's confidence and the amount of information underlying the forecasts (number of advisors and cues), the advisors' accuracy, and the distribution of cues over judges with special attention to the level of interjudge overlap in information. The chapter concludes with a general discussion of the theoretical and practical implications of the model and a road map for future work.

BACKGROUND

Practically every important decision involves integration of information[1] from multiple sources. Thus, it is not surprising that there is a

---

[1] The terms integration, aggregation, and combination (of information) are used interchangeably in this chapter.

327

voluminous literature that addresses various facets of the information aggregation process (see Armstrong, 2001; Clemen, 1989; Ferrel, 1985; Flores & White, 1988; Hogarth, 1977; Rantilla & Budescu, 1999, for partial reviews of this body of work). Interest in the aggregation problem cuts across disciplinary and paradigmatic boundaries, although it takes different forms and focuses on distinct questions in the various cases (Budescu, 2001; Rantilla & Budescu, 1999).

The normative approach that guides much of the relevant work in decision theory, statistics, and operations research, is primarily concerned with the identification of *optimal aggregation rules* that can be implemented by formal models to predict single-valued outcomes (e.g., inflation rate in a given year, Dow Jones index on a given date, or amount of precipitation in given month at a particular location). The definition of optimality is derived from a set of compelling axioms (e.g., Bordley, 1982; Morris 1983), from Bayesian reasoning (e.g., Clemen & Winkler, 1993), or is grounded in traditional statistical approaches (e.g., Walz & Walz, 1989; Winkler & Makridakis, 1983). In social contexts the issue of opinion aggregation was approached from both normative and descriptive perspectives. From a normative perspective, social choice (e.g., Arrow, 1963) and voting research (e.g., Black, 1958) deal with the issue of *aggregation of individual preferences* to best represent the elusive collective preference. In the same spirit, social psychologists modeled and studied empirically the performance of *small interactive groups*, including juries and committees striving to reach jointly common decisions regarding intellective or judgmental problems (e.g., Davis, 1973, 1996; Hastie, 1993; Laughlin & Ellis, 1986). Finally, the cognitive and behavioral decision literature includes many studies of the processes that govern, and factors that affect, the way *single DMs combine information form multiple sources* in the service of estimating numerical quantities (e.g., probabilities) or making predictions and decisions. The work by Fiedler (1996), Fischer & Harvey (1999), Soll (1999), Yaniv & Kleinberger (2000), as well as research by Sniezek and her colleagues (Kuhn & Sniezek, 1996; Sniezek, 1992; Sniezek & Buckley, 1995; Trafimow & Sniezek, 1994) on the judge-advisor system (JAS), exemplifies various facets of this paradigm.

The reasons individuals, small-groups, societies, and models integrate multiple sources of information are quite similar: They seek to (a) maximize the amount of information available for the decision, estimation, or prediction task; (b) reduce the potential impact of extreme or aberrant sources that rely on faulty, unreliable, and inaccurate information; and (c) increase the credibility and validity of the aggregation process by making it more inclusive and ecologically representative. All these factors should also increase the confidence in the final aggregate. The present chapter focuses on the confidence of individual DMs in the aggregate of probability judgments of various advisors.

AGGREGATION BY INDIVIDUAL DMs AND CONFIDENCE IN THE AGGREGATE

We call on advisors[2] for advice in mundane problems (e.g., tuning in to weather forecasters to determine what to wear on a given day) as well as important decisions (e.g., asking several financial advisors how to invest one's savings). Decision makers are most likely to seek advice in the presence of uncertainty about the possible outcomes and/or their relative likelihoods, and when they believe that the advisors have access to relevant information and have the expertise necessary to interpret it properly. In most cases DMs solicit advice from multiple advisors. For example, when facing the possibility of a major medical procedure it is common practice, and common sense, to seek advice from several specialists at different medical centers. Similarly, when deciding whether to fund a certain research proposal, granting agencies collect opinions from several reviewers with different backgrounds and qualifications. This chapter is concerned with various aspects of this process: I discuss how people aggregate the opinions of multiple advisors and how confident they are in the process and in its final outcome.

A great deal of research has been devoted to the identification of the various aggregation rules (normative, statistical, heuristic, or intuitive) used by DMs and to the comparison of their relative merits. These comparisons have focused primarily on the accuracy of the rules. A general conclusion from this work is that some form of averaging describes very well the observed behavior in most cases (e.g., Anderson, 1981; Clemen, 1989; Fischer & Harvey, 1999; Rantilla & Budescu, 1999; Wallsten et al., 1997).[3] Related questions are whose, and how many, opinions to aggregate. The emerging consensus is that it is best to combine opinions of accurate and conditionally uncorrelated advisors. The accuracy of the aggregate increases with every advisor added to the pool, but at a diminishing rate that depends on the interjudge correlation (e.g., Ariely et al., 2000; Clemen & Winkler, 1985; Hogarth, 1978; Johnson, Budescu, & Wallsten, 2001; Wallsten & Diederich, 2001).

The primary focus of this chapter is the DMs' confidence, which is defined as the strength of belief in the goodness, accuracy, and appropriateness of the judgment or decision in question (Salvadori, van Swol, & Sniezek, 2001). Confidence is an important determinant and a useful predictor of postdecisional attitudes and actions such as willingness to endorse, support, and persevere in pursuing the chosen course of action (Budescu & Rantilla, 2000; Sniezek, 1986). Having the correct (or the most accurate) answer, or the best algorithm to generate that answer, is

---

[2] The terms advisors, forecasters, and judges are used interchangeably in this chapter.
[3] Under certain special circumstances averaging is also *normatively* optimal.

useless if the DM does not trust, and is not willing to rely on, this aggregate. A robust empirical finding is that the advisors' confidence in their
opinions, regardless of actual accuracy or ability at the task, is one of the
primary determinants of the DM's confidence and his or her likelihood
to take action consistent with the advice (e.g., Salvadori et al., 2001; van
Swol & Sniezek, 2002). In other words, judges weight heavily opinions of
confident advisors, regardless of how accurate those advisors are, and are
more confident when adopting them.

I am interested in the situation in which a single DM integrates probabilistic opinions (forecasts) from J distinct judges, regarding a unique target
event. Once their opinions are communicated to the DM, he or she has to
combine them to (a) generate the best estimate of the probability of the
target event and (b) express his or her confidence in this estimate. I assume
that all parties involved (the DM and the advisors) are properly motivated
and wish to optimize the quality of communication and of the eventual
decision. It is also assumed that the advisors operate separately and do not
communicate with each other.

The judges' forecasts are based on an information pool consisting of N
distinct "diagnostic cues." A cue is diagnostic if it is correlated with the target event, so one can use it to predict the occurrence of the event. In Bayesian
terms a cue is diagnostic if $Pr(Cue|Event) \neq Pr(Cue|No\ Event)$. Such
cues are "ecologically valid" in the Brunswikian sense (e.g., Gigerenzer,
Hoffrage, & Kleinbölting, 1991; Juslin, Olsson, & Bjorkman, 1997). Assume that the jth judge ($j = 1, \ldots, J$) has direct access to n_j of these cues
($1 \leq n_j \leq N$), and that each of the cues is accessed by *at least* (but, possibly, more than) one judge. Let $n_{jj'}$ be the number of common cues seen by
judges j and j' (where $j \neq 'j' = 1, \ldots, J$ and $0 \leq n_{jj'} \leq n_j, n_{j'}$).

The DM does not have direct access to the information pool and/or lacks
the expertise required to interpret the cues and form his or her personal independent estimate of the event's probability. However, the DM knows the
amount of information that was available to each judge, as well as the level
of overlap between judges. Thus, the DM's final estimate depends only on
the forecasts provided by the judges (see work by Harvey & Harries, 2003;
Soll & Larrick, 2000; Yaniv & Kleinberger, 2000; Yaniv, 2004, for the case
where DMs adjust their own opinions in light of advice from others).

To fix ideas, consider a medical diagnosis scenario: A patient (the DM)
is examined by J physicians (the advisors) who provide separate diagnoses
of the type "*the probability that the patient has Disease A is X_j.*" Each advisor
derives this probability from a set of n_j diagnostic tests, and results of
some of the tests may be viewed by more than one physician. The patient
knows (a) how many physicians have examined him or her, (b) how many
tests were available to each of them, (c) the level of overlap between the
physicians (in terms of the number of tests they saw), and (d) the distinct
forecasts provided by the J advisors. The DM may also have information

about (i) the differential reliability and diagnostic value of the various tests and (ii) the quality (accuracy, reputation, etc.) of the various physicians. It is up to the patient to determine the probability that he or she has Disease A. Subsequently, depending on the patient's confidence in this estimate, the patient will have to decide on some course of action: look for additional information and advice, follow the treatment prescribed, reject it, etc.

This situation involves a two-level hierarchical aggregation process: First, each of the advisors has to combine the information from the various cues to form an estimate. Then the DM must combine these forecasts. There is a close resemblance between this problem and the structure underlying the BIAS model (Fiedler, 1996). In the next section I describe a simple model of this two-stage process, from the DM's perspective. The model reflects two basic assumptions:

(a) People (forecasters and DMs) combine information by applying simple averaging rules.

(b) The DM's level of confidence in the aggregate is a monotonically decreasing function of its perceived variance.

The first assumption is intuitively compelling and has considerable empirical support (e.g., Clemen, 1989; Fischer & Harvey, 1999; Wallsten et al. 1997), although it is not necessarily clear that averaging is a universally accepted principle (see Larrick & Soll, 2003). The second assumption is at the core of the present work. In a nutshell, this assumption implies that the DMs are sensitive to the factors that drive the variance of the aggregate. Of course, I do not mean to imply that DMs actually calculate this variance, or even that they recognize the validity of this statistical metric. However, I assume that those factors that increase (decrease) the variance of the final estimate (weighted average of the forecasts) are also those that decrease (increase) the DM's intuitive confidence. These factors are consistent with three general psychological principles that drive the DM's confidence, namely, sensitivity to amount of information, consensus among sources, and distinctiveness of information. More specifically, these principles predict that (1) confidence increases monotonically as a function of the total amount of information available to the DM (e.g., Einhorn & Hogarth, 1978) and (2) confidence increases monotonically as a function of the perceived agreement and consensus among sources (e.g., Kahneman & Tversky, 1973) and it peaks in the case of unanimity among sources. The third principle is less obvious and lacks solid empirical support at this time, in part because it has not been studied systematically and carefully, so I state it in the form of a conjecture. I believe that the impact of any piece of information of the DM's confidence is moderated by its perceived distinctiveness, so the impact of information on one's confidence increases to the degree to which it is perceived by the DM to be distinct, and not merely duplicating other sources. Naturally, this can attenuate the effects of the first two factors (total information and consensus). Note that the variance is sensitive to these factors in similar fashion, so this model provides a general framework for formulating hypotheses and evaluating the results of experiments in which one elicits direct confidence judgments. Some of these experiments are summarized and discussed in a later section of the chapter.

A MODEL OF THE AGGREGATION PROCESS AND CONFIDENCE

The first part of the model describes the way the DM perceived the judges and their estimates. Let π_i be the validity of the ith cue: the probability that the target event would materialize, conditional on the cue's presence. The DM assumes that when the jth judge sees the ith cue, he or she forms an overall (posterior) probabilistic opinion regarding the occurrence of the target event, X_{ij}, which can be expressed as $X_{ij} = \pi_i + e_{ij}$, where e_{ij} is a random variable with a zero mean and a finite standard deviation, σ_{ij}. This is a special case of the more general judgment model proposed by Erev, Wallsten, & Budescu (1994) and Budescu, Erev, & Wallsten (1997).[4] There are many possible sources of the variability of X_{ij}, including (but not restricted to) the perceived unreliability and imprecision of the cue and the imperfections in the advisor's processing of the cue owing to lack of knowledge or momentary distractions. It is impossible to distinguish between the various sources and their effects experimentally, but this is of no consequence for the present analysis. Assume that the e_{ij}s are uncorrelated (a) across various cues for each judge, (b) across distinct judges, and (c) with the actual validity, π_i.[5] In the interest of simplicity, also assume that (a) the magnitude of the variance is fixed for each judge across all the cues: all $\sigma_{ij} = \sigma_j$; (b) all cues are equally valid: all $\pi_i = \pi$; and (c) all cues are equicorrelated: all $\rho_{\pi_i \pi_{i'}} = \rho$. We propose that the DM considers the judge's final forecast, X_j, as communicated to him or her to be, simply, the average of all the cue-specific judgments presented to the judge. Formally,

$$X_j = \sum_{i=1}^{n_j} X_{ij}/n_j = \sum_{i=1}^{n_j} (\pi_i + e_{ij})/n_j. \tag{14.1}$$

The DM aggregates the J forecasts by taking an average weighted by the number of cues underlying each forecast:

$$\overline{P} = \sum_{j=1}^{J} n_j X_j \bigg/ \sum_{j=1}^{J} n_j. \tag{14.2}$$

---

[4] The model states that $X_{ij} = f(\pi_i, e_{ij})$, without assuming additivity of the two terms.
[5] A normal distribution of e_{ij} is not assumed. In fact, the requirement that $0 \le X_{ij} \le 1$ may impose constraints on the shape of this distribution.

The most general form of the variance of this aggregate can be obtained from the variances of, and covariances between, the judges' forecasts.[6] Specifically,

$$
\sigma_{\bar{P}}^2 = \left\{ \sum_{j=1}^{J} n_j \sigma_j^2 + \sigma_\pi^2 \left[\sum_{j=1}^{J} n_j (1 + (n_j - 1)\rho) \right. \right.
$$

$$
\left. \left. + \sum_{j=1}^{J} \sum_{j' \neq j=1}^{J} (n_{jj'} + (n_j n_{j'} - n_{jj'})\rho) \right] \right\} \bigg/ \sum_{j=1}^{J} n_j. \tag{14.3}
$$

In this formula the variance of the aggregate is expressed as a function of the (1) inaccuracy of the typical judge, σ^2, (2) the unpredictability of the target event from the typical cue, $\sigma^2 \pi$, (3) the average intercue correlation, ρ, (4) the number of cues seen by each judge, n_j, (5) the number of common cues seen by any pair of judges, $n_{jj'}$, and (6) the total number of judges, J.

At this point, it is convenient to distinguish between strictly symmetric and asymmetric advisors. Judges are considered to be symmetric when they are undifferentiated in terms of expertise, amount and quality of evidence they can access, or any other relevant distinguishing feature. In other words, any sensible DM would be indifferent if asked to choose one (or more) judge(s) from such a group. In contrast, judges are considered asymmetric if they vary along some of these dimensions to the degree that a reasonable DM would have good reason to prefer some of them to the others.

Budescu and Rantilla (2000) focused on the symmetric case where (a) all advisors see the same number of cues, $n_j = n = gN$, (where $0 < g \leq 1$) (b) the level of pairwise overlap for all advisor pairs is identical, $n_{jj'} = fn = gfN$ (where $0 \leq f \leq 1$), and (c) the DM considers the J advisors to be equally accurate ($\sigma_j = \sigma$). Under these simplifying assumptions, the variance is reduced to

$$
\sigma_{\bar{P}}^2 = \left\{ \sigma^2 + \sigma_\pi^2 [(1 + (J - 1)f) + (J Ng - 1)\rho - (J - 1)f\rho] \right\} / J Ng. \tag{14.4}
$$

For the case of asymmetric judges we relax the assumptions that $n_j = n$, and abandon the assumption of uniform pairwise overlap among advisors, but maintain the assumption that all J advisors are equally accurate ($\sigma_j = \sigma$). Thus, the flow of information from the advisors to the DM is much less constrained, and the same amount of information can be transferred to the DM in a variety of ways. Let $T = \sum n_j$ be the total number of "information pathways" between the N cues and the DM, mediated by the J advisors. For example, if there are $N = 4$ cues and $J = 3$ judges, T can vary between 4, when each cue is seen by only one advisor, to 12, where all advisors see all the cues. Note, however, that there are multiple patterns that have the

---

[6] A complete derivation of the model is provided in the appendix of Budescu et al. (2003).

same number of pathways. For example, if $T = 6$, the information can be distributed symmetrically between the judges (two cues presented to each), or in two asymmetric ways ($n_1 = 3, n_2 = 2, n_3 = 1$, or $n_1 = 4, n_2 = 1, n_3 = 1$) and each of these distributions can involve various levels of interjudge overlap. If all possible patterns of overlap are equally likely, the expected variance of the aggregate is given by (Budescu et al., 2003)

$$E\left\{\sigma_{\bar{P}}^2\right\} = \frac{\sigma^2}{T} + \frac{\sigma_\pi^2}{NT}\{N(1 - \rho) + T[1 + [N - 1]\rho\} - \frac{\sigma_\pi^2(1 - \rho)}{NT^2}\sum_{j=1}^{J} n_j^2.$$

(14.5)

The factors in the model can be classified into *natural and structural*. The former class includes all the factors that describe naturally occurring relationships between the cues and the target event. In this group I count the validity of each cue, the uncertainty associated with each cue, the correlation between the cues, and the accuracy of the judges. These factors are related to the consensus and distinctiveness principles discussed earlier. These values may vary across contexts and from one problem to another. I believe that in most cases these values are estimated or inferred *subjectively*. For example, the level of error is inferred from the variance in the forecasts of advisors who see the same evidence, and the level of intercue correlation is inferred from the agreement among judges who have access to distinct cues. It is difficult to communicate the levels of these factors to participants in experiments. This information can be easily misunderstood, distorted, or disregarded. For example, DMs may not believe that financial or medical advisors can, or fail to, achieve certain levels of accuracy. The most difficult factor to convey to DMs is the correlation between cues. Simply put, even sophisticated DMs do not know necessarily know how to deal with this information (e.g., Maines, 1996; Soll, 1999).

The second category consists of all those variables that define the structure and magnitude of the decision problem but are independent of its content. They are the total number of cues, the number of judges, the fraction of cues seen by each judge, and the fraction of pairwise overlap in the cues presented to the judges. These factors define the process by which information is sampled from the environment and communicated to the DM, via the J advisors, and they are related to the total information principle mentioned earlier. In experimental studies of aggregation, these factors are under the investigator's complete control and are both easy to manipulate and communicate precisely to the DMs.

Table 14.1 summarizes some of the model's predictions regarding the DM's confidence. As stated earlier, I assume that the factors that increase (decrease) the variance are those that decrease (increase) the DM's confidence. Because confidence is measured by subjective reports, and the

TABLE 14.1. *Predicted Changes in the DM's Confidence as a Function of Various Factors in the Model*

As Factor Increases	Confidence in the Aggregate
Number of Cues in the Pool ($N > 1$)	Increases
Number of Judges ($J > 1$)	Increases
Total Number of Information Pathways ($T \geq J$)	Increases
Asymmetry in Information among Judges	Increases
Inaccuracy of Experts (σ^2)	Decreases
Unpredictability of the Event $\{\sigma_\pi^2\}$	Decreases
Intercue Correlation ($\rho > 0$)	Decreases
Overlap in Cues Presented to Judges	Decreases
Structural Overlap	Increases

relationship between confidence and the variance of the aggregate is only ordinal, only predictions about the marginal effects of the individual factors are listed. These qualitative predictions are the most general one can derive without making stronger (and hard to justify) assumptions about the functional relation between confidence and variance. Thus, it is impossible to predict the presence of specific interactions from this form of the model. However, I expect that if such interactions are detected they would be internally consistent. Thus, interactions between factors with congruent marginal effects (for which all factors involved increase, or decrease, the confidence) will be synergistic – they will accentuate the common direction of the effects. Conversely, most interactions between factors with incongruent marginal effects (for which some are increasing and some decreasing) will be antagonistic – they will reduce the relevant effects.

Note that the marginal effects listed in the table are in line with the three psychological principles listed earlier: The predicted confidence boost associated with more cues, more judges, and a higher number of pathways is consistent with the total information effect. The reduced confidence predicted in the presence of low diagnostic cues and inaccurate judges is associated with the consensus effect. Finally, the predicted drop in confidence associated with highly correlated cues and high overlap in information captures the moderating effect of lack of distinctiveness.

EMPIRICAL TESTS OF THE MODEL

We have performed four experiments designed to validate the models' predictions. Complete descriptions of the studies can be found in two previously published empirical papers. Budescu and Rantilla (2000) describe two studies involving strictly symmetric advisors (which will be labeled

S1 and S2), and Budescu et al. (2003) present two studies involving asymmetric advisors (which will be labeled AS1 and AS2).

Although the experiments varied in many specific details, their basic methodology was sufficiently similar to allow me to summarize their results jointly. In all cases participants were presented with multiple (from 36 in S1 to 104 in AS1) scenarios that listed the opinions of J (between 3 to 8 in the various studies) hypothetical experts in specific domains (business, medicine, weather forecasting, etc.) with regard to well-defined unique events. All opinions were presented as probability judgments on a 0–100% scale (e.g., *Physician A estimates the probability that the patient has condition X is 73%*). Any given numerical forecast was associated with only one judge, eliminating the possibility that the DM could choose the modal forecast. We varied the distribution of the judges' opinions by controlling its mean and range.

In each scenario the DMs knew both (a) the total number of cues, N, and (b) the number of cues underlying the various advisors' opinions, n_j, so they could infer the level of overlap (number of common cues) among the judges. For example, when DMs were informed that there were a total of $N = 12$ cues, and each of the $J = 4$ judges saw $n = 3$ cues, they could infer that the there was *no overlap between these symmetric judges*. However, if told that that there were a total of $N = 8$ cues, that the first judge saw $n_1 = 6$ cues, and that the second saw $n_2 = 4$, they could infer that the there were $n_{12} = 2$ *common cues presented to these asymmetric judges*.

The DMs were instructed to (a) estimate the probability of the target event on the same 0–100% scale, and (b) report how confident they were in their estimate. The confidence ratings were performed on seven-point scales but, prior to analyses, they were standardized to have a zero mean and unit variance within subject. This transformation eliminates individual differences and minimizes the effects of different response styles from the final analysis.

Budescu and Rantilla (2000) examined behavior in the case where the DMs were told that probabilistic information was supplied by *equally qualified* advisors who based their forecasts on *equal amounts of equally diagnostic information*. The scenarios varied in terms of (a) the number of advisors, (b) the total number of relevant cues, (c) the amount of overlap in the cues they saw, (d) the mean forecast, (e) the level of disagreement between the advisor opinions, and (f) the decision's context.

Two major findings stand out. First, the DMs averaged the J forecasts. Since the model uses the variance *of the mean* as a proxy for confidence in the aggregate, it is reassuring to confirm that DMs do, indeed, average. More interestingly, the DMs' confidence was sensitive to the structural factors manipulated. Consistent with the predictions of the model, DMs were more confident as (i) the degree of structural overlap increased, (ii) the number of cues in the database increased, (iii) the range of the opinions expressed by the judges (a proxy for σ^2) decreased, and (iv) the number of judges and/or cues available increased. Finally, (v) the results were invariant across the various contexts examined.

The most interesting interactions involved structural overlap and the range of opinions. Decision makers expect these two variables to operate in concert with high levels of overlap among the cues seen by judges expected to induce high levels of agreement between them. When this expectation is fulfilled, DM confidence peaks, but when this pattern is violated, that is, when judges who have seen all the information disagree, confidence drops dramatically. The anticipated correlation between the two factors is further reflected in their separate, but almost identical in form, interactions with the number of judges and of cues. These results lend further support to the notion that confidence is a direct function of the consensus among sources.

If a DM wishes to differentiate among various asymmetric advisors he or she needs to weigh their advice according to the relevant dimension(s) (e.g., Fischer & Harvey, 1999). The most natural way to introduce asymmetry and differentiate among the judges is to vary the amount of information that each advisor can access. The intuitive prediction is that confidence should peak when information is distributed evenly across advisors, and no differential weighting is required. However, the model (Eq. 14.5) predicts that the DM's confidence should be the highest when the information is distributed most unequally! Budescu et al. (2003) tested this prediction in AS1. Decision makers were told that *J equally qualified advisors* who based their forecasts on *unequal amounts of equally diagnostic information* supplied the forecasts. We compared the DM's responses to predictions derived from four plausible aggregation models:

(a) MC – the opinion of the single advisor who had access to the *largest number of cues*, disregarding all others;
(b) Mean – the *unweighted mean* of the three forecasts;
(c) Median – the *median* of the three forecasts; and
(d) WCMean – the *weighted mean* of the forecasts, with weights proportional to the number of cues.

For each participant and model we computed the mean absolute deviation (MAD) between the response and the model's prediction across all questions[7] and rank-ordered the four models, in terms of their fit. Figure 14.1 presents the distribution of ranks of each model across all participants. The weighted mean model best fits the results of 47 (of 81) participants and it was ranked either first or second in 78 out of 81 cases. As assumed, DMs were sensitive to the differentiating information available to the expert opinions, and they weighted their opinions accordingly. This

---

[7] In some cases two or more of the models' predictions coincide.

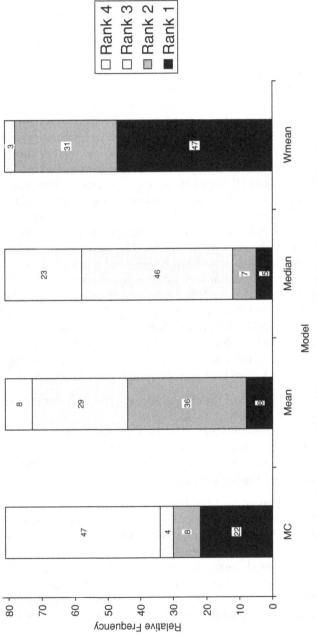

FIGURE 14.1. Distribution of closeness of fit of four aggregation models to the observed data (study AS1).

result also holds at the group level. The global MAD of the weighted mean across all questions and participants was considerably and significantly lower than the values for the other three models.

In the second study (AS2) we induced a more complex form of asymmetry. Decision makers were told that the forecasts were supplied by advisors with *different qualifications* (with DMs informed about each advisor's prior accuracy rates) who based their forecasts on *unequal amounts of equally diagnostic information*. We compared the DM's responses to predictions derived from a set of plausible aggregation models. In addition to MC, Mean, Median, and WCMean models, we also tested the following:

(e) MA – the opinion of the single *most accurate* advisor, disregarding all others;
(f) WAMean – the *weighted mean* of the forecasts, with weights proportional to the *judges' level of accuracy*; and
(g) WACMean – the *weighted mean* of the forecasts, with weights proportional to the product of the *number of cues and the judges' level of accuracy*.

We rank-ordered the seven models, from the best to the worst in terms of their MAD from the DM's responses across all items, for each subject separately. The distribution of the ranks of the seven models is displayed in Figure 14.2. The model that combines information about the number of cues and accuracy rate (WACMean) is the clear winner: It fits best 37 (of the 74) participants and it is in the top three for 69 respondents. The superiority of this model can be also observed at the group level where its MAD was significantly lower than any of its competitors. Thus, as expected, the DMs aggregation process was highly sensitive to the asymmetry factors among advisors: number of cues available and their level of accuracy.

Consistent with the model's predictions, confidence in these aggregates was higher (i) when multiple judges had access to a higher fraction of common cues, (ii) when there was higher interjudge agreement, (iii) when the judges were more accurate. The two asymmetric studies also confirmed that DMs were more confident (iv) when the cues were distributed unevenly across advisors. Moreover, the asymmetry of distribution of cues interacted synergistically with most other factors manipulated in the design. This pattern is illustrated in the three-way interaction of structural overlap, level of asymmetry, and the judges' accuracy, which is displayed in Figure 14.3. The plots trace the mean standardized confidence of DMs who receive $J = 3$ forecasts. On the abscissa we distinguish among the accuracy of the advisors (from three advisors with low accuracy, LLL, at one extreme, to three advisors with high accuracy, HHH, at the other end). The four curves represent groups where the advisors had access to equal, or unequal, number of cues with, or without, common cues. There is an obvious monotonic relation between the DM's confidence and the judges' accuracy,

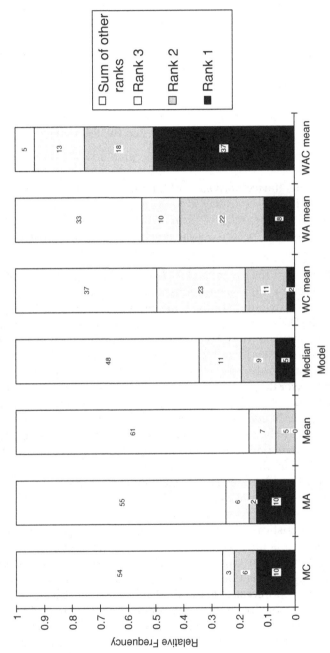

FIGURE 14.2. Distribution of closeness of fit of seven aggregation models to the observed data (study AS2).

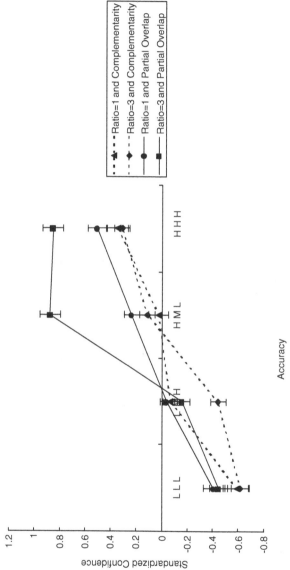

FIGURE 14.3. Interactive effect of structural overlap, information ratio, and accuracy on DMs' standardized confidence (study AS2).

TABLE 14.2. *Summary of the Model's Predictions Tested in Our Studies and of Their Principal Results*

Factor Increased	Prediction about Confidence	Tested in Experiments	Confirmed?
(Symmetric and Asymmetric)			
No. of Cues ($N > 1$)	Increases	S1, S2	Yes
No. of Judges ($J > 1$)	Increases	S1, S2	Yes
Inaccuracy of Judges (σ^2)	Decreases	S2	Yes
Structural Overlap	Increases	S1, S2, AS1, AS2	Yes
Context	No Change	S1, AS1, AS2	Yes
Mean Forecast	No Change	S1, S2, AS1	No
(Asymmetric Only)			
Asymmetry Among Judges	Increases	AS1, AS2	Yes
Accuracy of Judges	Increases	AS2	Yes

but the rate of increase is not uniform. The most salient feature of this interaction is the fact that the increase in confidence is most pronounced when the cues overlap (partially) and are divided unequally among the advisors.

To summarize, in a series of experiments involving over 300 participants who judged a large number of scenarios under a variety of contexts we found the following: (a) There is very strong evidence that DMs combine probabilistic opinions from multiple sources by averaging them, (b) in cases where the judges are asymmetric the DM's averaging process is highly sensitive to the sources of asymmetry, and (c) the DM's confidence in the aggregate is affected by the structural and natural factors, as predicted by our model. Table 14.2 summarizes the predictions tested in the various studies and the empirical results. The table indicates that, for the most part, the reported patterns were robust, consistent, and invariant across various contexts.

Most empirical results are (ordinally) consistent with the model's prediction, but a single remarkable violation of the model's prediction stands out. Normatively, DMs should be equally confident in any pair of complementary estimates (e.g., 30% and 70%) that are equally close to the point of equiprobability (50%) and the corresponding certainty points (0% and 100%, respectively). Contrary to this normative expectation we found that confidence increases as a function of mean probability: On average, DMs were more confident in higher estimates (with the median within-subject correlation between the aggregate and its confidence being 0.17 and being significant for about 70% of the DMs.). It is possible that the psychological concept of confidence is more complex than the model that we have proposed and, in particular, that it is sensitive to factors that do not affect the variance of the aggregate. It is also conceivable that the DM's judgments reflect some form of illusory correlation (Chapman & Chapman, 1967, 1969) between probabilities and confidence, possibly reflecting the "confidence heuristic" studied by Price and Stone (2004).

DISCUSSION OF THE MODEL THEORETICAL AND PRACTICAL IMPLICATIONS

The relevant information for the DM can be represented in a rectangular array that includes a row for each of the J advisors and a column for each of the N cues.[8] This array constitutes the sampling frame, and every decision problem can be described as a sample of entries from the frame. Every sample can be summarized by its unique pattern of binary entries: A value of 1 indicates that the advisor in that row has seen the cue in that particular column, and a value of 0 indicates that he or she has not seen it. The only constraints that apply to these samples are that they include at least one nonzero entry in each row and each column. This guarantees that each advisor has access to, at least, one cue and that each cue is presented to, at least, one advisor. In line with the book's focus on information sampling, I turn now to a discussion of the effects of amount and distribution of information on the DM's confidence.

As indicated in the exposition of the model, the process by which information is sampled from the environment and communicated to the DM through the judgments of the advisors is represented by the structural factors in the model. Evidently, increasing the sampling frame by simply adding more cues and/or more advisors increases the amount of information available to the DM and reduces the variance of the aggregate. The size of the frame has implications for the effective sample size, T, which is bounded from below by $Max(N, J)$, and from above by NJ. The model predicts, and the empirical results confirm, that increasing T also increases the DM's confidence. There are two interesting consequences of increasing the sample size: (1) It increases of the level of structural overlap between the judges, and (2) it increase the likelihood of asymmetric distributions across cues and/or judges. In the present studies all the cues were equally valid, and the DMs were only told how many cues were presented to each advisor without any specific cues being identified. Thus, asymmetry among cues was not studied. Next, I focus on the effects of asymmetry among judges and the structural overlap among advisors on the DMs' confidence.

The positive effects of asymmetry among judges on the DM's confidence seems to contradict the common intuition that symmetry is a natural state of affairs that is easier to process. The key psychological insight that accounts

---

[8] Typically, N, the number of cues, is larger than J, the number of advisors.

for this result is the realization that symmetry is associated with the highest level of fragmentation of information, so the integration of the forecasters' opinions requires the highest level of effort on the part of the DM, who needs to pay equal attention to all the forecasts and reconcile possible inconsistencies and disagreements among the judges. However, the more asymmetric the distribution of information, the easier it is to anchor one's judgment on a smaller subset of judges, and the level of effort involved can be reduced without sacrificing much in terms of accuracy (Payne, Bettman, & Johnson, 1993). In fact, in the extreme case where one judge has access to all N cues and the others ($J - 1$) have little information, the DM could simply adopt the single forecast, in a variation of the "Take the Best" heuristic (Gigerenzer & Goldstein, 1999). For example, in studies AS1 and AS2 in the symmetric case each of the three judges saw six cues (one-third of the information) and in the asymmetric case the three judges saw nine, six, and three cues (one-half, one-third, and one-sixth), respectively. The DM could rely on only two judges with relatively little loss of information, but with a considerable reduction in the amount of effort required. Another possibility is that the DMs may choose to ignore the opinions of the advisors with the least amount of information, especially when such opinions are unusual and/or extreme (see Yaniv, 1997). This may reduce the level of perceived disagreement among the advisors and, naturally, increase the DM's confidence.

The single factor studied in all four experiments (see Table 14.2) is the level of structural overlap among the J judges. The level of structural overlap is bounded from below and above by intuitive and commonly encountered cases: If every cue is seen by one, and only one, judge (thus $T = N$), there is no structural overlap. I refer to this case as *complementary*, to highlight the fact that the judges complement each other. At the other extreme, if all the judges see all the cues (so $T = J N$), the level of overlap is maximal. I label this case *redundant*, to highlight the fact that ($J - 1$) judges are redundant. In between these two qualitatively distinct cases there are

TABLE 14.3. *Mean Standardized Confidence (and Total Number of Information Pathways) as a Function of Number of Cues N, Number of Judges J, and Structural Overlap (Study S2)*

J	N	Structural Overlap			Mean
		Complementarity	Partial Overlap	Redundancy	
2	4	−.423 (4)	−.212 (6)	.445 (8)	−.063 (6)
2	8	−.161 (8)	−.229 (12)	.324 (16)	−.022 (12)
4	4	−.266 (4)	−.228 (12)	.511 (16)	.006 (10.67)
4	8	−.230 (8)	−.155 (24)	.625 (16)	.080 (21.33)
Mean		−.270 (6)	−.206 (13.5)	.476 (18)	.000 (12.5)

many cases of *partial overlap* where at least two judges see at least one common cue (and $N < T < J\,N$).[9] Budescu et al. (2003) showed that the model predicts that the variance of the aggregate should always peak in the complementary case and be smallest in cases of redundancy. Although no general proof is available, my conjecture is that the partial redundancy case would, typically, lead to intermediate levels. This prediction was, indeed, confirmed, with the results of all our experiments being consistent with the predicted ranking:

Confidence (Redundancy) > Confidence (Partial Overlap)
> Confidence (Complementarity).

The increase in confidence described by this inequality is consistent with the general principle that relates confidence to the total amount of information available to the DM. Indeed, for any fixed number of judges J and cues N, the effect of structural overlap coincides with this principle. However, if one compares cases with identical number of pathways T, obtained from various combinations of judges and cues, a more complicated picture emerges. Consider Table 14.3, which summarizes some results from study S2 (Budescu & Rantilla, 2000). For each of the twelve distinct cases defined by factorially crossing the number of judges J, the number of cues N, and the level of structural overlap, the table presents the mean standardized confidence, as well as the total number of pathways T. The table illustrates the increase in confidence as a function of each of the three factors: Confidence peaks in the case of maximal structural overlap (redundancy); it is minimal in the case of fragmented information (complementarity); and confidence grows as the number of judges and cues increases. However, the rank-order correlation between the mean confidence and the numbers of pathways, although positive and significant, is far from perfect (Kendall $\tau_b = 0.60$), indicating that the effect of structural overlap is not simply one of quantity (see Fiedler's, 1996, insightful discussion on the concept *information quantity*). The interactions among these factors indicate that the rate of change in confidence as a function of structural overlap is sharpest in the presence of more judges ($J = 4$), especially for the case of maximal information ($J = 4$, $N = 8$). In other words, higher levels of overlap boost the DM's confidence more substantially in the presence of larger amounts of information.

The previous interaction involves only structural factors that describe the nature and size of the aggregation problem. From a theoretical

---

[9] Consider, for example, $J = 2$ judges in an environment with $N = 6$ cues. If the first judge has access to the first two cues and the second judge sees the other four, the judges are *complementary*. If both judges see all six cues, they are *redundant*. In all cases where they see at least one common cue (e.g., the first judge sees the first four cues and the second judge sees the last four), we have *partial overlap*.

TABLE 14.4. *Mean Standardized Confidence as a Function of Structural Overlap and Range of Opinions (Study S2)*

Range of Opinions	Structural Overlap			
	Complementarity	Partial Overlap	Redundancy	Mean
High (.40)	−.545	−.388	−.004	−.312
Medium (.20)	−.240	−.264	.585	.027
Low (.10)	−.024	.033	.847	.285
Mean	−.270	−.206	.476	.000

perspective it is more intriguing to study the ways in which structural overlap interacts with natural factors that depend on the behavior of the people involved in the aggregation process. Table 14.4 summarizes data from the S2 study (Budescu & Rantilla, 2000). It shows the mean standardized confidence for all combinations of the three levels of structural overlap and the range of opinions of the *J* judges. On average, participants were more confident when all the experts saw all the cues and when the experts' estimates were in agreement (low range of opinions). The highest confidence was observed when all advisors saw all the information and agreed in their judgments, whereas the lowest confidence was found when the information was fragmented over judges and they disagreed in their forecasts. Confidence was reduced dramatically when participants could not reconcile the high structural overlap (redundancy) and the large differences in the judges' estimates.

In the forecasting literature it is recognized that correlations among experts in well-defined domains are quite high.[10] Morris (1986) explains the high interjudge correlations by noting, "*most experts have access to the same basic information and are basing their opinions on roughly the same body of data. Overlapping methodology may exist if experts in the field have similar academic and professional training*" (p. 143). The interaction between structural overlap and level of agreement uncovered in our studies suggests that our participants have similar mental models of experts, the source of their expertise, and the expected level of interexpert agreement.

What are some of the practical implications of these results? Sniezek (1986) argued that confidence is a critical postdecisional variable that determines the DM's strength of support and commitment to his or her decision, and Price and Stone (2004) confirmed the DMs' preference for confident judges. The model predicts, and our experiments confirmed, that increasing the number of judges, the number of cues, the number of pathways, the level of structural overlap, and the interjudge asymmetry could boost

---

[10] For example, consider the high correlations among the forecasts of professional sports writers (Winkler, 1971), executives at *Time* magazine (Ashton, 1986), emergency room physicians (Winkler & Poses, 1993), and econometric models (Clemen & Winkler, 1986).

the DMs' confidence. We have also demonstrated that confidence increases when judges are more accurate and tend to agree with each other. These findings provide additional support for the principles of total information and consensus, formulated in the introduction.

A clear implication of these results is that, everything else being equal, it is possible to induce higher (or lower) levels of confidence by structuring the situation and/or choosing judges such that the variance of the (weighted) average opinion is reduced (or increased). A DM who seeks advice from *many accurate* advisors who *share* unlimited access to *all (or most of) the cues* is likely to have the highest confidence in the aggregate. We have also shown that DMs are more confident when working with accurate advisors and that accuracy interacts synergistically with the amount of information. Thus, when more information is available to the most (least) accurate judges, the DM's confidence is likely to increase (decrease). We have also identified an interactive effect of the level of shared information, asymmetry of information, and the judges' accuracy. This implies that confidence should increase (decrease) when the more accurate judges are given access not only to more (fewer) cues, but also to more (less) common information. This list of effects and implications should not be interpreted as a practical guide to manipulating the DMs' confidence in their aggregates (though, clearly, in some situations one could attempt this), but rather it should be used to explain why the same information may induce different levels of confidence, in different settings.

I conclude this discussion by highlighting an intriguing similarity between the individual processes discussed in this chapter and certain robust results in the group decision-making literature (see the excellent review in the chapter by Mojzisch and Schulz-Hardt, 2004, in this volume). The so-called shared information effect refers to the finding that members of interacting groups tend to mention more often, discuss more, and overweight information that is shared by all (or most) members. In some cases, such as the "hidden profile" paradigm, these tendencies have detrimental effects (see Mojzisch & Schulz-Hardt, 2004; Wittenbaum & Stasser, 1996). Budescu and Rantilla (2000) speculated that one possible explanation for this phenomenon may be the confidence-boosting effect associated with higher levels of information overlap and redundancy. The interface between the group-level biases (Mojzisch & Schulz-Hardt, 2004) and the individual-level phenomena documented in this chapter should be studied carefully and systematically.

FUTURE DIRECTIONS

In all our studies we provided our participants with advice from hypothetical expert advisors in the context of scenarios that were embedded in the advisors' alleged domain of expertise (medicine, business, etc.). This approach is particularly useful to study, as we did, the effects of external *structural factors*. It is also convenient because it allows considerable control over the advice provided to the DMs. Thus, we tailored the forecasts by controlling key features of their distribution such as its mean and its range. However, to achieve this control we sacrificed a considerable degree of realism. The DMs may find the scenarios unnatural, some of the events may appear to be too abstract and/or vague, and DMs may doubt the veracity and credibility of the forecasts. It is also quite difficult to study the effects of the natural factors in the model (such as event predictability or interjudge correlation) under this paradigm, because this information cannot be communicated in a simple and direct fashion.

To address these issues, in future work I plan to use a different paradigm that is better suited for the study of the *natural factors* in models. In this paradigm multiple participants will run simultaneously. First they will all be trained and become familiar with the decision environment and the cues. Then, some of them will be designated to be advisors and others will serve as DMs. The advisors will have direct access to a subset of the cues for any given problem, whereas the DMs will only get the advisors' opinions and, as in the previous studies, will have to aggregate them and express their confidence in the final aggregate. The experimenter will maintain complete control over some structural factors (the number of judges, cues, etc.) but, evidently, will no longer control the distribution of forecasts, as this will depend on the advisors. However, the experimenter will control the validity of the cues, and their intercorrelations, and all participants will have an opportunity to experience and learn them, so we will not have to rely on (occasionally, difficult to understand) statistical descriptions of the diagnosticity of the cues and their correlations. This paradigm would allow rigorous tests of the model's predictions regarding the effects of the natural factors on confidence. An additional advantage is that it will reduce to a large degree the reliance on verbal descriptions of the situation (cues, judges, validity of cues, intercorrelation among cues, etc.), since the DMs will have direct first-hand experience with these factors.

Finally, this approach will provide a convenient vehicle to study a new facet of the process, namely, the selection of advisors. Unlike laboratory experiments, in real life people have the opportunity to pick and choose their sources of information. This choice is driven by external constraints (accessibility, budget, etc.) and intrinsic considerations of the *perceived quality* of the advisors. I hypothesize that there is a direct monotonic relation between perceived quality of an advisor and the DM's confidence in his or her opinion. Therefore, everything else being equal, the model predicts that DMs would select those advisors who inspire the highest confidence. In other words, the model suggests that the DM would pick advisors who (a) have access to many cues, (b) have access to the more discriminating (diagnostic) cues, and (c) are most accurate. Moreover, in those cases in which multiple advisors are selected, the model predicts that DM's would choose advisors (d) with access to overlapping information. These future studies in conjunction with the results summarized in this chapter would provide (at least partial) answers to important theoretical and practical questions such as the following: (a) What kind of, and how many, advisors should a DM consult, to achieve a desired level of confidence? (b) What factors make some advisors more, or less, attractive in certain types of decision problems? (c) How does one distribute information among advisors in a manner that is both cost effective and likely to increase the confidence of the DMs who depend on the advisors' opinions.

Acknowledgements

Preparation of the chapter was supported by a grant from the US National Science Foundation under Award No. NSF SES 02-41434. Many thanks to my collaborators Tzur Karelitz, Adrian Rantilla, and Hsiu-Ting Yu for their assistance in all stages and aspects of the empirical work, and for helping shape my thinking about this problem. Thanks also to Jack Soll and the editors of the book for their useful comments on earlier versions of the chapter.

References

Anderson, N. H. (1981). *Information integration theory*. Hillsdale, NJ: Lawrence Erlbaum.

Ariely, D., Au, W. T., Bender, R. H., Budescu, D. V., Dietz, C. B., Gu, H. Wallsten, T. S., & Zauberman, G. (2000). The effects of averaging subjective probability estimates between and within judges. *Journal of Experimental Psychology: Applied*, 6, 130–147.

Armstrong, J. S. (2001). Combining forecasts. In J. S. Armstrong (Ed.), *Principles of forecasting: A handbook for researchers and practitioners*. Norwell, MA: Kluwer Academic Publishers.

Arrow, K. (1963). *Social choice and individual values* (2nd edition). New Haven, CT: Yale University Press.

Ashton, R. H. (1986). Combining the judgments of advisors: How many and which ones? *Organizational Behavior and Human Decision Processes*, 38, 405–414.

Black, D. (1958). *The theory of committees and elections*. Cambridge, UK: Cambridge University Press.

Bordley, R. F. (1982). A multiplicative formula for aggregating probability assessments. *Management Science*, 28, 1137–1148.

Budescu, D. V. (2001). *Aggregation of probabilistic forecasts and opinions*. Presidential address at the annual meeting of the Society for Judgment and Decision Making (SJDM), Orlando, FL.

Budescu, D. V., Erev, I., & Wallsten, T. S. (1997). On the importance of random error in the study of probability judgment, Part I: New theoretical developments. *Journal of Behavioral Decision Making*, 10, 157–172.

Budescu, D. V., & Rantilla, A. K. (2000) Confidence in aggregation of expert opinions. *Acta Psychologica, 104,* 371–398.

Budescu, D. V., Rantilla, A. K,Yu, H., & Karelitz, T. M. (2003). The effects of asymmetry among advisors on the aggregation of their opinions. *Organizational Behavior and Human Decision Processes, 90,* 178–194.

Chapman, L. J., & Chapman, J. P. (1967). Genesis of popular but erroneous psychodiagnostic observations. *Journal of Abnormal Psychology, 72,* 193–204.

Chapman, L. J., & Chapman, J. P. (1969). Illusory correlation as an obstacle to the use of valid psychodiagnostic signs. *Journal of Abnormal Psychology, 74,* 271–280.

Clemen, R. T. (1989). Combining forecasts: A review and annotated bibliography. *International Journal of Forecasting, 5,* 559–583.

Clemen, R. T., & Winkler, R. L. (1993). Aggregating point estimates: A flexible modeling approach. *Management Science, 39,* 501–515.

Clemen, R. T., & Winkler, R. L. (1985). Limits for the precision and value of information from dependent sources. *Operation Research, 33,* 427–442.

Clemen, R. T., & Winkler, R. L. (1986). Combining economic forecasts. *Journal of Business and Economic Statistics, 4,* 39–46.

Davis, J. H. (1973). Group decision and social interaction: A theory of social decision schemes. *Psychological Review, 80,* 97–125.

Davis, J. H. (1996). Group decision making and quantitative judgments: A consensus model. In E. H. Witte & J. H. Davis (Eds.), *Understanding group behavior (Vol 1): Small group processes and interpersonal relations.* Hillsdale, NJ: Lawrence Erlbaum.

Einhorn, H. J., & Hogarth, R. M. (1978). Confidence in judgment: Persistence of the illusion of validity. *Psychological Review, 85,* 395–416.

Erev, I., Wallsten, T. S., & Budescu, D. V. (1994). Simultaneous over- and underconfidence: The role of error in judgment processes. *Psychological Review, 101,* 519–527.

Ferrel, W. R. (1985). Combining individual judgments. In G. Wright (Ed.), *Behavioral decision making*(pp. 111–145). New York: Plenum.

Fiedler, K. (1996). Explaining and simulating judgment biases as an aggregation phenomenon in probabilistic multiple-cue environments. *Psychological Review, 103,* 193–214.

Fischer, I., & Harvey, N. (1999). Combining forecasts: What information do judges need to outperform the simple average? *International Journal of Forecasting, 15,* 227–246.

Flores, B. E., & White, E. M. (1988). A framework for the combination of forecasts. *Academy of Marketing Science, 16,* 95–103.

Gigerenzer, G., & Goldstein, D. G. (1999). Betting on one good reason: The take the best heuristic. In G. Gigerenzer, P. M. Todd, & the ABC Research Group (Eds.), *Simple heuristics that make us smart* (pp. 75–96). New York: Oxford University Press.

Gigerenzer, G., Hoffrage, U., & Kleinbölting, H. (1991). Probabilistic mental models: A Brunswikian theory of confidence. *Psychological Review, 98,* 506–528.

Harvey, N., & Harries, C. (2003). Effects of judges' forecasting on their later combination of forecasts for the same outcomes. *International Journal of Forecasting,* in press.

Hastie, R. (1993). *Inside the juror: The psychology of jury decision-making.* Cambridge, UK: Cambridge University Press.

Hogarth, R. M. (1977). Methods for aggregating opinions. In H. Jungermann & G. de Zeeuw (Eds.), *Decision making and change in human affairs* (pp. 231–255). Boston: Riedel.

Hogarth, R. M. (1978). A note on aggregating opinions. *Organizational Behavior and Human Performance, 21,* 40–46.

Johnson, T. R., Budescu, D. V., & Wallsten, T. S. (2001). Averaging probability judgments: Monte Carlo analyses of asymptotic diagnostic value. *Journal of Behavioral Decision Making, 14,* 123–140.

Juslin, P., Olsson, H., & Bjorkman, M. (1997). Brunswikian and Thurstonian origins of bias in probability assessment: On the interpretation of stochastic components of judgment. *Journal of Behavioral Decision Making, 10,* 189–210.

Kahneman, D., & Tversky, A. (1973). On the psychology of prediction. *Psychological Review, 80,* 237–251.

Kuhn, K. M., & Sniezek, J. A. (1996). Confidence and uncertainty in judgmental forecasting: Differential effects of scenario presentation. *Journal of Behavioral Decision Making, 9,* 231–247.

Larrick, R. B., & Soll, J. B. (2003). *Lay intuitions about combining quantitative judgments.* Paper presented at the workshop on Information Aggregation in Decision Making, Silver Spring, MD.

Laughlin, P. R., & Ellis, A. L. (1986). Demonstrability and social combination processes on mathematical intellective tasks, *Journal of Experimental Social Psychology, 22,* 177–189.

Maines, L. (1996). An experimental examination of subjective forecast combination. *International Journal of Forecasting, 12,* 223–233.

Mojzisch, A., & Schulz-Hardt, S. (2004). Information sampling in group decision making: Sampling biases and their consequences. This volume.

Morris, P. A. (1983). An axiomatic approach to expert resolution. *Management Science, 29,* 24–32.

Morris, P. A. (1986). Comment on Genest and Zideck's "Combining probability distributions: A critique and annotated bibliography." *Statistical Science, 1,* 141–144.

Payne, J. W., Bettman, J. R., & Johnson, E. R. (1993). *The adaptive decision maker.* New York: Cambridge University Press.

Price, P. C., & Stone, E. R. (2004). Intuitive evaluation of likelihood judgment producers: Evidence for a confidence heuristic. *Journal of Behavioral Decision Making, 17,* 39–57.

Rantilla, A. K., & Budescu, D. V. (1999). Aggregation of advisor opinions. In *Proceedings of the 32nd Annual Hawai'i International Conference on Systems Sciences.*

Salvadori, L., van Swol, L. M., & Sniezek, J. A. (2001). Information sampling and confidence within groups and judge advisor systems. *Communication Research, 28,* 737–771.

Sniezek, J. A. (1986). The role of variable labels in cue probability learning tasks. *Organizational Behavior and Human Decision Processes, 38,* 141–161.

Sniezek, J. A. (1992). Groups under uncertainty: An examination of confidence in group decision making. *Organizational Behavior and Human Decision Processes, 52,* 124–155.

Sniezek, J. A., & Buckley, T. (1995). Cueing and cognitive conflict in judge-advisor decision making. *Organizational Behavior and Human Decision Processes, 62,* 159–174.

Soll, J. B. (1999). Intuitive theories of information: Beliefs about the value of redundancy. *Cognitive Psychology, 38,* 317–346.

Soll, J. B., Larrick, R. (2000). *The 80/20 rule and the revision of judgment in light another's opinion: Why fo we believe ourseleves so much?* Paper presented at the annual meeting of the Society for Judgment and Decision Making (SJDM), Los Angeles.

Trafimow, D., & Sniezek, J. A. (1994). Perceived expertise and its effect on confidence. *Organizational Behavior and Human Decision Processes, 57,* 290–302.

van Swol, L. M., & Sniezek, J. A. (2002). *Trust me, I'm an expert: Trust and confidence and acceptance of expert advice.* Paper presented at the 8th Conference on Behavioral Decision Research in Management (BDRM), Chicago.

Wallsten, T. S., Budescu, D. V., Erev, I., & Diederich, A. (1997). Evaluating and combining subjective probability estimates. *Journal of Behavioral Decision Making, 10,* 243–268.

Wallsten, T. S., & Diederich, A. (2001). Understanding pooled subjective probability estimates. *Mathematical Social Sciences, 18,* 1–18.

Walz, D. T., & Walz, D. D. (1989). Combining forecast: Multiple regression versus a Bayesian approach. *Decision Sciences, 20,* 77–89.

Winkler, R. L. (1971). Probabilistic prediction: Some experimental results. *Journal of the American Statistical Association, 66,* 675–685.

Winkler, R. L., & Makridakis, S. (1983). The combination of forecasts. *Journal of the Royal Statistical Society: Series A, 146,* 150–157.

Winkler, R. L., & Poses, R. M. (1993). Evaluating and combining physician's probabilities of survival in an intensive care unit. *Management Science, 39,* 1526–1543.

Wittenbaum, G., & Stasser, G. (1996). Management of information in small groups. In J. L. Nye & A. M. Brower (Eds.), *What's social about social cognition? Research on socially shared cognition in small groups.* (pp. 3–28). Thousand Oaks, CA: Sage.

Yaniv, I. (1997). Weighting and trimming: Heuristics for aggregating judgments under uncertainty. *Organizational Behavior and Human Decision Processes, 69,* 237–249.

Yaniv, I. (2004). Receiving other people's advice: Influence and benefit. *Organizational Behavior and Human Decision Processes, 93,* 1–13.

Yaniv, I., & Kleinberger, E. (2000). Advice taking in decision-making: Egocentric discounting and reputation formation. *Organizational Behavior and Human Decision Processes, 83,* 260–281.

15

Self as Sample

Joachim I. Krueger, Melissa Acevedo,
and Jordan M. Robbins

The popular imagination owes many stimulating images to the science fiction series *Star Trek*. Among the more disturbing ones are the Borg, an alien life form operating "under a collective consciousness, whereby the thoughts of each drone are interconnected with all others in what is referred to as the 'Hive Mind,' eliminating any sense of individuality."[1] The image of the Borg evokes familiar human concerns. For decades, sociologists and social psychologists sought to understand the psychology of the masses or crowds. Crowds came to be feared both for what they could do to others (e.g., lynch them) and for what they could do to the people within them (e.g., lead them toward self-destruction). Members of crowds were thought to be depersonalized as a result of the homogenization of their thoughts and behaviors (Le Bon, 1895).

In modern parlance, the idea that crowds have minds of their own is an argument regarding an emergent property. Emergent properties are irreducible; they cannot be understood with reference to the properties of their constituent elements. Social scientists studying crowds would be wasting their time trying to understand the behavior of the collective from the behavior of a sample of individuals. Nonetheless, this is what F. H. Allport (1924) proposed. He suggested that the analysis of collective behavior should begin with an analysis of individual behavior because "the individual in the crowd behaves just like he would behave alone, *only more so*" (p. 295, emphasis in the original). This view implied that members of a crowd can rely on their own behavioral tendencies to predict collective behavior. They can, in essence, sample their own ideas and inclinations to understand the crowd. Allport thereby suggested that processes of social projection link the experience of individuals with their perceptions of what the group is like or what it is likely to do.

---

[1] This quote was retrieved from www.locutus.be/locutus.php.

In this chapter, we are concerned with the similarities people see between themselves and others in the group. From the collectivist perspective, perceptions of similarity arise from processes of *self-stereotyping*. Self-stereotyping means that people have well-formed ideas about what the group is like, and under certain circumstances, they expect to be like or act like the typical group member. Like any other type of stereotyping, self-stereotyping is a form of inductive reasoning in which knowledge of the properties of a class is sampled or retrieved from memory, and then applied to specific members of the class.

From the individualist perspective, perceptions of similarity arise from processes of *social projection*. Social projection means that people have well-formed ideas about what they themselves are like and that they generalize their own properties or behaviors to others. Social projection is also a form of inductive reasoning. Here it is the knowledge of specific instances that is generalized to the whole class. When conceptualized as a sequence of sampling and inference, social projection shares much in common with other processes of inference and judgment discussed in this volume. Some of the same questions arise. How, for example, is one to know whether a sample is biased, and how is one to correct for such a bias? In the case of self-based predictions, this question turns out to be difficult to answer because the self constitutes only a sample of one. If such a sample is random, the question of whether it is biased has no answer, much like there is no answer to the question of whether the number six is random when a die is cast only once. Nonetheless, it is possible to approach the question of sample bias in less direct ways, as we shall see.

We first review the history of ideas leading to the self-stereotyping and the social-projection hypotheses. This review is essential because both hypotheses lay claim to explaining perceptions of self-group similarity. Then, we review recent studies with relevant evidence. Thus far, both hypotheses have been examined in relative isolation. When perceptions of similarity are observed, investigators often accept such data as evidence for whichever of the two hypotheses they are inclined to favor. From the outset, we note that a comparative review is not likely to refute one or the other hypothesis conclusively. Rather, the boundary conditions of both may come into focus, and questions for future research may emerge.

COLLECTIVISM AND INDIVIDUALISM IN SOCIAL PSYCHOLOGY

In various continental European philosophies of the nineteenth century, the whole was regarded as more than the sum of its parts. Durkheim's (1901) structural-functional sociology and Wundt's (1920–1929) work on folk psychology reflected this legacy. The Romance school of crowd psychology made the radical assumption that, in a crowd, individual minds are surrendered to the emergent group mind (Le Bon, 1895; Sighele, 1892;

de Tarde, 1895). Influential writers such as Freud (1921), McDougall (1920), and Reich (1933) extended and popularized this idea. Elwood (1920) wrote that "all human social groups elaborate what may be called a 'group will' [by acting] to a certain extent as individuals, and develop a sort of individuality or quasi-personality [which] means that each individual must, to some extent, lose his personality in his group" (pp. 114–115). This early definition reveals a conceptual difficulty. Although the group mind was considered a qualitatively distinct and emergent property, it could only be defined metaphorically with recourse to individual minds, which, one might surmise, are easier to understand.

Allport's (1924) critique of the group-mind thesis was influenced by the empiricism of Hume, Locke, and Mill. These philosophers were concerned with the problem of induction. Although they failed to find a logical proof of induction,[2] they justified it pragmatically (Reichenbach, 1951). Induction involves learning from experience. Learners stake their bets for the future on observations they made in the past. In the past this tended to work, and so they expect it to continue to work. In Allport's time, behaviorist models of animal and human learning came to dominate experimental psychology. One way or another, these models were concerned with the learning of covariations: Conditioned stimuli predict unconditioned stimuli, behaviors predict consequences, and so forth. To Allport, the individual stood at the center of analysis, and so it was important to note that the behavior of individuals predicted the behavior of the group.

Over the following decades, the concept of the group mind was questioned by psychologists (Allport, 1962; Asch, 1952; Hofstätter, 1957; Turner et al., 1987), sociologists (Berger & Luckmann, 1966), and philosophers of science (Hayek, 1952; Popper, 1957; Watkins, 1952). Allport's insistence that the individual person be the primary target of analysis was a reflection of his methodological individualism, which Popper described as "the quite unassailable doctrine that we must try to understand all collective phenomena as due to the actions, interactions, aims, hopes, and thoughts of individual men, and as due to traditions created by individual men (Popper, 1957, pp. 157–158). Yet, Allport's (1924) concern that the "ghost [of the group mind] has been exceedingly difficult to lay" (p. 60) was prophetic. Weaker forms of the group-mind hypothesis have been proposed. Their core assumption is that, under certain conditions, the contents of individual minds are shaped by what these individuals *perceive* to be the contents of the group mind. By this conceptual device, a group mind could be brought into play while simultaneously denying its existence.

Questioning the ability of purely individualist theories to account for the complexity of self and social perception, Tajfel & Turner (1986)

---

[2] As Bertrand Russell humorously remarked, there is a special place in hell reserved for those who thought they did (Russell, 1955).

suggested that people understand themselves not only in terms of their unique and personal attributes but also in terms of their membership in important groups. Turner and collaborators took the idea of group-based self-concepts further. Their theory of self-categorization claims a middle ground between individualism and collectivism by "rejecting both the fallacy of the group mind and the reduction of group life" (Turner et al., 1987, p. 5). Nonetheless, self-categorization theory incorporates the assumption that the collective self enjoys primacy over the individualist self. On the one hand, "the collective self is a precursor to the emergence of a personal self; there would be no personal self in the absence of a higher order 'we' that provides the context for self-other differentiation in terms of person-specific attributes" (Onorato & Turner, 2002, p. 165). On the other hand, the collective self can undo its own creation, the individual self, through the process of depersonalization. "The self can become defined almost exclusively in social identity terms in intergroup contexts" (Onorato & Turner, 2002, p. 162).

SELF-STEREOTYPING

The process leading to depersonalization is self-stereotyping. "Once some specific social identification is salient, a person assigns to self and others the common, typical or representative characteristics that define the group as a whole. [Thus, they come to] perceive themselves as relatively interchangeable with other in-group members" (Brown & Turner, 1981, p. 39). Inasmuch as stereotyping affects perceptions of "self and others," self-perception is not regarded as being unique (Hogg & Abrams, 1988). Indeed, according to a recent review, self-stereotyping is obligatory. "Self-categorization as an in-group member *entails* assimilation of the self to the in-group category prototype and enhances similarity to other in-group members" (Hewstone, Rubin, & Willis, 2002, p. 578, emphasis added).

We hesitate to accept the full force of this claim. The strength of stereotypic inferences tends to be constrained by the amount of individuating or person-specific information available (Krueger & Rothbart, 1988; Locksley et al., 1980). Arguably, such specific information is more readily available with regard to the self than to others (G. W. Allport, 1950; Krueger, 2003). If so, self-stereotyping may be rather difficult to demonstrate. Indeed, self-categorization theory predicts self-stereotyping to occur only under certain conditions.

A PsychInfo search conducted on February 24, 2002, yielded thirty-one relevant entries since 1984. Of these, twelve articles were published in major journals (four in the *Journal of Personality and Social Psychology*, three in the *British Journal of Social Psychology*, two in the *European Journal of Social Psychology*, and one each in the *Personality and Social Psychology Bulletin*, the *Social Psychology Quaterly*, and *Group Processes & Intergroup Relations*). We set aside studies using a single-item measure of perceived similarity (e.g., Simon, Pantaleo, & Mummendey, 1995; Spears, Doosje, & Ellemers, 1997). Single-item measures tend to be psychometrically labile, yielding only low correlations with multiple-item measures of profile similarity (Krueger & Stanke, 2001; Pickett, Bonner, & Coleman, 2002). Having no content, global measures of perceived similarity cannot capture the "cognitive redefinition of the self – from unique attributes and individual differences to shared social category memberships and associated stereotypes" (Turner et al., 1987, p. 528). Moreover, these measures cannot effectively separate self-stereotyping from social projection.

Our review focuses on studies examining self-stereotyping on specific personality traits. These studies exemplify the three primary conditions under which self-stereotyping is assumed to occur: when a self-category is salient (Hogg & Turner, 1987), when the attributes in question are positive (Biernat, Vescio, & Green, 1996), or when the self-concept is threatened (Pickett et al., 2002).

Category Salience

The salience of social categories is the foundational requirement for self-stereotyping. In a classic study, Hogg & Turner (1987) had male and female participants read a description of a debate in which either one man and one woman or two men and two women argued opposite sides of a controversial issue. Participants were then asked to rate how a typical person of their sex would describe him- or herself. Later, the participants found themselves in the situation they had evaluated. After the debate, they rated themselves "as you see yourself *now*, and not in terms of enduring personality attributes" (p. 329, emphasis in original). Thus, the experimental manipulation was powerful, categorization by sex was salient, and the stereotypes being tapped were situation-specific. In short, participants were forced to self-categorize. Hogg & Turner predicted that postexperimental self-ratings would be more similar to preexperimental stereotype ratings when categorization by gender was salient (i.e., when a pair of men debated a pair of women). A marginally significant trend emerged for positive traits ($r = .22$), but it did not consistently replicate (see also Mussweiler & Bodenhausen, 2002).

Recently, Onotrato & Turner (2004) followed up this work by asking groups of female and male participants to reach consensus on the applicability of various personality traits (e.g., dominance, aggressiveness) to women and men. Participants then tended to endorse traits typical of their gender as being descriptive of themselves. This effect held regardless of whether participants had described their individual selves in primarily feminine of masculine terms during the initial screening phase.

The interpretation of this finding as evidence for self-stereotyping is complicated by two critical design features. First, participants were asked to think of themselves "as a woman [man], and...to think about the characteristics you have as a woman [man] compared to men [women]" (Onorato & Turner, 2004, p. 263). This feature demanded rather than tested self-stereotyping. Second, participants were told that "if you think you have each characteristic compared to men [women], please respond by pressing the US key. If you think men [women] have each characteristic in contrast to you, please press the THEM key" (p. 263). This feature confounded judgments of the self with judgments of the ingroup. A woman, for example, could have pressed the US key for the trait of dependence because her self-concept was assimilated to her stereotype of women, or because she indicated her knowledge of that stereotype.

Attribute Valence

Some investigators speculated that self-stereotyping might be limited to either positive or negative attributes. Simon, Glässner-Bayerl, & Stratenwerth (1991) found that, compared with members of the straight majority, gay men were more likely to ascribe negative attributes of the gay stereotype to themselves. Biernat et al. (1996), however, suggested that negative self-stereotyping is paradoxical because the "acceptance of negative stereotypes work[s] against the general goal of self-enhancement" (p. 1,194). Following Hogg & Turner (1987), these authors suggested that self-stereotyping might more readily occur for positive attributes, where there is less of a conflict between accepting a collective self and seeking personal self-enhancement. In an empirical study, sorority women responded to positive and negative attributes that were stereotypic of sororities. As expected, descriptions of proximal in-groups (i.e., own sororities) were more favorable than descriptions of the more general in-group (all sororities) or descriptions of the superordinate in-group (university students). From the point of view of the self-stereotyping hypothesis, it was curious "that in these studies that placed great emphasis on group membership, the self emerged as superior" (Biernat et al., p. 1,206).

The social-projection hypothesis offers an alternative interpretation of the results. This hypothesis makes three critical assumptions: First, most people hold favorable self-images; second, ratings of in-groups are imperfect reflections of people's self-views; third, the strength of projection diminishes as social target groups become larger, more heterogeneous, or otherwise psychologically remote (e.g., out-groups; Krueger & Clement, 1996). These assumptions are sufficient to account for the finding that self-ratings are more favorable than ratings of proximal in-groups, which in turn are more favorable than ratings of more distant in-groups. There is no paradox to be solved nor is there a need to deny negative attributes.

Correlations between self-ratings and group ratings provide a more direct test. If self-stereotyping were selective, similarity correlations should be larger when computed across positive attributes than when computed across negative ones. Analysis of the tabulated average ratings provided by Biernat et al., 1996, (Table 1, p. 1,198) shows, however, that perceptions of similarity were high regardless of attribute valence (for own sorority, both $r = .87$; for all sororities, $r = .63$ [positive] and .66 [negative]; for university students, $r = .78$ [positive] and .66 [negative].[3]

Threat to Self

Following Brewer (1991), Pickett et al. (2002) proposed that self-stereotyping can restore a sense of optimal distinctiveness. According to this view, a person can satisfy the need to belong to a group by seeing him- or herself as a typical member. Likewise, a person can satisfy the need to be unique by seeing both the self and the in-group as differentiated from a salient out-group. To test these ideas, honors students were asked to complete a self-concept inventory. For some, a need for assimilation was activated by telling them that their personal score differed from the average score obtained by honors students. For others, a need for differentiation was activated by telling them that their scores were average. Still others were told that they had scored like the average honors student but different from the general student population. All participants then described themselves on a series of positive trait adjectives, some of which were stereotypical of honors students. As expected, self-descriptions were more stereotypical when a need for assimilation or a need for differentiation was activated $(M[r] = .26)$.[4] In a second study, this effect emerged as a trend among participants who were highly identified with the group. In a third study, highly identified sorority women showed a trend of self-stereotyping on positive traits, and significant self-stereotyping on two out of three measures on negative traits. Like the findings reported by Hogg & Turner (1987), these results point to conditions under which self-stereotyping yields perceptions of similarity that cannot be attributed to social projection. Future replications of these findings may lead to firm conclusions as to when and under which conditions self-stereotyping occurs.

---

[3] Participants who were not sorority members (Study 2) yielded no clear correlational picture (sorority: $r = .27$ [positive] and .67 [negative]; students in general: $r = .93$ and .78). These correlations were so high, in part, because of the prior aggregation of ratings across participants. For Study 3, no separate correlations for positive and negative attributes were computed because the means in Table 4 (Biernat et al., 1996, p. 1204) were averages across the salience conditions.

[4] For Study 1, the effect size was estimated by deriving r from t and df (Rosenthal & DiMatteo, 2001, p. 72). For Study 2, the effect sizes could not be estimated because only a p value was reported.

SOCIAL PROJECTION

The social-projection hypothesis suggests that the sampling of self-related evidence precedes and affects inferences about the group. Because this sequence is assumed to involve a causal relation, changes in the former should bring about changes in the latter. We now consider evidence suggesting that self-related knowledge enjoys temporal primacy over group knowledge and that projected self-ratings systematically increase the variability of group ratings.

Response Time

The social-projection hypothesis suggests that self-related knowledge is more accessible than group-related knowledge. Consistent with this view, participants in recent studies made self-ratings faster than group ratings ($M[r] = .42$; Cadinu & De Amicis, 1999; Clement & Krueger, 2000; Krueger & Stanke, 2001). In other studies, response latencies were examined separately for those stimulus items for which self-ratings or group ratings were the same (i.e., matched) and those for which they were different (i.e., mismatched). Here, matched responses were made faster than mismatched responses (Coats et al., 2000; Smith, Coats, & Walling, 1999; Smith & Henry, 1996). This effect has been taken to mean that self-images merge with perceptions of the group (Onorato & Turner, 2002; Tropp & Wright, 2001).

Cadinu & De Amicis (1999) examined response latencies as a function of both the target of judgment (self versus group) and the type of item (matched versus mismatched responses). This full-design analysis revealed that responses were faster for self than for group, and faster for matched than for mismatched items. [See Figure 15.1 for a replication with data from Krueger & Stanke (2001); data from Clement & Krueger (2000) also yielded the same pattern.] The matched traits are those on which self-group similarities are perceived. Because the latencies of self-ratings remained shorter than the latencies of group ratings for this subset of items, it seems that group ratings were matched with self-ratings rather than vice versa. The shorter latencies of group ratings on matched items than on mismatched items suggested that self-ratings facilitated group ratings when similarities were perceived (Dunning & Hayes, 1996; Mussweiler & Bodenhausen, 2002).

To test the facilitation hypothesis further, the order of self- and group ratings was varied in one study, and participants also made nonsocial control ratings about the traits (e.g., "Does the trait word contain the letter S?"; Clement & Krueger, 2000). The facilitative effect of self-ratings on group ratings was the degree to which group ratings were sped up when preceded by self-ratings instead of control ratings; the facilitative effect of group ratings was the degree to which self-ratings were sped up when preceded

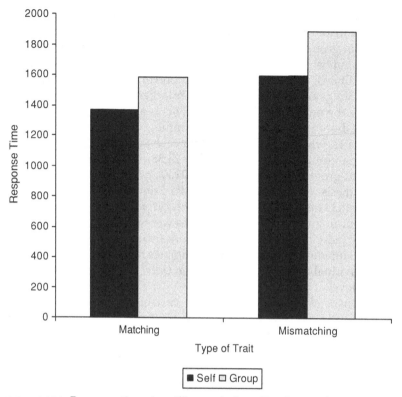

FIGURE 15.1. Response times in milliseconds for self-ratings and group ratings on matching and mismatching traits (calculated from Krueger & Stanke, 2001).

by group ratings instead of control ratings. Both facilitation effects were observed, but the facilitative effect of self-ratings was larger than the facilitative effect of group ratings ($r = .16$). Self-reports also revealed that participants experienced less difficulty when rating the self than when rating the group ($r = .62$), and that they felt more confident about the accuracy of their self-ratings ($r = .57$). In addition, participants described themselves with greater consistency over time than they described groups ($r = .48$; Krueger & Stanke, 2001).

Response Variability

Another way to state the social-projection hypothesis is to say that the variability of group ratings reflects, in part, the variability of self-ratings. If people did not project, one source of variability for group ratings would be disabled, and group ratings should become more homogeneous (Dawes, McTavish, & Shaklee, 1977).

To test this hypothesis, we analyzed data from two studies, in which participants made twenty self-ratings and consensus estimates (in percent) for an in-group or an out-group created in the laboratory (Acevedo & Krueger, unpublished; Clement & Krueger, 2002). Across thirteen experimental conditions, the idiographic projection correlations, computed between self-ratings and group estimates, were larger for the in-group ($M[r] = .47$) than for the out-group ($M[r] = .11$). This is a common and expected finding (see discussion below). When examining the standard deviations of the estimates, which were computed for each participant separately, we found that in-group estimates ($M[SD] = 23.38$) were indeed more variable than out-group estimates ($M[SD] = 21.77$, $p = .02$). For a second test of the hypothesis, we correlated the standard deviations of the estimates with the participants' projection coefficients. When participants made estimates about in-groups, these correlations were positive ($M[r] = .27$). In other words, individuals who projected more also generated more variable consensus estimates. The variability of estimates for out-groups was unrelated to individual differences in projection ($M[r] = .05$), perhaps because the overall level of projection was low.

As consensus estimates become more variable, they tend to become more extreme. As estimates depart from the 50% mark, they imply greater perceived group homogeneity for a particular attribute. When the variability of the estimates across attributes is at its largest (i.e., when only estimates of 0% or 100% are made), the assumed variability of the group on any particular attribute is zero. Inasmuch as social projection is selectively addressed to in-groups, it can contribute to perceptions of in-group homogeneity. Perceiving groups as homogeneous is also a facet of stereotyping. Self-categorization theory suggests that, under certain conditions, in-groups are perceived to be more homogeneous than out-groups (Haslam, Oakes, & Turner, 1996). The present analysis shows, however, that projection can yield the same effect in the absence of stereotypes.

Self-categorization theory further assumes that self-stereotyping "makes sense only if people are assumed to share similar mental representations" (Turner & Onorato, 1999, p. 26).[5] In other words, the group ratings made by different observers should be highly intercorrelated. We identified a data set in which this condition was met and asked whether social projection could still be identified. Data came from 174 Brown University students who rated themselves and their fellow collegiates on a series of twelve personality-descriptive terms (Robbins & Krueger,

---

[5] Without shared stereotypes, the meaning of depersonalization dissolves. If, for example, Per holds an idiosyncratic stereotype of Swedes, his self-image remains unique even with a massive increase in self-stereotyping. Per may be depersonalized in his own mind, but not in the estimation of others. Thus, the effect is not a collective one, though it should be, according to self-categorization theory.

unpublished). The idiographic correlations between self- and group ratings were in the typical range ($M[r] = .47$). We then computed the average group rating for each trait as an index of the group's autostereotype. This stereotype was held with considerable consensus, as suggested by the correlations between these averages and individual respondents' group ratings ($M[r] = .76$). We then isolated the unique component of perceived self-group similarity by computing the correlations between self- and group ratings while controlling for the average group ratings (Kenny & Winquist, 2001). The result ($M[r] = .22$) suggested that perceptions of self-group similarity went beyond the shared component of the social stereotype. We suspect that this unique idiographic component of perceived similarity was the result of projection. In line with the variability hypothesis discussed in the foregoing, we submit that when people who differ in their personal self-descriptions project to the group, the variability of their group ratings systematically increases and covaries with their self-ratings. The more this happens, the weaker the social or consensual nature of the group stereotype will become (Ames, 2004).

SAMPLING THE SELF IN THE LABORATORY

Most studies on perceived self-group similarities are correlational, thus allowing only indirect assessments of the relative contributions of stereotyping and projection. Experimental procedures permit more direct tests when participants learn new information about themselves or about others. Early work on the effects of arbitrary feedback to the self was interpreted as showing an illusion of uniqueness, or a perception of *dissimilarity*. In what became known as the Barnum effect, participants accept personality feedback as highly accurate when in fact the presented sketch is of the generic one-size-fits-all variety (Forer, 1949). Compared with participants who only read the sketch, but who are not led to believe that it describes them, recipients estimate that the sketch is descriptive of a larger segment of the population (Clement & Krueger, 2002). They thus do not feel particularly unique. Once a new piece of personal information is attached to the self-concept, it tends to be projected to others (Agostinelli et al., 1992; Schimel, Greenberg, & Martens, 2003).

The inverse effect has, however, also been found. One of the recurring themes of social-psychological theory and research is that the self-concept is not monolithic but is responsive to social influence. Classic work by Asch (1956) and Sherif (1936) attests to the malleability of individuals' responses when in the presence of others whose responses they witnessed. By a variety of processes, group settings increase the homogeneity of behavior so that the responses of individual group members become intercorrelated. The changes in individuals' responses are, however, rather modest and perceptions of others remain anchored on the self-concept even when a considerable amount of sampling information has accumulated (Alicke & Largo, 1995; Krueger & Clement, 1994; Kulig, 2000).

When a lot of information is available about the probability of a certain trait or behavior in a group, it can – and should – be used to predict what a sampled individual is like. Statistically, it is less uncertain to predict the features of an individual from knowledge of the feature's distribution in the group than vice versa (Dawes, 1990). Because of this asymmetry, evidence for processes such as self-stereotyping or conformity should be easier to obtain than evidence for social projection. When, however, both types of inference are examined in the same study, inferences from the self to the group tend to be stronger than the reverse. The processes accounting for this difference appear to be egocentric because there is no comparable difference in the strength of the inferences when information about some other individual group member is involved (Cadinu & Rothbart, 1996).

In a complex experiment reported by Brewer and Weber (1994), participants were categorized into laboratory groups of different sizes, and each was shown a video clip of either an academically competent or incompetent student. When the incompetent student belonged to the minority group, members of that group rated themselves less favorably compared with the baseline ratings of participants who belonged to no group. When the incompetent student belonged to the majority, members of that group rated themselves more favorably. These findings were consistent with the idea that people seek optimal levels of distinctiveness. For the purpose of the present analysis, it is instructive to also consider the ratings of the videotaped student. Changes in these ratings were more than twice as large as changes in the self-ratings. Regardless of their majority–minority status, participants rated the incompetent in-group student more positively, and thus more similar to themselves, than did participants who belonged to a different group or to no group at all.[6]

In experiments on perceptions of similarity, sampling information is, by definition, not random. The question of whether people overuse self-related sample information or underuse other-related information still awaits a conclusive answer (Engelmann & Strobel, 2000). There is, however, an emerging literature with studies in which participants are deliberately exposed to a sample of self-related information that is known to be biased. Here, people either do not recognize the nature of the bias or they are unable to correct it. These phenomena of egocentric overprojection are variously known as the "spotlight effect" or the "illusion of transparency" (Gilovich & Savitsky, 1999). Their common denominator is that the target person's attention is focused on an emotionally significant aspect of the self. In the case of the spotlight effect, this might be an embarrassing piece

---

[6] No separate means were reported for the competent in-group target.

of clothing; in the case of the illusion of transparency, it might be a state of stage fright. Either way, participants overestimate the degree to which an audience detects or denigrates the source of the person's own anguish.

Overprojection occurs when self-related information is biased in the sense that it is not drawn from the same sampling space as the audience's judgments that it is meant to predict. When the self's perspective involves unique visual information or access to private emotional cues, it is not interchangeable with the perspective of an audience. To perceive strong similarities between the audience's and one's own responses is, in a sense, a case of overprojection to a social group that does not include the self. This raises the general question of how and under what conditions social categorization moderates the strength of social projection.

SOCIAL CATEGORIZATION

Any account of sampling and inference needs to address the question of how the sampled information is related to the population about which inferences are made. The standard ideal is to specify a population first (e.g., all eligible voters) and then draw a random sample from it. When this ideal is met, the standard errors of the sample statistics are inversely related to sample size, and their validity converges on their reliability. Often, however, inferences need to be drawn about populations that have not been sampled. Most psychological research is conducted with undergraduate students, and most biomedical research is conducted with a small number of animal species. Yet, findings are generalized to other categories of humans or animals. On the one hand, it is recognized that the samples are biased with regard to these other target categories; on the other hand, certain a priori beliefs or other empirical data suggest that the sampled and unsampled categories are sufficiently similar to justify cross-category inferences.

In social perception, people assess the similarities between themselves and groups to which they belong (in-groups) and between themselves and groups to which they do not belong (out-groups). Both the social-projection hypothesis and the self-stereotyping hypothesis suggest that people see greater similarities between themselves and in-groups than between themselves and out-groups. To discriminate between the two hypotheses, it is useful to consider different kinds of category used in this research area.

In correlational studies, participants are grouped according to existing memberships in various social categories, such as race, gender, or academic affiliation. In experimental studies, participants are grouped according to controlled feedback they receive in the laboratory. Feedback often consists of scores presumably obtained on some test or task. This information is arbitrary, and the resulting groups are considered "minimal" (Rabbie & Horwitz, 1969; Tajfel et al., 1971). Compared with real social groups, minimal groups have two important properties: They do not comprise any preexisting social stereotypes that group members might use to "tune" their self-concepts, and there is no reason to believe that these groups differ in anything but their labels. Information sampled from one group might as well be generalized to the other.

A recent meta-analysis revealed that perceptions of similarity are stronger for minimal than for real in-groups ($M[r] = .57$ and $.40$, Robbins & Krueger, 2005). Projection can operate where self-stereotyping is blocked (i.e., in minimal groups). If self-stereotyping were a potent source of perceived similarities, correlations should be larger when real groups are involved. The attenuation of similarity correlations in real groups is also consistent with general processes of sampling and inference. In real social categories, people come across other group members whose attributes or behaviors may differ from their own. Taking this information into account, they make predictions that are less egocentric than their predictions in the laboratory where no other sampling information exists.

The meta-analytic data also showed that perceptions of similarity are weaker for minimal than for real out-groups ($M[r] = .10$ and $.15$). Again, the modest correlation for real out-group targets is consistent with the idea that people have *some* relevant sampling information even though they tend to be less familiar with out-groups than with in-groups (Fiedler, 2000; Linville, Fischer, & Salovey, 1989). The puzzle is why perceptions of similarity are so low for minimal out-groups. Here, people do not appear to "overproject" as they do in the case of the spotlight effect and the illusion of transparency. Instead, they appear to take the view of a conservative scientist who refrains from generalizing findings to unsampled populations. Because, however, they are asked to make *some* predictions as part of the experimental protocol, research participants oblige, but their predictions remain independent of the self-related sampling information they possess.

Many social-psychological theories assume that various cognitive or motivational processes contribute to the perception of intergroup differences. Social-identity theory and self-categorization theory stress the psychological need to differentiate the in-group from the out-group (Tajfel & Turner, 1986); optimal-distinctiveness theory assumes that out-groups are actively contrasted away from the self and the in-group (Brewer, 1991). The asymmetry between projection to in-groups and out-groups offers a different view, suggesting that in-groups and out-groups are perceived differently even in the absence of any specific comparative process.

Figure 15.2 shows two arrangements of the relevant trivariate relationships. In the left panel, the correlations between self-ratings (S) and group ratings (IG and OG, respectively, for in-groups and out-groups) are set to values that approximate empirical findings. Assuming that people make

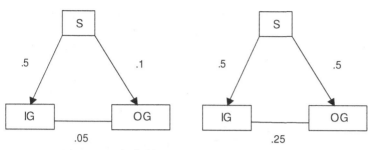

FIGURE 15.2. Asymmetric (left) versus symmetric (right) projection to in-groups and out-groups.

no direct assessment of in-group–out-group similarity, the correlation between the two sets of group ratings is the product of the two projection correlations. The low value of this correlation indicates the near independence of in-group and out-group perceptions. In contrast, the right panel shows that in-groups and out-groups would appear to be rather similar if projection to the two groups were equally strong. As social projection becomes more asymmetrical, perceptions of intergroup differences emerge more strongly.

A PROJECTION MODEL OF IN-GROUP BIAS

Since Sumner's (1906) speculations about the ubiquity of ethnocentrism, empirical work has consistently shown that people hold more favorable images of their in-groups than of out-groups. Many, but not all, out-groups are seen as neither positive nor negative, making the bias one of in-group favoritism rather than out-group derogation (Brewer, 1999). To explain in-group bias from a collectivist point of view, Turner & Onorato (1999) emphasized the role of motivation, suggesting that there is "a *psychological requirement* that relevant in-groups compare favorably with relevant out-groups ... a *motivation* producing a *drive* for in-group superiority [and] *pressures* for intergroup differentiation" (p. 18, emphases added). How this might happen in minimal groups is not clear. Here, it is necessary to assume that people treat minimal groups as if they were real, that is, as if they offered platforms for enduring and meaningful social identities, and as if they were endowed with behavioral norms or social stereotypes.

The ability of asymmetric social projection to yield perceptions of intergroup differences has an important consequence for how favorably in-groups and out-groups are described. Social projection accounts for in-group bias by using assumptions of inductive reasoning and self-love. The inductive assumption is that people project more to in-groups than to out-groups (i.e., $r_{S,IG} > r_{S,OG}$). The self-love assumption is that most people rate positive attributes as being more descriptive of themselves than negative attributes, that is, the correlation between self-ratings and ratings of the attributes' social desirability, $r_{S,SD}$, is high (Krueger, 1998). When no other systematic relationships are assumed, an estimate of in-group bias can be derived as follows. The perceived favorability of the in-group is the product of the favorability of the self and projection to the in-group (i.e., $r_{IG,SD} = r_{S,SD}r_{S,IG}$). Analogously, the perceived favorability of the out-group is the product of the favorability of the self and projection to the out-group (i.e., $r_{OG,SD} = r_{S,SD}r_{S,OG}$). In-group bias is the difference between the perceived favorability of the in-group and the perceived favorability of the out-group (i.e., $r_{IG,SD} - r_{OG,SD}$). It is useful to examine how the difference between in-group judgments and out-group judgments is correlated with attribute desirability. In-group bias is present if the in-group is rated more highly than the out-group inasmuch as the attribute in question is desirable (i.e., if $r_{IG-OG,SD} > 0$). This difference-score correlation can be computed as the ratio of overall in-group favoritism (i.e., $r_{IG,SD} - r_{OG,SD}$) over a term expressing the complement of intergroup differentiation (i.e., $\sqrt{2[1 - r_{IG,OG}]}$). For simplicity of exposition, it is assumed that all variables are standardized (for unstandardized variables see Asendorpf & Ostendorf, 1998).

Figure 15.3 represents this account of in-group favoritism. The top panel diagrams the associations among the four input variables (i.e., ratings of desirability, self, in-group, and out-group). Direct paths are shown as solid lines; indirect or mediated paths are shown as dashed lines. The three input correlations approximate empirical values (i.e., $r_{S,SD} = .8$, $r_{S,IG} = .5$, and $r_{S,OG} = .1$). The bottom panel shows the predictions of the model as a set of regression lines. Self-ratings, in-group ratings, and out-group ratings are regressed on attribute desirability, with the slope being steepest for self-ratings and flattest for out-group ratings. Thus, in-group-favoritism is strongest for the most desirable and the most undesirable attributes. The model also suggests that perceived differences between self and in-group are greatest for evaluatively extreme attributes. This latter prediction differs from the self-categorization perspective, which assumes that in-group favoritism and depersonalization go hand in hand.

To examine some predictions of the projection model, we conducted a study with minimal groups. Following an ostensible testing procedure, some participants were told that according to their cognitive style, they belonged to the category of "grounders," whereas others learned that they did not belong to this group. Perceptions of self-group similarity and desirability were computed from subsequent ratings of a set of personality-inventory statements (e.g., "I like poetry"; see Krueger & Clement, 1996). The idiographic correlations between group ratings and desirability ratings showed the expected in-group bias ($M[r] = .30$ and .04, for in-groups and out-groups, respectively). The asymmetry in the projection coefficients, however, surpassed the size of the in-group bias. The correlations between

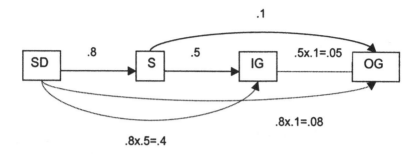

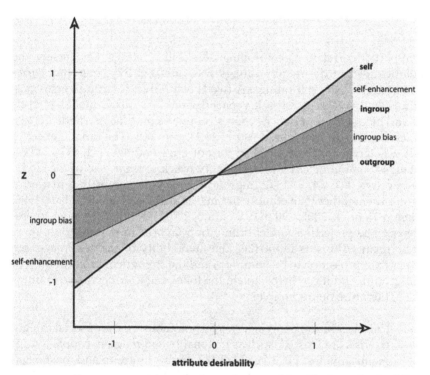

FIGURE 15.3. From asymmetric projection to in-group bias: The hypothesized flow of inferences (top) and the resulting differences in regressiveness scaled as z scores (bottom).

self-ratings and group ratings were much larger for the in-group ($M[r] =$ .71) than for the out-group perspective ($M[r] = $ .05). This pattern suggested that asymmetric projection could more effectively account for in-group bias than vice versa.

Table 15.1 presents a partitioning of the variance of the group ratings. The unique effect of self-ratings on group ratings was captured by

TABLE 15.1. *Predicting Group Ratings from Self- and Desirability Ratings: Variance Accounted for and Standard Deviations (in Parentheses)*

	In-group	Out-group
	r^2 (SD)	r^2 (SD)
Total Variance (Multiple r)	.52 (.22)	.23 (.22)
Effect of Self: r(self, group.[desirability])^2	.38 (.22)	.16 (.20)
Effect of Desirability: r(desirability, group.[self])^2	.04 (.04)	.07 (.11)
Shared Variance[a]	.10 (.15)	.003 (.05)

[a] The shared variance is the difference between the r^2 in the top row and the sum of the r^2 in the two center rows.

semipartial correlations controlling desirability ratings. Conversely, the unique effect of desirability ratings was captured by semipartial correlations controlling self-ratings. A two (Perspective: in-group versus out-group) by two (Predictor: self versus desirability) mixed-model analysis of variance with repeated measures on the second factor yielded a significant interaction, with $F(1, 84) = 24.23$, $p < .001$. The unique effect of self was larger for in-group than for out-group ratings, $F(1, 84) = 48.04$, whereas the unique effect of desirability was low regardless of the raters' perspective, $F(1, 84) = 1.26$. In-group bias was thus mostly a matter of egocentrism rather than ethnocentrism (see also Cadinu & Rothbart, 1996; Chen & Kenrick, 2002; Otten & Wentura, 2001; or Otten, 2002a,b, for reviews). The projection model of in-group favoritism requires only three basic relationships as input (i.e., the favorability of the self-image, the strength of projection to the in-group, and the strength of projection to the out-group), and it can help explain the following four pervasive findings in the literature on in-group bias.

1. *Cross-categorization*: When two dichotomous means of social categorization are crossed, such as nationality and religion, people's judgments are most positive about the double in-group and most negative about the double out-group (see Migdal, Hewstone, & Mullen, 1998, for a review and meta-analysis). The social-projection hypothesis predicts this result because a double in-group should receive the full benefit of projection, whereas groups that are in-groups according to one means of categorization and out-groups according to the other should receive less. Double out-groups should be seen as the least appropriate target of projection.

2. *Recategorization*: One way of eliminating in-group bias is to eliminate categorization altogether (Gaertner et al., 2001). The projection model states that changes in categorization beget changes in perceived similarity. People who gain membership in one group while losing membership in another increase their projection to

the new in-group while decreasing projection to the old in-group (Clement & Krueger, 2002). If, as this pattern suggests, projection is mainly a question of inclusion of others into the same social category, then others who join the in-group become targets of projection for people's positive self-concepts.

3. *Group size*: Members of minority groups show greater in-group bias than members of majority groups (Mullen, Brown, & Smith, 1992). Being a member of a minority group is a more salient experience and engenders higher levels of identification than being a member of a majority group (Leonardelli & Brewer, 2001). It may, however, be sufficient to assume that the strength of projection increases as groups become smaller. Although group size alone does not matter from the perspective of pure induction (Dawes, 1989), small groups also tend to be more homogeneous than large groups. Thus, inferences based on small samples tend to be less uncertain.

4. *Discrimination*: A central aim of social identity and self-categorization theories is to explain why people discriminate against out-groups. Both theories assume that preferential treatment of in-group members can occur in the absence of any rational assessment of individual self-interest. In contrast, Rabbie, Schot, & Visser (1989) proposed that people favor others inasmuch as they expect reciprocal benefits (see Gaertner & Insko, 2000, for empirical evidence). To the extent that people expect their actions to be reciprocated, they may rationally choose to bestow benefits onto others. As such projective expectations of reciprocity come to mind more readily when the other person is a member of the in-group instead of the out-group, a discriminatory behavioral gap opens (Acevedo & Krueger, 2004; Yamagishi & Kiyonari, 2000).[7]

CONCLUSION

In this chapter, we explored the processes underlying perceptions of self-group similarities. From a sampling-and-inference perspective, the hypotheses of self-stereotyping and social projection offer different accounts for such perceptions. Whereas the idea of self-stereotyping focuses on how different in-groups come to be salient sampling spaces for inferences about the self, the idea of social projection focuses on the sampling of self-related information, and on how this information is generalized to available in-groups.

---

[7] Tajfel originally speculated that participants "would assume others to behave as they themselves did, and that this assumption would in turn affect their behavior" (Tajfel et al., 1971, p. 175). This projective idea was subsequently no longer emphasized in theories of social identity or self-categorization.

As we noted at the outset, it was not expected that one hypothesis would entirely prevail to the exclusion of the other. Although at present, the social-projection hypothesis appears to be better supported by empirical data, the self-stereotyping hypothesis is too compelling to be discarded. The self-concept is, after all, responsive to real and perceived social influences, as well as perceptual contrast and assimilation effects (Mussweiler, 2003). A practical recommendation for future research is to develop experimental designs aimed at further clarification of the boundary conditions of either hypothesis and at shedding more light on how processes of self-stereotyping and social projection may complement each other.

Proponents of self-categorization theory recently offered the same recommendation (Onorato & Turner, 2004). At the same time, however, these authors noted that "depersonalization involves introjection [i.e., the assimilation of the self-concept to the group stereotype] and projection" (p. 275). Both mechanisms are "aspects of the same process of assigning group-level characteristics to self and others as the ingroup-outgroup level of self-categorization becomes salient" (p. 275). We worry that this formulation rather insulates the concept of depersonalization from empirical challenge, making self-categorization theory appear to be true a priori.

References

Acevedo, M., & Krueger, J. I. (unpublished). *Differential social projection accounts for ingroup bias in the minimal group paradigm.* Brown University.

Acevedo, M., & Krueger, J. I. (2004). Two egocentric sources of the decision to vote: The voter's illusion and the belief in personal relevance. *Political Psychology, 25,* 115–134.

Agostinelli, G., Sherman, S. J., Presson, C. C., & Chassin, L. (1992). Self-protection and self-enhancement biases in estimates of population prevalence. *Personality and Social Psychology Bulletin, 18,* 631–642.

Alicke, M. D., & Largo, E. (1995). The role of the self in the false consensus effect. *Journal of Experimental Social Psychology, 31,* 28–47.

Allport, F. H. (1924). The group fallacy in relation to social science. *Journal of Abnormal and Social Psychology, 19,* 60–73.

Allport, F. H. (1962). A structuronomic conception of behavior: Individual and collective. I. Structural theory and the master problem of social psychology. *Journal of Abnormal and Social Psychology, 64,* 3–30.

Allport, G. W. (1950). *The nature of personality.* Cambridge, MA: Addison-Wesley.

Ames, D. R. (2004). Inside the Mind Reader's Tool Kit: Projection and Stereotyping in Mental State Inference. *Journal of Personality and Social Psychology, 87,* 340–353.

Asendorpf, J. B., & Ostendorf, F. (1998). Is self-enhancement healthy? Conceptual, psychometric, and empirical analysis. *Journal of Personality and Social Psychology, 74,* 955–966.

Asch, S. E. (1952). *Social psychology.* New York: Prentice-Hall.

Asch, S. E. (1956). Studies of independence and conformity: I. A minority of one against a unanimous majority. *Psychological Monographs 70.*

Berger, P. L., & Luckmann, T. (1966). *The social construction of reality*. Garden City, NY: Doubleday.

Biernat, M., Vescio, T. K., & Green, M. L. (1996). Selective self-stereotyping. *Journal of Personality and Social Psychology, 71*, 1194–1209.

Brewer, M. B. (1991). The social self: On being the same and different at the same time. *Personality and Social Psychology Bulletin, 17*, 475–482.

Brewer, M. B. (1999). The psychology of prejudice: Ingroup love or out-group hate? *Journal of Social Issues, 55*, 429–444.

Brewer, M. B., & Weber, J. G. (1994). Self-evaluation effects of interpersonal versus intergroup social comparison. *Journal of Personality and Social Psychology, 66*, 268–275.

Brown, R. J., & Turner, J. C. (1981). Interpersonal and intergroup behaviour. In J. C. Turner & H. Giles (Eds.), *Intergroup behaviour* (pp. 33–65). Oxford: Blackwell.

Cadinu, M. R., & De Amicis, L. (1999). The relationship between the self and the ingroup: When having a common conception helps. *Swiss Journal of Psychology, 58*, 226–232.

Cadinu, M. R., & Rothbart, M. (1996). Self-anchoring and differentiation processes in the minimal group setting. *Journal of Personality and Social Psychology, 70*, 661–677.

Chen, F. F., & Kenrick, D. T. (2002). Repulsion or attraction? Group membership and assumed attitude similarity. *Journal of Personality and Social Psychology, 83*, 111–125.

Clement, R. W., & Krueger, J. (2000). The primacy of self-referent information in perceptions of social consensus. *British Journal of Social Psychology, 39*, 279–299.

Clement, R. W., & Krueger, J. (2002). Social categorization moderates social projection. *Journal of Experimental Social Psychology, 38*, 219–231.

Coats, S., Smith, E. E., Claypool, H. M., & Banner, M. J. (2000). Overlapping mental representations of self and in-group: Reaction time evidence and its relationship with explicit measures of group identification. *Journal of Experimental Social Psychology, 36*, 304–315.

Dawes, R. M. (1989). Statistical criteria for establishing a truly false consensus effect. *Journal of Experimental Social Psychology, 25*, 1–17.

Dawes, R. M. (1990). The potential nonfalsity of the false consensus effect. In R. M. Hogarth (Ed.), *Insights in decision making: A tribute to Hillel J. Einhorn* (pp. 179–199). Chicago: University of Chicago Press.

Dawes, R. M., McTavish, J., & Shaklee, H. (1977). Behavior, communication, and assumptions about other people's behavior in a commons dilemma situation. *Journal of Personality and Social Psychology, 35*, 1–11.

de Tarde, G. (1895). *Les lois de l'imitation; étude sociologique*. Paris: Alcan.

Dunning, D., & Hayes, A. F. (1996). Evidence for egocentric comparison in social judgment. *Journal of Personality and Social Psychology, 71*, 862–871.

Durkheim, E. (1901). *Les règles de la méthode sociologique*. Paris: Alcan.

Elwood, C. A. (1920). *An introduction to social psychology*. New York: Appleton.

Engelmann, D., & Strobel, M. (2000). The false consensus effect disappears when representative information and monetary incentives are given. *Experimental Economics, 3*, 241–260.

Fiedler, K. (2000). Beware of samples! A cognitive-ecological sampling approach to judgment biases. *Psychological Review, 107*, 659–676.

Forer, B. R. (1949). The fallacy of personal validation: A classroom demonstration of gullibility. *Journal of Abnormal and Social Psychology, 44,* 118–123.

Freud, S. (1921). *Massenpsychologie und Ich-Analyse.* Vienna: Internationaler Psychoanalytischer Verlag.

Gaertner, L., & Insko, C. (2000). Intergroup discrimination in the minimal group paradigm: Categorization, reciprocation, or both? *Journal of Personality and Social Psychology, 79,* 77–94.

Gaertner, S. L., Mann, J., Murrell, A., & Dovidio, J. F. (2001). Reducing intergroup bias: The benefits of recategorization. In M. A. Hogg & Abrams, D. (Eds.), *Intergroup relations: Essential readings* (pp. 356–369). Philadelphia: Psychology Press.

Gilovich, T., & Savitsky, K. (1999). The spotlight effect and the illusion of transparency: Egocentric assessments of how we are seen by others. *Current Directions in Psychological Science, 8,* 165–168.

Haslam, S. A., Oakes, P. J., & Turner, J. C. (1996). Social identity, self-categorization, and the perceived homogeneity of ingroups and out-groups: The interaction between social motivation and cognition. In R. M. Sorrentino & E. T. Higgins (Eds.), *Handbook of motivation and cognition,* Vol. 3: *The interpersonal context* (pp. 182–222). New York: Guilford.

Hayek, F. A. v. (1952). *The counter revolution in science.* Chicago: University of Chicago Press.

Hewstone, M., Rubin, M., & Willis, H. (2002). Intergroup bias. *Annual Review of Psychology, 53,* 575–604.

Hofstätter, P. R. (1957). *Gruppendynamik. Die Kritik der Massenpsychologie.* Hamburg: Rowohlt.

Hogg, M. A., & Abrams, D. (1988). *Social identifications: A social psychology of intergroup relations and group processes.* London: Routledge.

Hogg, M. A., & Turner, J. C. (1987). Intergroup behaviour, self-stereotyping and the salience of social categories. *British Journal of Social Psychology, 26,* 325–340.

Kenny, D. A., & Winquist, L. (2001). The measurement of interpersonal sensitivity: Consideration of design, components, and unit of analysis. In J. Hall & F. Bernieri (Eds.), *Interpersonal sensitivity: Theory and measurement* (pp. 265–302). Englewood Cliffs, NJ: Lawrence Erlbaum.

Krueger, J. (1998). Enhancement bias in the description of self and others. *Personality and Social Psychology Bulletin, 24,* 505–516.

Krueger, J. I. (2003). Return of the ego – self-referent information as a filter for social prediction: Comment on Karniol (2003). *Psychological Review, 110,* 585–590.

Krueger, J., & Clement, R. W. (1994). The truly false consensus effect: An ineradicable and egocentric bias in social perception. *Journal of Personality and Social Psychology, 67,* 596–610.

Krueger, J., & Clement, R. W. (1996). Inferring category characteristics from sample characteristics: Inductive reasoning and social projection. *Journal of Experimental Psychology: General, 125,* 52–68.

Krueger, J., & Rothbart, M. (1988). Use of categorical and individuating information in making inferences about personality. *Journal of Personality and Social Psychology, 55,* 187–195.

Krueger, J., & Stanke, D. (2001). The role of self-referent and other-referent knowledge in perceptions of group characteristics. *Personality and Social Psychology Bulletin, 27,* 878–888.

Kulig, J. W. (2000). Effects of forced exposure to a hypothetical population on false consensus. *Personality and Social Psychology Bulletin, 26*, 629–636.

Le Bon, G. (1895). *Psychologie des foules*. Paris: Alcan.

Leonardelli, G. J., & Brewer, M. B. (2001). Minority and majority discrimination: When and why? *Journal of Experimental Social Psychology, 37*, 468–485.

Linville, P. W., Fischer, G. W., & Salovey, P. (1989). Perceived distributions of the characteristics of in-group and out-group members: Empirical evidence and a computer simulation. *Journal of Personality and Social Psychology, 57*, 165–188.

Locksley, A., Borgida, E., Brekke, N., & Hepburn, C. (1980). Sex stereotypes and social judgment. *Journal of Personality and Social Psychology, 39*, 821–831.

McDougall, W. (1920). *The group mind, a sketch of the principles of collective psychology, with some attempt to apply them to the interpretation of national life and character*. New York: Putnam.

Migdal, M., Hewstone, M., & Mullen, B. (1998). The effects of crossed categorization on intergroup evaluations: A meta-analysis. *British Journal of Social Psychology, 37*, 303–324.

Mullen, B., Brown, R., & Smith, C. (1992). Ingroup bias as a function of salience, relevance, and status: An integration. *European Journal of Social Psychology, 22*, 103–122.

Mussweiler, T. (2003). Comparison processes in social judgment: Mechanisms and consequences. *Psychological Review, 110*, 472–489.

Mussweiler, T., & Bodenhausen, G. V. (2002). I know you are, but what am I? Self-evaluative consequences of judging in-group and out-group members. *Journal of Personality and Social Psychology, 82*, 19–32.

Onorato, R. S., & Turner, J. C. (2002). Challenging the primacy of the personal self: The case for depersonalized self-conception. In Y. Kashima, M. Foddy, & M. J. Platow (Eds.), *Self and identity: Personal, social, and symbolic* (pp. 145–178). Mahwah, NJ: Lawrence Erlbaum.

Onorato, R. S., & Turner, J. C. (2004). Fluidity of the self-concept: the shift from personal to social identity. *European Journal of Social Psychology, 34*, 257–278.

Otten, S. (2002a). "Me" and "us" or "us" and "them"? – The self as heuristic for defining novel ingroups. In: W. Stroebe & M. Hewstone (Eds.), *European Review of Social Psychology* (Vol. 13, pp. 1–33). Philadelphia: Psychology Press.

Otten, S. (2002b). I am positive and so are we: The self as a determinant of favoritism toward novel ingroups. In J. P. Forgas & K. D. Williams (Eds.), *The social self: Cognitive, interpersonal, and intergroup perspectives* (pp. 273–291. New York: Psychology Press.

Otten, S., & Wentura, D. (2001). Self-anchoring and in-group favoritism: An individual profiles analysis. *Journal of Experimental Social Psychology, 37*, 525–532.

Pickett, C. L., Bonner, B. L., & Coleman, J. M. (2002). Motivated self-stereotyping: Heightened assimilation and differentiation needs result in increased levels of positive and negative self-stereotyping. *Journal of Personality and Social Psychology, 82*, 543–562.

Popper, K. R. (1957). *The poverty of historicism*. New York: Harper & Row.

Rabbie, J. M., & Horwitz, M. (1969). Arousal of ingroup–out-group bias by a chance win or loss. *Journal of Personality and Social Psychology, 13*, 269–277.

Rabbie, J. M., Schot, J. C., Visser, L. (1989). Social identity theory: A conceptual and empirical critique from the perspective of a behavioral interaction model. *European Journal of Social Psychology, 19,* 171–202.

Reich, W. (1933). *Massenpsychologie des Faschismus.* Copenhagen: Verlag für Sexualpolitik.

Reichenbach, H. (1951). *The rise of scientific philosophy.* Berkeley: University of California Press.

Robbins, J. M., & Krueger, J. I. (unpublished). *Social projection among ethnic majorities and minorities.* Brown University.

Robbins, J. M., & Krueger, J. I. (2005). Social projection to ingroups and outgroups: A review and meta-analysis. *Personality and Social Psychology Review, 9,* 32–47.

Rosenthal, R., & DiMatteo, M. R. (2001). Meta-analysis: Recent developments in quantitative methods for literature reviews. *Annual Review of Psychology, 52,* 59–82.

Russell, B. (1955). *Nightmares of eminent persons.* New York: Simon & Schuster.

Schimel, J., Greenberg, J., & Martens, A. (2003). Evidence that projection of a feared trait can serve a defensive function. *Personality and Social Psychology Bulletin, 29,* 969–979.

Sherif, M. (1936). *The psychology of social norms.* New York: Harper.

Sighele, S. (1892). *La foule criminelle: Essai de psychologie collective.* Paris: Alcan.

Simon, B., Glässner-Bayerl, B., & Stratenwerth, I. (1991). Stereotyping and self-stereotyping in a natural intergroup context: The case of heterosexual and homosexual men. *Social Psychology Quarterly, 54,* 252–266.

Simon, B., Pantaleo, G., & Mummendey, A. (1995). Unique individual or interchangeable group member? The accentuation of intragroup differences versus similarities as an indicator of the individual self versus the collective self. *Journal of Personality and Social Psychology, 69,* 106–119.

Smith, E. R., Coats, S., & Walling, D. (1999). Overlapping mental representations of self, in-group, and partner: Further response time evidence and a connectionist model. *Personality and Social Psychology Bulletin, 25,* 873–882.

Smith, E. R., & Henry, S. (1996). An in-group becomes part of the self: Response time evidence. *Personality and Social Psychology Bulletin, 22,* 635–642.

Spears, R., & Doosje, B., & Ellemers, N. (1997). Self-stereotyping in the face of threats to group status and distinctiveness: The role of group identification. *Personality and Social Psychology Bulletin, 23,* 538–553.

Sumner, W. G. (1906). *Folkways.* Boston: Ginn.

Tajfel, H., Billig, M. G., Bundy, R. P., & Flament, C. (1971). Social categorization and intergroup behavior. *European Journal of Social Psychology, 1,* 1–39.

Tajfel, H., & Turner, J. C. (1986). The social identity theory of intergroup behavior. In S. Worchel & W. G. Austin (Eds.), *Psychology of intergroup relations* (pp. 7–24). Chicago: Nelson-Hall.

Tropp, L. R., & Wright, S. C. (2001). Ingroup identification as the inclusion of ingroup in the self. *Personality and Social Psychology Bulletin, 27,* 585–600.

Turner, J. C., Hogg, M. A., Oakes, P. J., Reicher, S. D., & Wetherell, M. (1987). *Rediscovering the social group: A self-categorization theory.* Oxford: Blackwell.

Turner, J. C., & Onorato, R. S. (1999). Social identity, personality, and the self-concept: A self-categorization perspective. In T. R. Tyler, R. M. Kramer, & O. P.

John (Eds.), *The psychology of the social self* (pp. 11–46). Mahwah, NJ: Lawrence Erlbaum.

Watkins, J. W. N. (1952). Ideal types and historical explanation. *British Journal for the Philosphy of Science, 3,* 22–43.

Wundt, W. M. (1920). 1929 *Völkerpsychologie. Eine Untersuchung der Entwicklungsgesetze von Sprache, Mythus und Sitte.* Leipzig: Engelmann.

Yamagishi, T., & Kiyonari, T. (2000). The group as the container of generalized reciprocity. *Social Psychology Quarterly, 63,* 116–132.

PART V

VICISSITUDES OF SAMPLING IN THE RESEARCHERS' MINDS AND METHODS

16

Which World Should Be Represented in Representative Design?

Ulrich Hoffrage and Ralph Hertwig

University of California, Berkeley campus, early 1940s. A man and a young woman are strolling across the university campus. Suddenly the man stops and says, "Now." The woman also stops, raises her arm, points to the building in front of her, and says, "The window on the second floor." After contemplating the scene for a moment, she says "Three-and-a-half feet." The man records this and other numbers. They begin walking again. A few moments later, the man stops once again and the same interaction repeats itself.
University of Constance, vision laboratory, mid-1980s. A young man enters a dark room. The experimenter welcomes him and then asks him to sit down in front of a large cubicle, to put his chin on a rest (thus making it impossible to move his head), and to look into the cubicle through two small holes. Its interior is completely dark. Suddenly a white square appears, only to disappear a few seconds later. Then, a tone is heard, which is followed by another white square to the right of the previous one. The young man says, "The left one." This episode frequently repeats itself over the course of the next half hour.

Although both scenes represent the same genre, namely, psychological studies on the perception of object size, they share few common features. As we shall see shortly, their differences can be traced back to the fact that the Berkeley study was conceived by Egon Brunswik (1944; for a replication, see Dukes, 1951), whereas a contemporary experimental psychologist conducted the Constance study. In Brunswik's study, the participant (Miss Johanna R. Goldsmith, a graduate student in psychology) was interrupted at randomly chosen intervals during the course of her daily activities, in various outdoor and indoor situations (on the street, on campus, in the laboratory, at home, etc.). Once interrupted, she was asked to "indicate which linear extension happened to be most conspicuous to her at the moment" (Brunswik, 1944; cited after Hammond & Stewart, 2001, p. 70). Brunswik summarized the sample of situations thus obtained as follows: Although it may "not be perfectly representative of 'life', there is no doubt that it is more representative and variegated with regard to size, distance, proportion, color, surrounding patterns of objects, and other characteristics than any laboratory design could hope to be" (Brunswik, 1944; cited in Hammond & Stewart, 2001, p. 70).

It seems fair to conclude that the study in the vision laboratory was not representative of "life" – indeed, this dimension was probably not even considered. Rather, the experimenter created highly constrained situations in which he had full control over variables, such as the size of the objects, their distance, and their colors. He also determined whether the participant's vision was binocular or monocular. (The participant was the first author, who was required to participate in a number of psychological studies as an undergraduate.) Whereas Brunswik aimed to sample situations that were representative of the actual demands of the environment in which the individual lives, the Constance study investigated situations that rarely occur outside the artificial world of the vision lab. For Brunswik, the key concern was being able to generalize the results to the person's natural habitat. Accordingly, he sampled situations directly from the person's habitat. In contrast, for many experimentalists – today as well as in Brunswik's time – the prime objective is to carefully disentangle possible variables that affect a person's perceptual and cognitive performance. Consequently, experimentalists create microworlds that may never occur in reality, but over which they exert full control.

The topic of this chapter is what we believe to be the constructive tension between these two paradigms in psychological experimentation. We begin with a short historical review of the notion of *representative design*, Brunswik's term for an experimental design that aims for a veridical representation of the environment in which organisms naturally perform. Then we ask whether representative design matters for the results obtained. Lastly, we identify a key conceptual difficulty of representative design, namely, the issue of how to define the reference class from which situations are sampled. We demonstrate how different reference classes may lead to different conclusions and discuss possible solutions for this problem.

BRUNSWIK'S CRITIQUE OF PSYCHOLOGISTS' WAY OF CONDUCTING BUSINESS

The methodological dictate that has ruled experimental practices since the birth of experimental psychology is *systematic design*. According to its rationale, experimenters select and isolate one or a few independent variables that are varied systematically, while holding all other variables constant or allowing them to vary randomly. Experimenters then observe the resulting changes in the dependent variable(s), thus hoping to identify cause–effect relationships. Systematic design bets on internal validity, that is, on the sound demonstration that a causal relationship exists between two or more variables, rather than on external validity, which refers to the generalizability of the causal relationship beyond the experimental context. After all, so the logic goes, if internal validity is not guaranteed, no conclusions can be drawn with confidence about the effects of the independent variable. This logic was espoused by the founding fathers of experimental psychology, for instance, by Hermann Ebbinghaus, and has frequently been propagated in textbooks on experimental psychology. Take, for example, Woodworth's (1938) classic textbook *Experimental Psychology*, also known as the *Columbia Bible*, which promoted systematic design as *the* experimental tool (see Gillis & Schneider, 1966, for the historical roots of systematic design).

Brunswik opposed psychology's default experimental method (Kurz & Tweney, 1997). He questioned both the feasibility of disentangling variables and the realism of the stimuli created in doing so. In what he considered to be the simplest variant of systematic design, the one-variable design, Brunswik (1944, 1955, 1956) pointed to the fact that variables were "tied," thus making it impossible to determine the exact cause of an observed effect. To illustrate this, he referred to classic experiments in psychophysics. Using the Galton bar, experimenters presented lines of different lengths and requested observers to estimate their size. The distance between the experimental and the observed stimuli is held constant. Consequently, physical and retinal size are artificially "tied" because a small object will project a smaller retinal size and a large object will project a larger retinal size. Of course, such ties can be disentangled by using a more sophisticated variant of systematic design, namely, factorial design. Here, the levels of one variable (e.g., physical size) are combined with the levels of another variable (retinal size), exhausting all possible combinations. Brunswik criticized factorial design on the grounds that even if two or more variables are untied, each variable still remains tied to many other factors that may affect the organism's achievement. In his view, complete systematic isolation of one variable as the crucial factor would involve the combination of this variable with a "very large, and in fact indefinite, number of originally tied situational variables" (Brunswik, 1955, p. 197). He also argued that factorial design, because it aims to control for all variables that are not being investigated, destroys the natural covariation among variables. Therefore, factorial design inhibits the researcher's ability to examine the level of achievement that an organism reaches within its habitat.

Probabilistic Functionalism and Representative Design

Brunswik's methodological convictions were closely intertwined with his theoretical outlook (Hammond & Stewart, 2001; in particular, Kurz &

Hertwig, 2001). To fully appreciate Brunswik's criticism, it is helpful to view it in combination with the theoretical framework that he advocated. In his theory of *probabilistic functionalism*, Brunswik (1943, 1952) argued that the real world is an important consideration in experimental research because psychological processes are adapted in a Darwinian sense to the environments in which they evolve and function (Hammond, 1996). To the extent that psychology's objective is to study the adjustment of organisms to the environment in which they actually live, any test that is implemented to study adjustment should, according to Brunswik (1944), ensure "that the habitat of the individual, group, or species is represented with all of its variables, and that the specific values of these variables are kept in accordance with the frequencies in which they actually happen to be distributed" (cited in Hammond & Stewart, 2001, p. 69).

The ecology that an organism adapts to is not perfectly predictable for the organism (Brunswik, 1943). A particular distal stimulus, for instance, does not always imply the same specific proximal effects. Sometimes a specific proximal effect does not occur despite the presence of the distal stimulus. Similarly, particular proximal effects do not always imply a specific distal stimulus because sometimes the same effect may be caused by other distal stimuli. Proximal cues are therefore only probabilistic indicators of a distal variable. Brunswik (1940, 1952) proposed measuring the ecological (or predictive) validity of a cue by the correlation between the cue and the distal variable. The proximal cues are themselves interrelated, thus introducing redundancy (or intra-ecological correlations) into the environment. This redundancy, in turn, is the basis for *vicarious functioning* – a principle that Brunswik (1952) considered as being the foundation on which the adaptive system is built. Since a cue is not always present, an adaptive system has to rely on multiple cues that can be substituted for each other. Systematic design, with its policy of isolating and controlling variables, risks destroying the causal texture of the environment to which an organism has adapted (Brunswik, 1944), and thus, ultimately, the process of various functioning. To keep the process intact, Brunswik (1956) thought it necessary to sample experimental situations that are representative of a defined population of stimuli in terms of their *number, values, distributions, intercorrelations,* and *ecological validities* of their variable components. Otherwise, the obtained results are no longer representative of the organism's actual functioning in its habitat.

Brunswik (1955) suggested three ways of achieving a representative design. The first and preferred way is by *random sampling* (also referred to as situational, representative, and natural sampling) of stimuli from a defined population of stimuli (reference class) to which the experimenter wishes to generalize the findings. This is one form of what we today call probability sampling, where each stimulus has an equal probability of being selected. Note that time sampling is not synonymous with random sampling because intervals may be sampled systematically (for a review of this sampling method, see Czikszentmilhalyi & Larson, 1992). Brunswik's study that we described at the beginning used time sampling; for another application of this method see Hogarth's (2005) chapter in the present book. The second way of achieving a representative design is through what Brunswik (1955, 1956) called the *canvassing* of stimuli and what today is known as nonprobability sampling, namely, stratified, quota, proportionate, or accidental sampling (Baker, 1988). However, these procedures only provide a "primitive type of coverage of the ecology" (Brunswik, 1955, p. 204). Perhaps the most desirable, yet least feasible, way of achieving a representative design aims for a *complete coverage* of the entire population of stimuli.

The fundamental shift in the method of psychology that Brunswik (1943) has called for did not occur. His contemporaries were united in their wish to maintain the status quo (Postman, 1955). Despite the fact that his ideas were published in leading journals, they were largely ignored, misunderstood, and treated with skepticism and hostility (Feigl, 1955; Hilgard, 1955; Hull, 1943; Krech, 1955; Postman, 1955). His colleagues' hostile response to the notion of representative design is not without irony when one considers that sampling (albeit the sampling of subjects, not of objects) has been considered a *sine qua non* of psychological experimentation. In fact, Brunswik (1943, 1944) called attention to the "double-standard" in the practice of sampling in psychological research, pointing out that the entire problem of generalization was thrown "onto the responder rather than onto the situation" (Brunswik, 1955, p. 195). To quote Hammond (1998):

Why, he wanted to know, is the logic we demand for generalization over the subject side ignored when we consider the input or environment side? Why is it that psychologists scrutinize subject sampling procedures carefully but cheerfully generalize their results – without any logical defense – to conditions outside those used in the laboratory. (p. 2)

It seems that many decades later, the same double-standard lives on, and the issue of generalization still defies solution by the conventional methods of experimental design. In their incisive critique of current social psychological experimentation, Wells & Windschitl (1999) point to the neglect of stimulus sampling as a "serious problem that plagues a surprising number of experiments" (p. 1115). Stimulus sampling, so they argued, is imperative whenever individual instances within categories (e.g., gender or race) vary from one another in ways that affect the dependent variable. For example, relying on only one or two male confederates to test the hypothesis that men are more courteous to women than to men "can confound the unique characteristics of the selected stimulus with the category," and "what might be portrayed as a category effect could in fact be due to the unique characteristics of the stimulus selected to represent that category" (p. 1116). Indeed, in his own studies on social perception, Brunswik advocated the importance of presenting respondents with a representative sample of "person-objects" (a phrase referring to the person to be judged) (Brunswik, 1945, 1956; Brunswik & Reiter, 1937; for work in social perception in the tradition of Brunswik, see Funder, 1995).

In conclusion, Brunswik stressed two major shortcomings of the systematic design that he hoped to remedy with the representative design: First, systematic design does not allow researchers to elicit and study the process of vicarious functioning – the defining mark of an adaptive system. Second, as a consequece, experimenters who rely on systematic design cannot generalize their findings to the organisms' natural ecology. Next, we discuss the extent to which sampling experimental objects is as crucial for the results obtained as Brunswik believed.

DOES SAMPLING OF EXPERIMENTAL STIMULI MATTER?

Although Brunswik (1956) spelled out the theoretical rationale for representative design, he could not point to evidence indicating that the way stimuli are sampled affects the results obtained. In the hope that some fifty years later such evidence would be available, Mandeep Dhami and ourselves compared the effects of systematic and representative designs in various lines of investigation, namely, policy-capturing research and research on overconfidence and on the hindsight bias (Dhami, Hertwig, & Hoffrage, 2004). In the following, we briefly summarize our main findings.

Do Judgment Policies Differ in Representative versus Systematic Designs?

The key goal of many studies in the tradition of *policy-capturing* research has been to pin down how people process (e.g., combine or weight) cues to judge real-world problems – for instance, the degree to which patients have a mental health problem, the amount of bail to be set on a number of cases, the quality of shopping centers, or whether it was safe for a car to cross an intersection. To this end, participants are typically required to make decisions on a set of either real or hypothetical cases each with a corresponding set of cues. A person's judgment policy is then inferred, traditionally, using a multiple linear regression analysis and is characterized, for instance, in terms of the number and weight of cues used to make judgments. In addition, achievement is frequently measured in terms of the correlation between a person's judgments and the criterion values and by comparing the person's policy with a model of the task (for overviews, see Cooksey, 1996; Stewart, 1988).

Many researchers who aim at capturing policies have adopted the Brunswikian approach to study judgment and decision making. They have expressed a commitment to the method of representative design, which, they argue, differentiates them from other researchers in cognitive psychology, in general, and judgment and decision making, in particular (e.g., see Brehmer, 1979; Cooksey, 1996; Hastie & Hammond, 1991, p. 498). In light of their explicit commitment to the method of representative design, we were surprised to find that a large proportion of studies (those that relied on formal situational sampling; see Hammond, 1966) often failed to represent the ecological properties toward which generalizations were intended. For instance, researchers rarely combined cues to preserve their intercorrelations – an essential condition for the operation of vicarious functioning.

Possibly, the most rigorous test of the effect of representative design is a within-subjects comparison of policies captured for individuals under both a representative and an unrepresentative condition. Not surprisingly, given the frequent failure to implement representative design, we found only two published studies that aimed for such a test. The first study examined how livestock experts judge the breeding quality of pigs. In an unrepresentative condition, Phelps & Shanteau (1978) asked experts to respond to descriptions of hypothetical cases comprising a fractional factorial combination of eleven cues indicative of the breeding quality of pigs. Two months later, now partaking in a representative condition, the same experts rated the overall quality of eight pigs in photographs and provided ratings on the eleven cues. The experts' inferred policy differed across conditions: In the unrepresentative condition, experts used markedly more cues (i.e., nine to eleven) than in the representative condition (i.e., fewer than three) as suggested by the number of statistically significant cues in the captured policies. In the second rigorous test of the impact of sampling procedure on the captured policy, Moore & Holbrook (1990) studied the car-purchasing policies of MBA students and found a significant difference between the weights attached to two cues in individuals' policies captured using representative and unrepresentative stimuli.

In addition, Dhami et al. (2004) found a small set of other studies that compared policies captured under representative and unrepresentative conditions, albeit less stringently. Overall, the findings in this set are mixed (for details, see Dhami et al., 2004, Table 5). Whereas some researchers found differences in the policies captured using representative and unrepresentative cases (Ebbesen & Konecni, 1975, 1980; Ebbesen, Parker, & Konecni, 1977; Hammond & Stewart, 1974), others concluded that there are no differences (Braspenning & Sergeant, 1994; Olson, Dell'omo, & Jarley, 1992; Oppewal & Timmermans, 1999). When differences were observed, they occurred on multiple dimensions including those that Phelps & Shanteau (1978) and Moore & Holbrook (1990) identified, that is, the number and weight of cues.

Because of the small number of studies, it is difficult to draw any definite conclusions regarding the effects of representative design in policy-capturing studies. Fortunately, however, in Dhami et al. (2004), we could turn to two other lines of research that have investigated the impact of sampling experimental stimuli, namely, research on the overconfidence effect and the hindsight bias. What caused researchers' interest in this issue?

How Rational Do People Appear in Representative and Systematic Designs? The Case of Overconfidence and Hindsight Bias

In Brunswik's understanding of psychology as a functionally oriented science, one should study the adjustment of organisms to the environments in which they actually live and the resulting level of the organisms' achievements. Although many contemporary psychologists would not necessarily share Brunswik's commitment to a functional perspective, akin to his program, they aim to describe and explain people's cognitive and behavioral achievements. However, the empirical results are typically not couched in terms of adjustment and achievement, but, for instance, in terms of people's *rationality* or lack thereof. For an illustration, take the heuristics-and-biases program (e.g., Kahneman, Slovic, & Tversky, 1982; Kahneman & Tversky, 1996), which is, on many accounts, the most influential research program within cognitive and social psychology over the past thirty years.

Since its inception in the early 1970s, the heuristics-and-biases program has produced a large and growing collection of findings demonstrating that human reasoning frequently departs from classic norms of rationality. These findings include insensitivity to sample size, base-rate neglect, misperceptions of chance, illusory correlations, overconfidence, and hindsight bias. Such "cognitive illusions" have been explained in terms of simple heuristics that people, being cognitively limited, need to rely on when they make inferences about an uncertain world (Tversky & Kahneman, 1974). Resting their judgment on the seemingly ubiquitous cognitive illusions, many researchers have arrived at a bleak assessment of human reasoning – as nothing more than "ludicrous," "indefensible," and "self-defeating" (see Krueger & Funder, 2004).

Like Brunswik, the heuristics-and-biases program has stressed that people need to function in an inherently uncertain world. In Brunswik's words, the "environment to which the organism must adjust presents itself as semierratic and that therefore all functional psychology is inherently probabilistic" (Brunswik, 1955, p. 193). If so, and this forms the core of representative design, then the statistical properties of the laboratory task need to represent the statistical properties of the ecology to which the results are to be generalized. Concerns about whether this has been true in studies of cognitive illusions have given rise to a Brunswikian perspective, first in research on the overconfidence effect and then in research on the hindsight bias.

The overconfidence effect plays a prominent role among the cognitive illusions catalogued by the heuristics-and-biases program (Kahneman & Tversky, 1996). It has received considerable attention both within psychology (for a recent review, see Hoffrage, 2004) and beyond [e.g., in economics – see Hertwig & Ortmann (2004); and in consumer decision making – see Alba & Hutchinson (2000)]. Studies in psychology that demonstrate the overconfidence effect typically present respondents with questions that test their general knowledge, such as "Which city has more inhabitants: Atlanta or Baltimore?" Participants are asked to select the correct option and then indicate their confidence that it is indeed correct. The frequently replicated finding is that among choices about which people say they are 100% confident, only about 80% are correct. Similarly, among choices about which people deem themselves 90% confident, only about 75% are correct, and so on. In quantitative terms, the overconfidence effect is usually defined as the difference between the mean confidence rating and the mean percentage correct across a series of such general knowledge questions.

Are people really as out of touch with the accuracy of their knowledge as the host of overconfidence studies suggests? Adopting a Brunswikian perspective, and thus paying special attention to how overconfidence researchers sampled general knowledge questions, Gigerenzer, Hoffrage, & Kleinbölting (1991) challenged this conclusion. According to their theory of *probabilistic mental models*, people respond to questions, such as which city has more residents, by constructing a mental model that contains probabilistic cues and their validities. For instance, when choosing between Atlanta and Baltimore, a person may retrieve the fact that Atlanta has one of the world's fifty busiest airports and that Baltimore does not, and that cities with such an airport tend to be larger than those without. Capitalizing on the knowledge of this cue, the person thus concludes that Atlanta has more inhabitants than Baltimore, and then states the cue's validity as her subjective confidence that this choice is correct.

At this point, the way in which experimenters sample questions becomes crucial. To illustrate this, let us assume that only one cue, the airport cue, can be retrieved, upon which basis the relative population size of U.S. cities is inferred. Among the fifty largest U.S. cities, the airport cue has an ecological validity of .6 (Soll, 1996). The ecological validity of a cue is defined as the percentage of correct choices rendered possible by this cue. If the participants' assessment of the validity of the cue approximates its ecological validity, then they should be well calibrated, that is, they do not overestimate the accuracy of their knowledge. This means that given a confidence category (e.g., 60%), the relative frequency of correct choices should equal this value (here, 60%). This, however, only holds true under two conditions: The first is that people's subjective cue validities approximate the ecological validities – an assumption that is consistent with a rich literature demonstrating that people seem to be keenly sensitive to environmental frequencies (e.g., Hasher & Zacks, 1984). The second condition is that the experimenter samples questions such that the cues' validities in the experimental sample of questions are preserved. Gigerenzer et al. (1991) suggested that overconfidence stems from the fact that the second condition typically is not preserved. Specifically, they argued that researchers do not sample general knowledge questions randomly but tend to overrepresent items in which cue-based inferences would lead to wrong choices. If so, then overconfidence would not reflect fallible reasoning processes but would be an artifact of the way the experimenter sampled the stimuli and ultimately misrepresented the cue-criterion relations in the ecology. Supporting this interpretation, Gigerenzer et al. (1991, Study 1) were able to show that people were well calibrated when questions included randomly sampled items from a defined reference class (here, German cities). Percentage correct and mean confidence amounted to 71.7% and 70.8%, respectively. They were also able to replicate the overconfidence effect in a nonrandomly selected set of items – here, percentage correct and mean confidence amounted to 52.9% and 66.7%, respectively; overconfidence was 13.8%. The same pattern of findings has been independently predicted and replicated by Peter Juslin and his colleagues (e.g., Juslin, 1993, 1994; Juslin & Olsson, 1997; Juslin, Olsson, & Björkman, 1997); for further effects that are consistent with a Brunswikian perspective, such as the confidence-frequency effect, see Gigerenzer et al. (1991).

One objection that has been raised against this Brunswikian interpretation of overconfidence is that the observed differences between the selected and representative set of questions are just another manifestation of the _hard–easy effect_ (e.g., Griffin & Tversky, 1992). The hard–easy effect is the observation that overconfidence covaries with item difficulty: Difficult item sets (i.e., percentage of correct answers about 75% or lower) tend to produce overconfidence, whereas easy sets (i.e., percentage correct about 75% or higher) tend to produce underconfidence. At first glance, the hard–easy effect seems to be confounded with the sampling procedure. Specifically, sets consisting of representatively drawn items tend to be easier and confidence judgments tend to be well calibrated. In contrast, selected sets are difficult (as items often have been selected to be difficult, or even misleading) and tend to yield overconfidence.

The differential impact of item difficulty and sampling procedure, however, can be teased apart empirically: In a meta-analysis, Juslin, Winman, & Olsson (2000) conducted a review of ninety-five independent data sets with selected items and thirty-five sets with representatively sampled items. Across all selected and representative item sets, overconfidence was 9% and 1%, respectively (with 95% confidence intervals for each of the two sampling procedures of ±2%). Having statistically controlled for item difficulty, the authors pointed out that this difference could not be explained by differences in percentage correct, as has been claimed by Griffin & Tversky (1992), based on three data points. Moreover, when they controlled for end effects of the confidence scale and linear dependence between percentage correct and the over-/underconfidence score (i.e., mean confidence minus percentage correct), the hard–easy effect nearly disappeared for the representative item sets.

The impact of the item-sampling procedure is not restricted to the overconfidence effect. It also matters, for instance, for the hindsight bias – the tendency to falsely believe, after the fact, that one would have correctly predicted the outcome of an event. [For a recent collection of papers on the hindsight bias, see Hoffrage & Pohl (2003).]

Akin to research on overconfidence, Winman (1997) presented participants with general knowledge questions, such as "Which of these two countries has a higher mean life expectancy, Egypt or Bulgaria?" Before they responded to the question, participants were told the correct answer, Bulgaria, and then were asked to identify the option they would have chosen had they not been told the correct answer. Winman (1997) presented participants with selected and representative sets of questions. In the latter, the countries involved were drawn randomly from a specified reference class. The differences in hindsight bias were striking: In the selected set, 42% of the items elicited the hindsight bias, whereas in the representative set only 29% did. Moreover, Winman observed that in the representative sample only three out of twenty participants reached a higher degree of accuracy in hindsight than in foresight (thus indicating the hindsight bias). In the selected sample, in contrast, fourteen out of twenty participants fell prey to the hindsight bias (in terms of higher hindsight accuracy); for more detailed analyses and additional data regarding the impact of sampling of items on the hindsight bias, see Winman and Juslin's (2005) chapter in the present volume.

In conclusion, in Brunswik's view, psychology is the study of the adjustment of organisms to environments and the resulting degree of achievement. A prominent research program in contemporary psychology, the heuristics-and-biases program, also focuses on achievement, measured in terms of the degree to which the cognitive system is able to reason in accordance with the laws of statistics and probability theory. For Brunswik, the experimental stimuli used to measure the organisms' degree of achievement need to be selected so that the sample is representative of the actual demands that the environment makes upon the organism. Recent research on the overconfidence effect and the hindsight bias has demonstrated that people's cognitive achievement, here in terms of the veridical assessment of their knowledge, depends strongly on how the experimental stimuli are sampled. Thus, experimenters' conclusions about how much or little people achieve in their uncertain inferences, and how rational or irrational they are, appear to depend also on how experimenters select the stimuli that constitute the reality of the laboratory.

REPRESENTATIVE DESIGN AND SIZE OF THE REFERENCE CLASS

Sampling of experimental stimuli matters. Overconfidence, or the lack thereof, for instance, appears to be a function of whether experimental stimuli are randomly drawn from a specified reference class. In the following, we examine the boundary condition of this observation. Specifically, we examine whether the overconfidence phenomenon is robust across different sizes of the reference class. According to the theory of probabilistic mental models (Gigerenzer et al., 1991), people are well adapted to their natural environments, and they are able to estimate cue validities with a reasonable degree of accuracy. But what if cue validities change as a function of different sizes of the reference class, and if so, which reference class is the one to which people are adapted?

Cue validities can depend on the size of the reference class. To appreciate this, consider the hypothetical environment of objects in Figure 16.1. In this environment, there are six objects with their cue values on one dichotomous cue: Objects A and C have a cue value of "1," and Objects B, D, E, and F have a cue value of "0". Cue values are coded such that in a comparison between two objects with values of "1" and "0," the object with the value of "1" is more likely to be larger. The validity of the cue (as computed in the complete paired comparison) is 87.5% (i.e., seven out of eight inferences are correct). How would this validity be affected if we reduced the size of

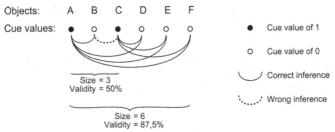

FIGURE 16.1. Six fictitious objects, ordered according to a numerical criterion, with A being the largest and F the smallest. Black circles represent a cue value of "1" (e.g., a particular feature is present), and white circles represent a cue value of "0" (e.g., a particular feature is absent). The validity of this cue depends on the size of the reference class, that is, on the number of objects larger than a specified threshold. Two objects between which the cue does not discriminate (and thus does not allow for an inference) are not connected; such a pair does not enter the computation of cue validity.

TABLE 16.1. *Validities of Twelve Cues in the Set of German Cities (Table Taken from Hoffrage, 1995)*

	Threshold			
	100,000	150,000	200,000	300,000
Size of reference class[a]	65	42	31	16
Resulting number of pairs	2080	861	465	120
Cues and their validities				
University	70	75	69	64
Intercity station	77	75	76	0
Airport	89	80	86	78
Soccer team in national league	90	82	81	65
Industrial belt	57	51	42	44
Government	84	72	76	73
License plate	92	90	84	65
Zip code	80	93	83	85
Symbol in road map	99	97	99	100
Court ("Oberlandesgericht")	80	79	86	83
National pension offices	76	71	62	87
Area (square kilometers)	82	87	85	89

[a]The size of the reference class is the number of cities with more inhabitants than the threshold.

the reference class from which objects are sampled to a subset of the largest *n* objects? (Whenever we use the term *size of the reference class*, here and in the following, we refer to the number of objects with a criterion value that is higher than a specific threshold.) It turns out that the validity depends on the threshold; specifically, for the reference classes of size 6, 5, 4, 3, and 2, the validities are 7/8, 5/6, 3/4, 1/2, and 1, respectively (see Figure 16.1). Thus, except for the reference class of size 2, validity decreases as the size of the reference class decreases.

Figure 16.1 illustrates a hypothetical environment, but what about real-world environments? Clearly, in many environments the issue of the size of the reference class will not matter. For instance, the reference class of all Nobel laureates, or nations that partook in the last Olympic Games, is well defined. Other reference classes, however, have fuzzy borders. Take the class of German cities. Though there is a strict bureaucratic definition of what counts as a city, people's subjective reference classes may take on very different sizes. If so, how would different sizes impact on cue validity? Gigerenzer et al. (1991) used all German cities with more than 100,000 inhabitants (as of 1988). Although 100,000 is a salient number, other thresholds might have been used. Indeed, as Table 16.1 shows, the cue validities in this environment depend on this threshold, that is, on the minimum number of inhabitants a city must have to be included in the set.

Across four different thresholds, cue validities varied widely: For one of the twelve cues, the validity dropped from 77% to 0%; for the others, the average absolute difference between the validities among all cities with more than 100,000 inhabitants and those among all cities with more than 300,000 was 10.3% (when we designed and published the studies reported in Gigerenzer et al., 1991, we were not aware of this dependency). The average Pearson and Spearman correlations between the cue validities (first computed across the twelve cues within a given pair of thresholds, and then averaged across the six possible pairs of thresholds that can be constructed from the four thresholds displayed in Table 16.1) were .66 and .62, respectively.

The observation that cue validities depend on the specific size of the reference class as determined by the experimenter can matter for the interpretation of empirical results and for the selection of experimental materials. In the following, we illustrate these points in the contexts of overconfidence research (Study 1) and of a computer simulation of different inference heuristics (Study 2).

Study 1: Over-/Underconfidence Depends on the Size of the Reference Class

The first study was designed both as a test of the recognition heuristics and to investigate the effect of different sizes of the reference class on overconfidence. The recognition heuristic is an inference heuristic that can be applied to infer, for instance, which of two cities has more inhabitants. Its policy is as follows: "If one of two objects is recognized and the other is not, then infer that the recognized object has the higher value with respect to the criterion" (Goldstein & Gigerenzer, 2002, p. 76).

Austrian students (from the University of Salzburg, $N = 60$) were asked to decide for 100 pairs of U.S. cities which of two cities has more residents (Hoffrage, 1995, Study 4). For ten participants, these pairs were created by combining cities that were *randomly* sampled from the set of all ($n = 72$) cities with more than 200,000 inhabitants (as of 1988, henceforth the *large* set); for another ten participants, the cities were randomly sampled from the set of those ($n = 32$) cities with more than 400,000 inhabitants (henceforth the *small* set). After making their decisions and judging their confidence, participants were asked to state the following for each city:

(i) whether they had any knowledge about the city beyond mere name recognition (henceforth denoted by K, for knowledge);

(ii) whether they recognized the city's name, but had no more knowledge about it (R, for recognition); or

(iii) whether they had never heard of the city (U, for unknown).

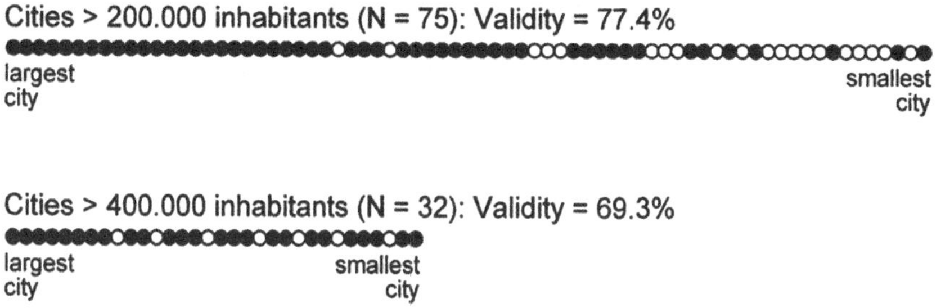

Cities > 200.000 inhabitants (N = 75): Validity = 77.4%

largest
city

smallest
city

Cities > 400.000 inhabitants (N = 32): Validity = 69.3%

largest
city

smallest
city

FIGURE 16.2. Austrian participants' recognition values of U.S. cities. Thirty participants provided recognition judgments about the largest seventy-five cities (top) and another thirty participants about the largest thirty-two cities (bottom). Black circles represent cities that at least half of the participants recognized (regardless of whether they also had some knowledge beyond mere name recognition); white circles represent cities that less than half of the participants recognized. The validity of the recognition cue was first computed for each participant separately and then averaged across participants.

Because the experimenter did not control which objects (of the large set and of the small set) were paired together, these two groups were tested in an approximation to a representative design. For the other participants, the experimenter systematically manipulated how the cities were paired. These forty participants first stated their recognition knowledge for each city: twenty participants for all cities from the large set and another twenty for all cities from the small set. For half of each group of twenty, the comparisons were constructed such that the recognition heuristic often discriminated; for the other half they were constructed such that the heuristic rarely discriminated.[1]

Figure 16.2 represents the knowledge of the average participant about the American cities. Specifically, the black circles represent cities that at least half of the participants recognized irrespective of whether they also had some knowledge beyond mere name recognition; the white circles represent the cities that less than half of the participants recognized. As one can see, the more inhabitants a city has, the higher the likelihood is that the name is recognized. The validity of the recognition knowledge is computed among all possible combinations of two cities where one is

[1] For each participant who was given the comparisons such that the recognition heuristic often discriminated, the types of comparisons (and their frequencies) were the following (types in which the recognition heuristic discriminated are in italics): K–U (25), R–U (30), K–R (30), K–K (5), R–R (5), and U–U (5). For participants who were given the comparisons such that the recognition heuristic rarely discriminated, the composition of pairs was as follows: K–U (5), R–U (20), K–R (20), K–K (5), R–R (30), and U–U (20). Note that in addition to the cases in which the recognition heuristic discriminated (K–U and R–U), knowledge about cue values could also lead to discrimination (potentially in K–R and K–K comparisons).

TABLE 16.2. *Mean Confidence, Percentage of Correct Inferences, and Over-/Underconfidence as a Function of the Size of the Reference Class and the Discrimination Rate of the Recognition Cue*

Discrimination Rate of Recognition Cue	Mean Confidence	Percent Correct	Over-/Under-Confidence
Large set (all cities > 200,000)			
Chance	70.7	72.3	−1.6
High	67.4	69.4	−2.0
Low	64.2	63.9	0.3
Small set (all cities > 400,000)			
Chance	73.6	61.5	12.1
High	74.9	65.5	9.4
Low	68.1	58.2	9.9

recognized and the other is not. The validity is the percentage of pairs in this set for which the recognized city has more inhabitants than the one that is not recognized (i.e., black–white combinations where the black circle is to the left). This validity was computed for each individual participant. Averaged across all participants who received the large set, the validity was 77.4%. For the small set, in contrast, this value was only 69.3%.

Did the participants' percentage of correct inferences show the same relationship? Yes. For each of the three discrimination conditions (cities randomly sampled irrespective of their recognition values, high discrimination rate, low discrimination rate), not only the validity of the recognition cue (Figure 16.2) but also the percentage of correct inferences (Table 16.2) was higher for the large set.

What about confidence and, thus, overconfidence? Were participants aware that the validity of the recognition cue (and probably also of the other cues they used) depends on the size of the reference class and did they adjust their confidences accordingly? In fact, mean confidences were different for the two sizes of the reference class. However, the difference points in the opposite direction: For each discrimination rate, confidences were higher for the small set (Table 16.2). That is, as the proportion of recognized cities increased, the validity of the recognition cue and the percentage correct decreased, whereas mean confidence increased. Taking mean confidence and percentage correct together, we can see that overconfidence disappeared for the large set, whereas there was substantial overconfidence for the small set (Table 16.2). Note that the effect of the size of the reference class on overconfidence was most pronounced in the chance set where the discrimination rate of the recognition cue had not been manipulated. However, also note that the effect could still be observed in the two conditions where the discrimination rate had been controlled for, suggesting that the effect in the chance condition was not due to only different frequency distributions of comparison types (e.g., a higher percentage of cases in which both cities were recognized in the small set compared to the corresponding percentage for the large set). For another example of different degrees of over-/underconfidence attributed to different reference classes, see Juslin, Olsson, & Winman (1998).

A closer look at the data (not shown in Table 16.2) revealed that once the type of comparison (K–U, K–R, etc.) was controlled, the participants in the same size of reference class condition did not differ with respect to mean percentage correct, mean confidence, and overconfidence. In other words, the effect of the discrimination rate (which was either quasi-experimentally observed or experimentally manipulated) could be fully accounted for by the different frequency distributions of the types of comparisons. Thus, the results support the hypothesis that participants used the recognition heuristic: Manipulating how often recognition discriminates between two cities affects overall performance on the group level.[2]

In summary, whether people appear overconfident in a study is a function of not only the sampling procedure (randomly versus selected) but also the size of the reference class from which the experimental stimuli are randomly drawn. In the next study, we turn to a daunting problem any experimenter faces when aiming to determine which of several cognitive policies people use. In doing so, one often realizes that it is difficult to discriminate among different policies because they frequently predict the same behavior. To illustrate this problem, let us return to the recognition heuristic.

Study 2: Policy Capturing and the Size of the Reference Class

If a person has to decide which of two cities is larger, and if the only information at hand is whether or not the person recognizes one of the cities, then that person can do little better than rely on partial ignorance, choosing recognized cities over unrecognized ones. Both the aforementioned study as well as Goldstein & Gigerenzer's (2002) studies show that people appear to rely on this judgment policy – a policy that works well when recognition

---

[2] The data allow for an even stronger test. Remember that for each city the recognition value was elicited, either before or after the comparisons were performed. In the vast majority of cases, participants decided in favor of the city they knew more about. In particular, for comparisons of the types K–U, K–R, and R–U , the percentages of decisions that were made in favor of the city with a higher recognition value were 91.1%, 79.3%, and 79.9%, respectively. Moreover, participants obtained a markedly better performance when they chose the city suggested by the recognition heuristic: The percentages of correct choices were 83.6%, 74.9%, and 65.3%, respectively. In contrast, if they decided in favor of the city they knew less about, the performance for each comparison type was even below chance level, namely, 21.2%, 48.5%, and 39.2%, respectively.

is correlated with the criterion that needs to be inferred. To explain how an association between recognition and a criterion may develop, Goldstein & Gigerenzer (2002) proposed that there are "mediators" in the environment that reflect the criterion and, at the same time, are accessible to the decision maker. For example, an American citizen may have no direct information about the population size of a German city, say, Mannheim. However, Mannheim's population size may be reflected in how often Mannheim is mentioned in U.S. daily newspapers. Frequency of mentions, in turn, is correlated with recognition: The more frequently the name of a city appears in the newspaper, the more likely it is that a reader will encounter its name. In this sense, the newspaper can serve as a mediator between recognition and the criterion (here, population size). In line with this view, Goldstein & Gigerenzer (2002) found that the ecological correlation – that is, the correlation between how often the names of German cities (with more than 100,000 residents) are mentioned in a major U.S. newspaper, the *Chicago Tribune*, and their actual populations – was .82.

As with any heuristics, the recognition heuristic is a judgment policy of limited scope: It cannot be applied when both cities are either recognized or not recognized. Another heuristic, the *fluency heuristic*, however, is applicable both when the recognition heuristic is applicable and when both cities are recognized and the recognition heuristic is thus not applicable. Like the recognition heuristic, it relies on only one reason: the fluency with which the objects are reprocessed. Fluency has been shown to function as a cue across a range of judgments (e.g., Begg, Anas, & Farinacci, 1992; Jacoby et al., 1989). In the context of a two-alternative choice, such as the city-size task, Schooler & Hertwig (2005) defined the heuristic task as follows: *If one of two objects is more fluently reprocessed, then infer that this recognized object has the higher value with respect to the criterion.*

To study the performance of both the recognition heuristic and the fluency heuristic, Schooler & Hertwig (2005) implemented them within a well-known cognitive architecture, namely, ACT-R (e.g., Anderson & Lebierre, 1998). Here, we will not be concerned with the details of this implementation (for such details, see Schooler & Hertwig, 2005). The bottom line is that fluency is a function of a city's activation, and activation within ACT-R is a function of two factors: the objects' environmental frequencies, such as mentions in the newspaper, and recency of occurrence (e.g., when a city was mentioned in the newspaper).[3] Using the

---

[3] Within ACT-R, the system cannot inspect activation levels directly; that is, it cannot simply read off the activation of a record. However, Schooler & Hertwig (2005) proposed that by taking advantage of the one-to-one mapping between activation and retrieval time, the speed of retrieval could be used as a proxy for activation. Rather than assuming that the system can discriminate between minute differences in any two retrieval times, however, they allowed for limits on the system's ability to do this. Specifically, if the retrieval times of the two alternatives were within a just noticeable difference of each other, then the system guessed.

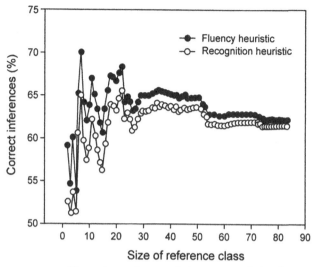

FIGURE 16.3. Percentage of correct inferences made by the recognition heuristic and the fluency heuristic as a function of the size of the reference class. Adapted from Schooler & Hertwig (2005).

environmental frequencies (i.e., mentions in the *Chicago Tribune*) and the recognition rate that Goldstein & Gigerenzer (2002) observed for the largest eighty-three German cities in a sample of students from the University of Chicago, Schooler & Hertwig (2005) calculated the cities' activations. This was carried out to ensure that performance based on the recognition rate of the model and empirical recognition was in agreement. Having thus ensured this correspondence, they could now analyze the performance of the recognition heuristic and the fluency heuristic, respectively.

Recall the policies of the two heuristics in question: For all possible pairs of cities within a reference class, the recognition heuristic chose the recognized city when only one of the cities in a given pair was recognized, and otherwise it guessed. In contrast, the fluency heuristic chose the city that was more fluently reprocessed as long as both cities were above the recognition threshold; otherwise it guessed. Figure 16.3 shows the performance of the two heuristics as a function of an increasing size of the reference class – the dimension that concerns us here. When the reference class and thus the pair comparison only include the ten largest German cities, then the fluency heuristic clearly outperforms the recognition heuristic. Whereas the fluency heuristic scores 70.1% correct inferences, the recognition heuristic scores 62%. This picture, however, changes as the reference class becomes more encompassing. The performance edge drops to a 5% difference when the twenty largest cities are included (69% versus 64.6% correct inferences), and it shrinks to a 1% advantage when all eighty-three largest German cities are included.

Why does the performance accuracy level of the heuristics converge with more encompassing reference size? The relative performance edge the fluency heuristic has over the recognition heuristic stems from the frequency of pairs in which both objects are recognized. It is only here that the two heuristics arrive at different judgments – when one object is recognized and the other is not, they behave identically, and when neither object is recognized, both heuristics need to guess. The likelihood, however, that both cities are recognized is particularly high for the very large cities (whose names appeared relatively frequently in the *Chicago Tribune*) and shrinks as the size of the reference class increases.

This finding highlights the possible performance dependence of the two heuristics on the reference class upon which they are tested. Had their performance been tested by relying only on the ten or twenty largest cities, then the fluency heuristic would have been judged to be markedly more accurate than the recognition heuristic. Indeed it is – but merely in the small set of large to very large German cities. Across a less restricted set of German cities (e.g., all cities with more than 100,000 residents), the strategies' performances are almost indistinguishable and there is no clear winner. This outcome thus demonstrates that the performance of heuristics, as well as ultimately the experimenter's conclusion, also depends on the size of the reference class.

Trying to distinguish between heuristics can be a daunting endeavor. To appreciate this, consider two experimenters who try to discover which of the two heuristics – the recognition heuristic or the fluency heuristic – people tend to use. To do so, the first experimenter uses all eighty-three German cities (with more than 100,000 residents), thus composing 3,403 pair comparisons. In this case, to find out whether a person's judgment policy is akin to the recognition heuristic or the fluency heuristic, respectively, will be extremely difficult, because in the majority of the 3,403 comparisons, both heuristics will arrive at the same predictions. The proportion of discriminatory cases will be minuscule, and the performance level of users of the recognition heuristic and users of the fluency heuristic will be hardly distinguishable. The second experimenter, in contrast, uses only the ten largest German cities, thus composing only 45 comparisons. Among them, the proportion (though not the absolute number) of discriminatory cases will be much larger, and the level of accuracy reached by users of the recognition heuristic and the fluency heuristic, respectively, will be markedly different.

This thought experiment suggests that the sampling procedure from the reference class is not the only thing that matters for the results obtained. The size of the reference class may also have drastic implications for, for instance, the observed accuracy level of heuristics and the likelihood with which an experimenter is able to distinguish among users of different heuristics.

DISCUSSION

We have discussed several conceptual and methodological implications of different sampling procedures in psychological experiments. In the first section, we briefly reviewed the historical and methodological foundations of representative design. In the second section, we introduced three lines of research – on policy capturing, overconfidence, and hindsight bias – in which empirical evidence has been accumulated indicating that the sampling procedure matters for the results obtained. Specifically, it matters whether the experimental stimuli are randomly sampled from a defined reference class, or whether the experimenter samples the objects systematically, thus causing the frequency distribution and informational structures in the sample and the population to diverge. After having shown that the *how* of sampling experimental stimuli matters, in the third section we turned to an issue that has rarely been addressed in the literature on representative design: Apart from the sampling procedure, the size of the reference class from which stimuli are sampled also matters for the results obtained. Clearly, both parties in an experiment – the participants and experimenters – cannot help but to settle on a reference class. The question is whether they will settle on the same one. We conclude with a discussion of the issues raised.

Selection of the Reference Class: A Time-Honored and Ubiquitous Problem

The problem of choosing the adequate reference class is neither trivial nor is it new. It is, for instance, fundamental to the frequentistic interpretation of probabilities (for the historical routes and interpretations of probabilities, see Gigerenzer et al., 1989). Conceptualizing probabilities in terms of relative frequencies requires specifying the reference class within which we count the objects or the events in question. Logic demands that the specification of the reference class precedes the counting of designated objects within it. Or, in the words of the great probability theorist Richard von Mises (1957), "We shall not speak of probability until a collective has been defined" (p. 18).

Take, for illustration, the problem of calculating the probability of a person dying within, say, the next ten years. Insurance companies need to estimate such probabilities to calculate a person's premium. But which of the person's innumerable properties – age, sex, profession, health, income, eating habits, family life, to name only a few – should be used to construct a reference population? Each of these and many other properties (as well as combinations thereof) could be used to define the reference class, and in all likelihood, many of the resulting reference classes would yield different statistics and thus different estimations for mortality risks, leaving open the question of which is the correct one (von Mises, 1957; for an example of different types of reference classes for probabilities, see Gigerenzer et al., 2005). To the best of our knowledge, neither probability theorists nor insurance companies have yet been able to provide a solution to this problem.

Selection of the Reference Class in Psychological Theory and Experimental Practice

The theory of probabilistic mental models predicts that representative sampling of experimental stimuli from a specified reference class results in well-calibrated confidence judgments (Gigerenzer et al., 1991). In this chapter, we have shown that this prediction may overlook an important additional condition for good calibration. Specifically, we have illustrated that cue validities can vary markedly as a function of the inclusiveness of the reference class. To the extent that confidence judgments rest on cue validities, they are only well calibrated when participants are sensitive to the inclusiveness of the reference class. Arguably, this is not a very plausible expectation. Thus, for overconfidence to be eliminated in psychological experiments, two conditions need to be met: (1) representative sampling and (2) representative sampling from *the* reference class from which people's knowledge of cues and cue validities stem. Therein, of course, lurks the problem: Which reference class is chosen, and what is its size? Moreover, how can the experimenter know the "right" reference class and its size?

Frankly, we do not have answers to these questions. Yet, let us suggest some directions in which one may search for answers. First and foremost, let us stress that we do not believe that the problem of the reference class is unique to the theory of probabilistic mental models. Rather, the problem of the "right" reference class permeates all of psychology. In principle, any psychological theory has to provide an answer as to which reference class of objects, stimuli, and situations it is meant to apply. For instance, any theory of classification and object perception ought to be explicit about the reference class of objects and properties in the world and in people's minds to which it applies. In reality, to the best of our knowledge, hardly any psychological theory tackles the reference class issue. Perhaps, the reason is that there seems to be no good answer – for purely logical reasons. For each reference class in relation to which a sampling process is representative, there exists a superordinate or a subordinate reference class relative to which the outcome of the sampling process is simply biased. For instance, sampling representatively from the 83 largest German cities results in a sample that is unrepresentative with respect to the 500 largest German cities or to the 10 largest German cities.

Does this mean that representative design is a chimera – a creature that exists in the fantasy of some experimenters, but for all practical purposes is nonviable? We do not think so. We believe that there are possible pragmatic routes toward a "good enough solution." Under some circumstances, experimenters may circumvent the problems that result from fuzzy reference classes either by selecting one that is small, finite, and complete (e.g., all African states) or by creating microworlds (e.g., Fiedler et al., 2002) over which experimenters have full control. They thus can determine participants' exposure to these worlds and make sure that the intended reference class and the participants' reference class converge.

Another route to determining the size of the reference class is to explore its boundaries empirically, for instance, by analyzing environmental frequencies. Anderson & Schooler (1991) examined a number of environmental sources (e.g., the *New York Times* and electronic mail) to show that the probability that a memory for a particular piece of information will be needed shows reliable relationships to frequency, recency, and patterns of prior exposure. Such an analysis of environmental statistics could also be conducted in the context of overconfidence research. For the German city task, for instance, it may show that people are much more likely to encounter larger cities, such as *Frankfurt* and *Stuttgart* (ranked 5 and 8), than smaller cities, such as *Leinefeld* and *Zeulenroda* (ranked 345 and 403). To the extent that environmental frequencies are also indicative of how often an object is the subject of an inference, we suggest that decision makers rarely need to make comparisons among items drawn from the lower tail of a criterion distribution.

In the context of social cognition, an empirical approach to determining the "right" size of *social* reference classes may capitalize on empirical studies indicating that the size of people's "sympathy" group is in the order of 10–15 people (i.e., number of friends and relatives whom they contact at least once a month), that the typical size of the network of acquaintances is around 135–150, and that the number of people whose faces one can attach names to is around 1,500–2,000 (see Dunbar, 1996; Hill & Dunbar, 2003). On the basis of these and similar quantities, researchers investigating social cognition infer plausible estimates of the size of people's social reference classes.

Yet another way to determine the "right" size of people's reference classes is to transfer the task of sampling experimental stimuli from the experimenter to the participants. In his aforementioned investigation of size constancy, Brunswik (1944) took exactly this approach (see also Hogarth, 2005, this volume). Specifically, he randomly interrupted the perceiver in her flow of daily activities at randomly chosen intervals, thereby letting the perceiver, the environment, and chance determine which environmental stimuli are designated to become experimental ones.

EPILOGUE

Participants in psychological experiments typically respond to stimuli selected by the experimenter. Each of these stimuli in and of itself constitutes reality. However, is this slice of reality in the experiment a sample of the reality to which generalization is intended? According to Brunwik, the answer is often "no." Owing to psychologists' high esteem of internal validity, experimental studies, according to Brunswik, often end up investigating phenomena at the fringe of reality. To remedy this situation, he advocated an entirely different approach to the sampling of stimuli: an approach in which experimenters apply the same standards to the sampling of stimuli that they espouse when they sample participants. We believe Brunswik's criticism of experimental practices and his proposal for reform are as timely today as they were back then. Although Brunswik's representative design is by no means convenient, nor without its own problems, it provides us with a valuable if preliminary framework to conceptualize stimulus sampling in experiments. Clearly, as experimenters we cannot help but sample stimuli, and it is up to us to strive for even better ways of doing so.

References

Alba, J. W., & Hutchinson, J. W. (2000). Knowledge calibration: What consumers know and what they think they know. *Journal of Consumer Research, 27*, 123–156.

Anderson, J. R., & Lebierre, C. (1998). *The atomic components of thought.* Mahwah, NJ: Lawrence Erlbaum.

Anderson, J. R., & Schooler, L. J. (1991). Reflections of the environment in memory. *Psychological Science, 2*, 396–408.

Baker, T. L. (1988). *Doing social research.* New York: McGraw-Hill.

Begg, I. M., Anas, A., & Farinacci, S. (1992). Dissociation of processes in belief: Source recollection, statement familiarity, and the illusion of truth. *Journal of Experimental Psychology: General, 121*, 446–458.

Braspenning, J., & Sergeant, J. (1994). General practitioners' decision making for mental health problems: Outcomes and ecological validity. *Journal Clinical Epidemiology, 47*, 1365–1372.

Brehmer, B. (1979). Preliminaries to a psychology of inference. *Scandinavian Journal of Psychology, 20*, 193–210.

Brunswik, E. (1940). Thing constancy as measured by correlation coefficients. *Psychological Review, 47*, 69–78.

Brunswik, E. (1943). Organismic achievement and environmental probability. *Psychological Review, 50*, 255–272.

Brunswik, E. (1944). Distal focussing of perception: Size constancy in a representative sample of situations. *Psychological Monographs, 56*, 1–49.

Brunswik, E. (1945). Social perception of traits from photographs. *Psychological Bulletin, 42*, 535.

Brunswik, E. (1952). The conceptual framework of psychology. In *International Encyclopedia of Unified Science* (Vol. 1, No. 10, pp. iv–102). Chicago: University of Chicago Press.

Brunswik, E. (1955). Representative design and probabilistic theory in a functional psychology. *Psychological Review, 62,* 193–217.

Brunswik, E. (1956). *Perception and the representative design of psychological experiments.* Berkeley: University of California Press.

Brunswik, E., & Reiter, L. (1937). Eindrucks-Charaktere schematisierter Gesichter. *Zeitschrift für Psychologie, 142,* 67–134.

Cooksey, R. W. (1996). *Judgment analysis: Theory, methods, and applications.* San Diego: Academic Press.

Czikszentmihalyi, M., & Larson, R. (1992). Validity and reliability of the experience sampling method. In M. W. de Vries (Ed.), *The experience of psychopathology: Investigating mental disorders in their natural settings* (pp. 43–57). New York: Cambridge University Press.

Dhami, M., Hertwig, R., & Hoffrage, U. (2004). The role of representative design in an ecological approach to cognition. *Psychological Bulletin, 130,* 959–988.

Dukes, W. F. (1951). Ecological representativeness in studying perceptual size-constancy in childhood. *American Journal of Psychology, 64,* 87–93.

Dunbar, R. (1996). *Grooming, gossip, and the evolution of language.* Cambridge, MA: Harvard University Press.

Ebbesen, E. B., & Konecni, V. J. (1975). Decision making and information integration in the courts: The setting of bail. *Journal of Personality and Social Psychology, 32,* 805–821.

Ebbesen, E. B., & Konecni, V. J. (1980). On the external validity of decision-making research: What do we know about decisions in the real world? In T. S. Wallsten (Ed.), *Cognitive processes in choice and decision behavior* (pp. 21–45). Hillsdale, NJ: Lawrence Erlbaum.

Ebbesen, E. B., Parker, S., & Konecni, V. J. (1977). Laboratory and field analyses of decisions involving risk. *Journal of Experimental Psychology: Human Perception and Performance, 3,* 576–589.

Feigl, H. (1955). Functionalism, psychological theory, and the uniting sciences: Some discussion remarks. *Psychological Review, 62,* 232–235.

Fiedler, K., Walther, E., Freytag, P., & Plessner, H. (2002). Judgment biases in a simulated classroom – A cognitive-environmental approach. *Organizational Behavior and Human Decision Processes, 88,* 527–561.

Funder, D. C. (1995). On the accuracy of personality judgment: A realistic approach. *Psychological Review, 102,* 652–670.

Gigerenzer, G., Hertwig, R., van der Broek, E., Fasolo, B., & Katsikopoulos, K. V. (2005). "A 30% chance of rain tomorrow:" How does the public understand probabilistic weather forecasts? *Risk Analysis, 25,* 623–629.

Gigerenzer, G., Hoffrage, U., & Kleinbölting, H. (1991). Probabilistic mental models: A Brunswikian theory of confidence. *Psychological Review, 98,* 506–528.

Gigerenzer, G., Switjink, Z., Porter, T., Daston, L., Beatty J., & Krüger, L. (1989). *The empire of chance.* Cambridge, UK: Cambridge University Press.

Gillis, J., & Schneider, C. (1966). The historical preconditions of representative design. In K. R. Hammond (Ed.), *The psychology of Egon Brunswik* (pp. 204–236). New York: Holt, Rinehart and Winston.

Goldstein, D. G., & Gigerenzer, G. (2002). Models of ecological rationality: The recognition heuristic. *Psychological Review, 109,* 75–90.

Griffin, D., & Tversky, A. (1992). The weighing of evidence and the determinants of confidence, *Cognitive Psychology, 24*, 411–435.

Hammond, K. R. (1966). Probabilistic functionalism: Egon Brunswik's integration of the history, theory, and method of psychology. In K. R. Hammond (Ed.), *The psychology of Egon Brunswik* (pp. 15–80). New York: Holt, Rinehart and Winston.

Hammond, K. R. (1996). Upon reflection. *Thinking & Reasoning, 2*, 239–248.

Hammond, K. R. (1998). Representative design. http://www.brunswik.org/notes/essay3.html.

Hammond, K. R., & Stewart, T. R. (1974). *The interaction between design and discovery in the study of human judgment.* Report No. 152. University of Colorado Institute of Behavioral Science.

Hammond, K. R, & Stewart, T. R. (Eds.) (2001). *The essential Brunswik: Beginnings, explications, applications.* Oxford: Oxford University Press.

Hasher, L., & Zacks, R. T. (1984). Automatic processing of fundamental information: The case of frequency of occurrence. *American Psychologist, 39*, 1372–1388.

Hastie, R., & Hammond, K. R. (1991). Rational analysis and the lens model. *Behavioral and Brain Sciences, 14*, 498.

Hilgard, E. R. (1955). Discussion of probabilistic functionalism. *Psychological Review, 62*, 226–228.

Hill, R. A., & Dunbar, R. I. M. (2003). Social network size in humans. *Human Nature, 14*, 53–72.

Hertwig, R., & Ortmann, A. (2004). The cognitive illusion controversy: A methodological debate in disguise that matters to economists. In R. Zwick & A. Rapoport (Eds.), *Experimental business research* (pp. 361–378). Boston: Kluwer.

Hoffrage, U. (1995). *Zur Angemessenheit subjektiver Sicherheits-Urteile. Eine Exploration der Theorie der probabilistischen mentalen Modelle [The adequacy of subjective confidence judgments: Studies concerning the theory of probabilistic mental models].* Doctoral thesis, University of Salzburg.

Hoffrage, U. (2004). Overconfidence. In R. F. Pohl (Ed.), *Cognitive illusions: A Handbook on fallacies and biases in thinking, judgement and memory* (pp. 235–254). Hove, UK: Psychology Press.

Hoffrage, U., & Pohl, R. F. (Eds.) (2003). Hindsight Bias (Special Issue). *Memory, 11*, 329–504.

Hogarth, R. (2006). Is confidence in decisions related to feedback? Evidence from random samples of real-world behavior. This volume.

Hull, C. L. (1943). The problem of intervening variables in molar behavior theory. *Psychological Review, 50*, 273–291.

Jacoby, L. L., Kelley, C. M., Brown, J., & Jasechko, J. (1989). Becoming famous overnight: Limits on the ability to avoid unconscious influences of the past. *Journal of Personality and Social Psychology, 56*, 326–338.

Juslin, P. (1993). An explanation of the "hard-easy effect" in studies of realism of confidence in one's general knowledge. *European Journal of Cognitive Psychology, 5*, 55–71.

Juslin, P. (1994). The overconfidence phenomenon as a consequence of informal experimenter-guided selection of almanac items. *Organizational Behavior and Human Decision Processes, 57*, 226–246.

Juslin, P., & Olsson, H. (1997). Thurstonian and Brunswikian origins of uncertainty in judgment: A sampling model of confidence in sensory discrimination. *Psychological Review, 104*, 344–366.

Juslin, P., Olsson, H., & Björkman, M. (1997). Brunswikian and Thurstonian origins of bias in probability assessment: On the interpretation of stochastic components of judgment. *Journal of Behavioral Decision Making, 10*, 189–209.

Juslin, P., Olsson, H., & Winman, A. (1998). The calibration issue: Theoretical comments on Suantak, Bolger, and Ferrell (1996). *Organizational Behavior & Human Decision Processes, 73*, 3–26.

Juslin, P., Winman, A., & Olsson, H. (2000). Naive empiricism and dogmatism in confidence research: A critical examination of the hard-easy effect. *Psychological Review, 107*, 384–396.

Kahneman, D., Slovic, P., & Tversky, A. (1982). *Judgment under uncertainty: Heuristics and biases* (pp. 493–508). Cambridge, UK: Cambridge University Press.

Kahneman, D., & Tversky, A. (1996). On the reality of cognitive illusions. *Psychological Review, 103*, 582–591.

Krech, D. (1955). Discussion: Theory and reductionism. *Psychological Review, 62*, 229–231.

Krueger, J. I., & Funder, D. C. (2004). Towards a balanced social psychology: Causes, consequences and cures for the problem-seeking approach to social behavior and cognition. *Behavioral and Brain Sciences, 27*, 313–327.

Kurz, E. M., & Hertwig, R. (2001). To know an experimenter. In K. R. Hammond & T. R. Stewart (Eds.), *The essential Brunswik: Beginnings, explications, applications* (pp. 180–186). New York: Oxford University Press.

Kurz, E. M., & Tweney, R. D. (1997). The heretical psychology of Egon Brunswik. In W. G. Bringmann, H. E. Lueck, R. Miller, & C. E. Early (Eds.), *A pictorial history of psychology* (pp. 221–232). Carol Stream, IL: Quintessence.

Moore, W. L., & Holbrook, M. B. (1990). Conjoint analysis on objects with environmentally correlated attributes: The questionable importance of representative design. *Journal of Consumer Research, 16*, 490–497.

Olson, G. A., Dell'omo, G. G., & Jarley, P. (1992). A comparison of interest arbitrator decision-making in experimental and field settings. *Industrial and Labor Relations Review, 45*, 711–723.

Oppewal, H., & Timmermans, H. (1999). Modeling consumer perception of public space in shopping centers. *Environment and Behavior, 31*, 45–65.

Phelps, R. H., & Shanteau, J. (1978). Livestock judges: How much information can an expert use? *Organizational Behavior and Human Performance, 21*, 209–219.

Postman, L. (1955). The probability approach and nomothetic theory. *Psychological Review, 62*, 218–225.

Schooler, L., & Hertwig, R. (2005). How forgetting aids heuristic inference. *Psychological Review, 112*, 610–628.

Soll, J. B. (1996). Determinants of overconfidence and miscalibration: The roles of random error and ecological structure. *Organizational Behavior and Human Decision Processes, 65*, 117–137.

Stewart, T. (1988). Judgment analysis: Procedures. In B. Brehmer & C. R. B. Joyce (Eds.), *Human judgment: The SJT view* (pp. 41–74). North-Holland: Elsevier.

Tversky, A., & Kahneman, D. (1974). Judgment under uncertainty: Heuristics and biases. *Science, 185*, 1124–1131.

von Mises, R. (1957). *Probability, statistics, and truth.* New York: Dover.

Wells, G. L., & Windschitl, P. D. (1999). Stimulus sampling and social psychological experimentation. *Personality and Social Psychology Bulletin, 25,* 1115–1125.

Winman, A. (1997). The importance of item selection in "knew-it-all-long" studies of general knowledge. *Scandinavian Journal of Psychology, 38,* 63–72.

Winman, A., & Juslin, P. (2006). "I'm m/n confident that I'm correct": Confidence in foresight and hindsight as a sampling probability. This volume.

Woodworth, R. (1938). *Experimental psychology.* New York: Holt.

17

"I'm *m/n* Confident That I'm Correct"

Confidence in Foresight and Hindsight as a Sampling Probability

Anders Winman and Peter Juslin

Occasionally the development of a scientific discipline recapitulates the trajectory of individual development from novice to expert and, particularly, that the early stages of a science coincides with the current naïve or novice understanding of phenomena. For example, people's lay theories of physical phenomena recapitulate many of the conceptions that characterized Aristotelian physics (McCloskey, 1983). Likewise, the animistic metaphysics of prescientific "primitive cultures," attributing spirits and subjectlike qualities to (what are today considered) inanimate objects like stones and trees, is recapitulated in the young child's understanding of the environment (Piaget, 1951).

The early development of probability theory was largely linked to the *classical (m/n) definition* of probability: The probability of an event equals the ratio between the number *m* of favorable outcomes relative to the number *n* of equally possible outcomes (Hacking, 1975). Although frequency and degree of belief were separate aspects, for a long time they were considered two associated and relatively unproblematic sides of the same basic probability concept. Since then classical probability has been replaced by the two competing schools of frequentists and subjectivists (Hacking, 1975; Oaks, 1986). Whereas relative frequencies naturally preserve the extensional properties that originally constrained the development of probability theory, such as subset–set inclusions and the property of being a real number in the interval [0, 1], the interpretation as a subjective degree of belief is premised on the assumption (or hope) that the judge's beliefs conform to these extensional rules (Savage, 1954).

For decades psychological research has been dominated by the idea that probability judgments are guided by useful but error-prone heuristics triggered by intensional, subjective aspects of stimuli, thereby contributing to cognitive illusions (Gilovich, Griffin, & Kahneman, 2002; Kahneman, Slovic, & Tversky, 1982). People who rely on similarity or representativeness thus overestimate the probability of a conjunction relative to the conjuncts (Tversky & Kahneman, 1983) and ignore base rates (Kahneman & Tversky, 1982). Hence, in this view probability judgments are not only subjective but fundamentally at variance with the extensional properties that originally constrained probability theory. In this chapter we will review research based on the idea that people's subjective probability judgments are based on their assessment of a *sampling probability* that closely mirrors the classical m/n probability concept. Probability judgments are thus often guided by considering n possible states or outcomes, m of which satisfy a specific event. In principle, the n possible outcomes may be produced by imagination, but in general we expect them to be a sample of previous events in the environment that are retrieved from long-term memory. The probability judgment is then made *as if* the event of concern has been sampled randomly among the n possible outcomes (e.g., Gigerenzer, Hoffrage, & Kleinbölting, 1991). Although sophisticated in some respects, this intuitive conception of probability does not encompass the distinction between subjective and frequentistic conceptions of probability: In this sense, the mind is neither a "subjectivist" nor a "frequentist" (cf. Gigerenzer et al., 1991); rather, it performs as a naïve intuitive statistician (Chapter 1) working with the classical interpretation of probability.

The implication is that the environment affords and defines a reference class of situations that may be encountered and the subjective probability (or confidence) judgment reflects the probability of making a correct choice when situations are randomly selected from this reference class (e.g., Brunswik, 1956; Gigerenzer et al., 1991; Juslin, 1994). This means that probability judgments cannot be addressed without attention to this reference class. Specifically, we will illustrate how this simplistic hypothesis has the power to relate and explain two previously disparate phenomena in the literature: *overconfidence* and *hindsight bias* (Winman, 1997b; Winman, Juslin, & Björkman, 1998).

The key concept that relates these two phenomena is Egon Brunswik's (e.g., 1956) notion of *representative design*. If people conceive of probability judgment as a sampling probability the appropriateness of the judgments will be disclosed only when a task that is representative of the participant's natural environment is used. The hypothesis suggests that people in many ways act as intuitive statisticians, albeit naïve ones (see Chapter 1 of this volume): As in the enlightenment era, they hold an eclectic notion of probability that blurs the conceptual distinction between frequentist and subjectivist interpretations of probability.

We first discuss probability judgment in foresight that generally is elicited by assessment ("What is the probability that your answer to this general knowledge item is correct?"). Thereafter, we address probability judgments made in hindsight, with the correct alternative indicated. The latter judgments may be elicited by a retrospective assessment of confidence ("How certain in your answer would you have been if the correct answer had not been indicated?") and/or by analyzing the retrospectively produced sequence of correct and wrong answers across a set of general knowledge items.

OVER- AND UNDERCONFIDENCE IN FORESIGHT

Trust in One's Reasoning

Although already implicit in Brunswik's probabilistic functionalism (see Hammond & Stewart, 2001, for an introduction) the notion that probability judgments derive from assessment of a sampling probability surfaced most clearly in research on overconfidence in general knowledge, and particularly in the theory of probabilistic mental models (Gigerenzer et al., 1991; see also Björkman, 1994; Juslin, 1994; Soll, 1996). In overconfidence studies the participants are commonly presented with sets of selected two-alternative general knowledge items; for example, "Which country has the larger population: (a) Brazil, or (b) Indonesia?" For each question, the task is to select an answer and thereafter to assess confidence in the selected answer as a subjective probability between .5 (pure random guess) and 1.0 (perfectly confident that the correct answer is selected).

Overconfidence refers to observation of proportions correct lower than the stated confidence levels with, for example, only 70% correct when 90% confident (see Figures 17.1A and B for the selected item samples). Whereas the early interpretations emphasized information processing biases (e.g., Koriat, Lichtenstein, & Fischhoff, 1980) or positive self-serving illusions (Taylor & Brown, 1988), the theory of probabilistic mental models (Gigerenzer et al., 1991) and the other Brunswik-inspired research (Björkman, 1994; Juslin, 1994) conceived of the inference problem facing the participants as a *sampling problem*. People have knowledge of the statistical structure of real environments, allowing them to make inferences of probabilistic or statistical validity – which are true for most situations that may be encountered, but not all – and confidence is the approximate expression of the sampling probability of attaining a correct answer when applying the knowledge to a specific situation.

Thus, knowledge that out of n cases that can be encountered in the environment most (m) yield an outcome guides people to estimate the probability of a correct answer to m/n.

For example, if you do not know the population of Brazil and Indonesia, you may capitalize on knowing that Asian countries in general have large populations and guess that the Asian country has a larger population. Although this will not produce the correct answer for all n pairs of Asian and South-American countries that can be encountered, it will for most of them. In terms of psychological theory (for example, PMM-theory;

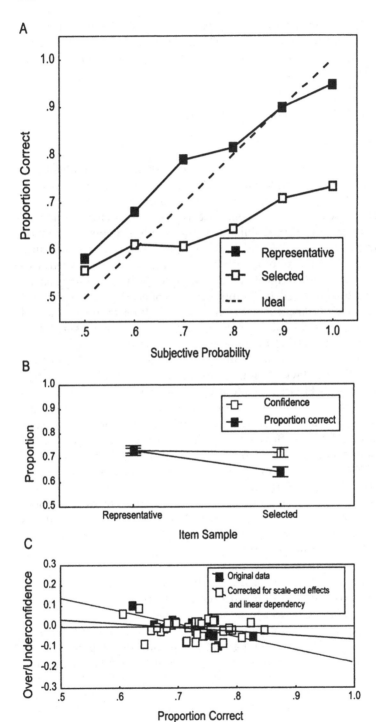

Gigerenzer et al., 1991) knowledge of the continent is said to be used as a *cue* to the target variable (population). Probabilistic cues may vary in *cue validity*; they may work more or less well. If from your previous experience you have some way of estimating the proportion m/n of cases where the correct answer will be selected this cue can be used to estimate the probability that you are correct. The suggestion by the Brunswik-inspired research on overconfidence was that confidence – partially or wholly – is an expression of such a sampling probability.

The key implication was that the way in which the experimental tasks (general knowledge items) are selected should be of crucial importance for the observed degree of realism in the confidence judgments (Gigerenzer et al., 1991; Juslin, 1994). Accurate judgments can only be observed in humans when the sampling of items in the experimental tasks does not systematically deviate from the way these humans have sampled the information in the environment. In Juslin (1994) two samples of general knowledge items were presented to the experimental participants: a *representative item sample*, where the objects of judgment (world countries) were sampled randomly from a natural environment (the countries of the world), thus implementing the sampling scheme that was hypothesized to underlie the inferences made by the participants, and a *selected item sample*, in which the items were specifically selected to be good knowledge-discriminating items, a procedure known to overrepresent the frequency of misleading and unusual facts. Twelve item-selectors naïve to the experimental hypothesis generated the selected sample. The calibration curves from twenty participants who responded to both item samples are presented in Figure 17.1A. For the representative item sample there was good calibration, but for the selected item set the proportions correct are too low, a manifestation of the overconfidence phenomenon (see Winman, 1997, and Kleitman & Stankov, 2001, for successful replications of this experiment).

To evaluate this sampling argument more generally we performed a meta-analysis (Juslin, Winman, & Olsson, 2000) based on ninety-five studies with selected item samples (e.g., Lichtenstein & Fischhoff, 1977; Ronis & Yates, 1987) and thirty-five studies with representative item samples (e.g., Doherty, Brake & Kleiter, 2001; Gigerenzer et al., 1991; Griffin & Tversky, 1992). In Figure 17.1B we see that – as predicted by

---

FIGURE 17.1. (A) Proportions of correctly selected answers to two-alternative general knowledge items plotted as a function of confidence for a representative item sample and a selected item sample (see text for explanations). Adapted from Juslin (1994). (B) Means with 95% confidence intervals from a meta-analysis of ninety-five data sets with selected item samples and thirty-five representative item samples. (C) The negative linear relationship between proportion correct (task difficulty) before and after correcting the data for linear dependency and scale-end effects. Both Panels B and C are adapted from Juslin, Winman, & Olsson (2000).

the sampling argument – mean confidences for selected and representative samples coincide and the proportion correct for representative item samples agrees with mean confidence, resulting in a close to zero over/underconfidence score. In contrast, there is substantial overconfidence for the selected item samples.

We further controlled for difficulty by a simple matching procedure for item samples with proportion correct below .7 (hard samples). For each representative item sample with a proportion correct below .7, we entered all selected item samples with the same proportion correct (i.e., as judged by two decimals). The mean over/underconfidence score with selected hard items was .10 (95% confidence interval, ±.02; proportion correct .65; $N = 29$), and the mean over/underconfidence for representative hard items was .05 (95% confidence interval, ±.02; proportion correct .65; $N = 12$). The difference between selected and representative samples is therefore observed, even if we control for proportion correct. Moreover, as illustrated in Figure 17.1C, the correlation between difficulty and over/underconfidence after the data have been controlled for artifacts (linear dependency and regression) is modest (Juslin et al., 2000).

Trust in One's Senses

So far, we have discussed beliefs in general knowledge. We may be more or less convinced that a certain fact is true by reliance on reasoning processes. Another, completely different source of uncertainty that we may experience is through the way our senses inform us of what is going on in the surroundings. Most often our senses function in a foolproof way and there is no doubt about how to interpret the information. Under scarce conditions, however, there might be an uncertainty as to what was, say, the source of a distant noise or a blurred object rapidly passing by in the dark. Psychological studies of *sensory discrimination* in the uncertainty zone are as old as experimental psychology itself. A common method is sensory discrimination with pair comparisons. A participant is, for example, presented with two almost similar weights and is asked "Which of these two presented weights is heavier: Weight a, or Weight b?" When participants in the first experimental studies reported that they had no confidence in their answers and that they were merely guessing, psychologists expected them to have equal proportions of correct and wrong responses, but this was not the case. Instead, the proportions of correct responses were consistently greater than the expected 50% (Peirce & Jastrow, 1884). Thus, it appeared that people were better than they thought they were at fine sensory discriminations. This phenomenon, which is often found for pair comparisons of sensory stimuli, has been labeled an *underconfidence bias*: People are more accurate than expected from their confidence judgments (Björkman, Juslin, & Winman, 1993; but see Baranski & Petrusic, 1994, Suantak, Bolger & Ferrell, 1996, for alternative conclusions). Figure 17.2A presents a

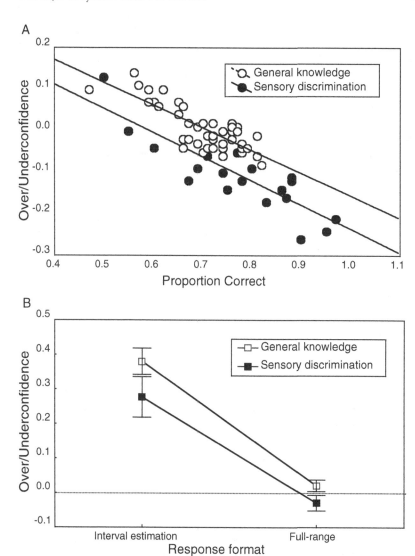

FIGURE 17.2. (A) Over/underconfidence scores (i.e., mean confidence minus proportion correct) plotted as a function of proportion correct in a meta-analysis comparing general knowledge tasks to sensory discrimination tasks. Adapted from Juslin, Olsson, & Winman, A. (1998). (B) Mean over/underconfidence scores with 95% confidence intervals for probability judgments with the interval format and the full-range format for general knowledge and sensory discrimination tasks. Adapted from Juslin, Winman, & Olsson (2003).

meta-analysis of general knowledge and sensory discrimination tasks with pair comparisons (Juslin et al., 1998). Over/underconfidence bias is plotted as a function of proportion correct (difficulty) separately for sensory and the epistemic tasks. There is a clear and significant difference between the two sorts of tasks with more underconfidence bias for the sensory discrimination tasks.[1]

This difference between general knowledge and sensory tasks has been replicated with other assessment formats like full-range assessment (e.g., "What is the probability that the left line/country is longer/more populous than the right?") and interval assessment (e.g., "Provide the smallest interval within which you are 90% confident that the true length/population of this line/country falls."). As illustrated in Figure 17.2B, interval assessment is generally more prone to overconfidence (Juslin, Wennerholm, & Olsson, 1999; Klayman et al., 1999, Chapter 11 of this volume), but there is also a clear main effect, with less overconfidence for sensory discrimination (Juslin et al., 2003).[2]

In 1997 Juslin and Olsson presented SESAM,[3] a model that accounts for the underconfidence bias in sensory discrimination. Although a detailed description of the SESAM model is beyond the scope of this chapter it is interesting to note that the processes in the model also imply that the probability judgment is treated as *if it were* a sampling probability. A sensory discrimination is assumed to produce a sample of n passing impressions held in memory at one moment. The decision about which of two presented stimuli, say weight A and weight B, has a larger magnitude (i.e., is heavier) is determined by the proportion of impressions that suggest each of the two outcomes. If most (m) of the n impressions suggest that weight A is heavier the decision favors weight A.

Confidence in the decision is based on the proportion of impressions that support the decision taken. For example, if a participant has sampled ten impressions ($n = 10$) and experienced weight A to be heavier for eight of these ($m = 8$), he or she will judge the probability that weight

---

[1] There is also a negative correlation between proportion correct and over/underconfidence – the so-called hard–easy effect (Lichtenstein & Fischhoff, 1977) in Figure 2A. The true magnitude of this relationship is, however, difficult to estimate from these data sets because neither data set has been corrected for the linear dependency between proportion correct (c) and the over/underconfidence score ($x - c$) or scale end effects. See Juslin, Winman, & Olsson (2000) for an extensive discussion of these problems.

[2] The sensory discrimination tasks in this study (Juslin, Winman, & Olsson, 2003) were marginally more difficult than the general knowledge tasks in terms of proportion of correct decisions.

[3] Vickers and Pietsch (2001) have recently criticized the processing assumptions of the sensory sampling model (Juslin & Olsson, 1997), in part on valid grounds. The discussion here, however, concentrates on the basic mechanism for explaining the underconfidence bias observed in sensory discrimination that was implemented in the model, rather than the specific processing assumptions made in Juslin & Olsson (1997).

B is heavier to be 80% (m/n). In effect, the judgment treats the n sensory inputs from the perceptual system as n possibilities and confidence is the probability of sampling a signal suggesting that weight A is heavier. Although the representations are different from those considered in the case of inferences for general knowledge items, SESAM suggest that there might be a common conception of probability that is shared in the two tasks.

Unfortunately, in this specific case, allowing confidence to be determined by the sampling probability for individual sensations is a mistake. The actual probability of being correct is not given by the probability of sampling individual impressions but by that of the sampling distribution for the whole collection of n sensations used to make the decision (i.e., the sampling distribution when n impressions are repeatedly sampled). Because the sampling distribution for a statistical aggregate is more precise than suggested by the distribution of the individual sensations, this will introduce a tendency to underestimate the probability of a correct answer: an underconfidence bias (for details see Juslin & Olsson, 1997).

In summary, the hypothesis that people interpret confidence as a sampling probability predicts that the way in which general knowledge items are selected has a profound effect on the observed over/underconfidence in foresight. On the assumption that people in a sensory discrimination task (inappropriately) use the sampling probability for passing impressions to estimate the probability of correct answer we expect an underconfidence bias. In general, these predictions are consistent with the data collected on confidence in foresight.

OVER- AND UNDERESTIMATION IN HINDSIGHT

Research on hindsight bias emerged in 1975 with Baruch Fischhoff's seminal paper that became the foundation for a tremendous amount of research and still continues to generate new ideas. Fischhoff (1975) was interested in the question of how the receipt of outcome feedback affects subjective probability judgments. The participants were presented with information about a presumably unpredictable event with several possible outcomes, in this case the Anglo–Nepalese War (1814–1816) over Nepal's southward and Britain's northward expansion into India (the British actually defeated Nepal). The experimental groups were given different endings, or outcomes, and were asked to rate the probability for the different outcomes that they would have assessed, *had they not been presented with the correct alternative*. These ratings were compared with those of a control group that was not presented with an outcome. The finding, which has proven to be robust, is that participants overestimate the predictability of the presented outcome relative to the control group: They act as if they *knew all along* what was going to happen.

Fischhoff (1975) explained the phenomenon by assuming that the information about the outcome is immediately assimilated with the information in long-term memory, which thus is modified, impairing or eradicating the memory for the original knowledge state. In the following we will accordingly refer to this hypothesis as the *memory impairment hypothesis*. As such, this hypothesis has apparently far-reaching theoretical implications for our understanding of human memory as well as large practical consequences. In the following, however, we will suggest an alternative to the memory impairment hypothesis in which the information in long-term memory is essentially unaffected by the feedback and left intact. Again, this alternative view is premised on the assumption that people interpret confidence and subjective probability as a sampling probability, in retrospect attempting to generate response patterns that are appropriate in the light of their belief about this sampling probability.

Consider the following example of a hindsight general knowledge item:

Which of these two countries has a larger area?
(a) Australia
(b) Brazil
(The correct answer is Brazil.)

This question is followed by the decision task:

Which alternative would you have chosen, had the correct answer not been indicated?

Or, alternatively, it is followed by the assessment task:

What estimates of probability would you have given to both alternatives, had the correct answer not been indicated?

This general knowledge item occurs in the studies presented in the following. In a control group not presented with the correct alternative only 15% of the participants select the correct answer. In hindsight, however, 65% of the participants respond that they would have chosen the correct alternative. This question is thus an example of an item that is associated with strong hindsight bias. We define hindsight bias as a change in subjective probability caused by feedback and occurring in spite of explicit instructions to ignore the feedback, presumably accompanied by ignorance that a change has taken place.

As in the original study by Fischhoff (1975), the persistent finding with general knowledge items has been that in retrospect people overestimate how much they knew before they were presented with the facts – they act as if they knew the correct answer all along. [See reviews by Christensen-Szalanski & Willham (1991), Fischhoff (1975, 1982a), and Hawkins & Hastie (1990); see also Hoffrage & Pohl (2003) for a recent journal issue dedicated to the topic.] For instance, the participants may indicate in retrospect that they would have selected the correct answers for 90% of the items, but when presented with the same items without the correct answer they actually select the correct answers for only 60% of the items, or alternatively, probability assessments shift in the direction indicated by the feedback. This hindsight bias has been explained by both motivational factors, such as the ambition to appear knowledgeable to oneself or others, and by cognitive factors, with most recent research favoring cognitive explanations (see Hawkins & Hastie, 1990).

The conceptual relation between overconfidence and hindsight bias has sometimes been acknowledged in the literature (Hawkins & Hastie, 1990; Hoch & Loewenstein, 1989). In general the relation has been interpreted as causal, with overconfidence bias originating from hindsight bias (but see Winman, 1997a,b, 1999). That one in retrospect experiences that one would have been able to predict things is assumed to create an inflated feeling of confidence in one's predictive and judgmental abilities: "I knew all along that it was going to rain today; therefore I am confident that I will be able to predict tomorrow's weather...." Both phenomena are presumed to have detrimental consequences; for example, overconfidence might lead to nonoptimal decision making in economic and other circumstances and hindsight bias to an inability to benefit from experience.

Why Is Representative Design Essential to Studies of Hindsight Bias?

The *accuracy-assessment model* (Winman, 1997b; Winman et al., 1998) proposes that both foresight confidence and hindsight responses emanate from the same process – in effect, the judgment of a sampling probability that a correct answer is produced. In both cases this probability refers to the probability m/n of identifying the correct answer when sampling from a reference class of n situations (e.g., items) that may be encountered in the environment. In this view neither phenomena causes the other: They are two manifestations of the same underlying factor – a misjudgment of this sampling probability.

Central to the notion of our model is that we must distinguish between two qualitatively different strategies that participants may adopt when confronted with a hindsight task. We refer to these separate approaches as *process simulation* and *accuracy assessment*, respectively. Although we propose that both of these processes operate, we will focus on the accuracy assessment, a process that is central in our approach and lies at the heart of the model.

Process simulation refers to an attempt to simulate the specific cognitive process that one would have used in foresight to arrive at an answer. This may involve considering what facts were known prior to receiving the outcome feedback and guessing or reconstructing what inferences would have been made in foresight. The traditional explanations of hindsight bias – such as the memory impairment hypothesis (Fischhoff, 1975) – emphasize that this reconstructive process is biased by knowledge of the outcome. Accuracy assessment involves the participants attempting to reproduce the accuracy that is reasonable in consideration of the task difficulty, that is, whether or not they would have been correct. In doing this, they consider the sampling probability of making a correct response when responding to items of this level of difficulty, given their assessed knowledge of the content of the item. For item content about which one has considerable knowledge, one expects to get most answers right; for content about which one is ignorant, one expects to be off the mark quite often. Although cognitive explanations of hindsight bias have highlighted process simulation, accuracy assessment has been neglected. To fully understand what is going on in a hindsight task we believe that it is necessary to acknowledge that both strategies are both relevant and used.

To clarify, consider an example illustrating process simulation. Let us, for instance, assume that a participant has been presented with the following task:

Which of these countries has a higher mean life expectancy?
(a) Bulgaria
(b) Egypt
(The correct answer is Bulgaria.)

A participant relying on process-simulation may reason along the following lines: "I would probably have answered 'Bulgaria,' since I would have known that Egypt is an African country, and I believe that these generally tend to have lower standards of living and accordingly probably lower mean life expectancies." The participant thus tries to reconstruct *the cognitive process* by means of which he or she would have solved the task in foresight (i.e., an inductive inference concerning life expectancies of African countries) and arrives at the answer that the simulated process would have produced (Bulgaria). The general idea in cognitive models of the hindsight bias is that these process simulations, automatically and largely unconsciously, tend to become biased by knowledge of the correct answer (Hawkins & Hastie, 1990). Process simulation may be more or less available to the participants depending on the saliency of the relevant foresight cognitive processes.

Without denying the importance of process simulation we propose that it is complemented also by considerations of accuracy assessment. The accuracy-assessment model implies that participants try to assess the probability with which they would have been correct. Instead of answering the question "Which alternative would I have chosen?" they ask themselves the question "Would I have been correct?" In Brunswikian terms (Brunswik, 1956) one could say that process simulation and accuracy assessment involves focusing on different distal variables. One example of accuracy assessment would be the following line of thought: "I don't really know how I would have reasoned, but since I don't have a lot of information about either of the countries I'm quite likely to have been wrong." In essence, if participants judge the probability of selecting the correct answer within a reference class of similar items to be m/n, in hindsight they will indicate that they would have been m/n certain and that they would have picked the correct answer for m/n of these general knowledge items.

The plausibility of accuracy assessment is evident if one considers that, across tasks deemed to be extremely difficult (e.g., the probability of selecting the correct answer for a randomly selected such item approaches .5), the participants can hardly expect to get every answer correct and this insight may overrule the outcome of a more or less fallible and biased process simulation. Both accuracy assessment and process simulation may indeed operate in parallel, often with process simulation as the primary process. In Winman et al. (1998) we distinguished between a strong and weak version of the accuracy-assessment hypothesis. The strong version implies that for *each specific* item participants "randomize" their answer as correct with a probability equal to the judged task difficulty (i.e., compare with probability matching). The weak version assumes that accuracy assessment acts as a constraint on the process simulation, checking that the response pattern appears reasonable given the task difficulty. The distinction between process simulation and accuracy assessment may seem subtle, but, as we will see, with further assumptions it generates powerful predictions.

Figure 17.3 summarizes the main conceptual relations and predictions of the accuracy-assessment model (for a detailed mathematical derivation, see Winman et al., 1998). At the center we have a belief about the difficulty of some reference class of tasks, defined as the probability of selecting a correct answer when sampling items from this reference class. For example, the participant may believe that a specific task item belongs to a difficulty level such that randomly selecting items from this reference class will produce 75% correct answers in the long run. In a foresight confidence judgment task the participant reports this probability as discussed in the section on over/underconfidence bias (see the right side of Figure 17.3).

If participants have reasonably correct beliefs about this sampling probability and the items are randomly selected from this reference class, the proportion correct is indeed .75 in the long run and the participants will be calibrated (e.g., the representative samples in Figure 17.1). However, if – for some reason – the task is unexpectedly difficult and only 60% of the answers are correct, the result is overconfidence bias of +.15 (i.e., the bottom of Figure 17.3). Correspondingly, if the task proves to be unexpectedly easy, allowing 90% of correct answers, the result is an underconfidence bias of −.15 (i.e., the upper part of Figure 17.3).

The accuracy-assessment model maintains that, at least in part, the responses produced in hindsight derive from the same belief about task

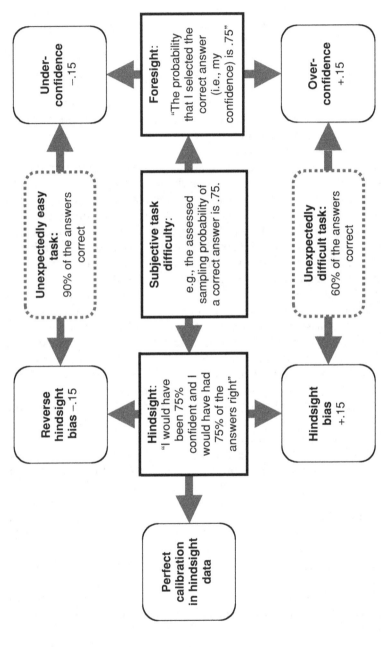

FIGURE 17.3. A schematic summary of the conceptual relations and the empirical predictions by the accuracy-assessment model.

difficulty (left side of Figure 17.3). When participants attempt to produce a response pattern appropriate in view of their belief about the task difficulty (.75) they will indicate that their foresight confidence would have been .75 and that they would have selected the correct answer for 75% of these items. After all, this is the response pattern that is appropriate with task difficulty .75. This brings us to a number of – insofar as we know – unique predictions by the accuracy-assessment model:

> *Prediction 1.* If calibration analysis is performed on data produced in hindsight – comparing the retrospectively stated confidence to the retrospectively indicated proportion of correct answers – the result should converge on perfect calibration (agreement).

In our simplistic example, both the retrospective confidence judgment and the indicated proportion of correct answers in hindsight should be .75. The degree to which this occurs depends on how successful the participants are in executing the accuracy-assessment strategy. Note further that if the participants have accurate beliefs about the task difficulty and the tasks are randomly selected from the appropriate reference class, they would indeed have had .75 confidence and 75% correct when performing the task in foresight:

> *Prediction 2.* If the participants are well calibrated in foresight (e.g., confidence and proportion correct .75 in foresight) they will produce no hindsight bias for the same material (i.e., they will also indicate confidence and proportion correct .75 in hindsight).

This presumes that the belief about task difficulty has not changed systematically between the foresight and the hindsight conditions, something that is approximated in familiar tasks and within-subjects hindsight designs (Winman, 1997a,b; Winman et al., 1998). The robust hindsight bias will be abolished in the global analysis of these data.

> *Prediction 3.* If participants misjudge the task difficulty and the task is unexpectedly difficult (lower part of Figure 17.3) this task will produce overconfidence bias in a foresight confidence study and ordinary hindsight bias in a hindsight study.

In our example in Figure 17.3 where the overconfidence bias is +.15 (.75 − .60), in retrospect, the participants in the hindsight condition will indicate that they would have been 75% confident (and correctly so), but they would erroneously claim that they would have gotten 75% of the answers correct, which amounts to an ordinary hindsight bias because the true proportion correct is only .6. To the extent that previous hindsight research with general knowledge items has relied on selected item samples that produce consistent overconfidence bias, hindsight bias should also appear a robust phenomenon.

Prediction 4. Correspondingly, if the task is unexpectedly easy (upper part of Figure 17.3) the participants will produce underconfidence bias in a foresight confidence study and a *reverse hindsight bias* in a hindsight condition with this task.

In Figure 17.3 where the underconfidence bias is −.15 (.75 − .90), in retrospect, the participants will indicate that they would have been 75% confident but erroneously claim that they would have gotten 75% of the answers correct, a reverse hindsight bias because the true proportion correct is .9. The typical participant will erroneously conclude that "I would never have known that," when in fact he or she would have! A reverse hindsight bias is inconsistent with all previous cognitive and motivational accounts. Together Predictions 2, 3, and 4 define the *confidence–hindsight mirror effect* that allows us to quantitatively predict the direction and magnitude of the hindsight bias from data on foresight over/underconfidence (for the detailed derivations and measures, see Winman et al., 1998). The accuracy-assessment model predicts a corresponding effect also in regard to global assessments in hindsight:

Prediction 5. When participants are overconfident in their local item-by-item confidence judgments but calibrated in their global judgments of *how many* items they have solved correctly ("the confidence–frequency effect"; see Gigerenzer et al., 1991), we expect a corresponding dissociation for hindsight bias (e.g., *How many of theses items would you have solved had the correct answers not been indicated?*). If the participants have a correct belief about the difficulty of selected item samples (e.g., in Figure 17.1) when probed in global terms, this should appear also in global assessments made in hindsight.

Prediction 6. The cognitive process and the origin of the uncertainly should not affect the hindsight bias; only the extent to which the task difficulty is misjudged.

This prediction contrasts the claim by alternative cognitive models suggesting that inferential and reconstructive cognitive processes by their very nature should be particularly prone to hindsight bias. The accuracy-assessment model implies that, if the foresight over/underconfidence bias is the same in two conditions that tax different cognitive processes, the hindsight bias should also be the same.

"I Never Would Have Known That": A Reversal of the Hindsight Bias

Perhaps the boldest prediction is the reversal of the hindsight bias in tasks that elicit underconfidence bias in foresight. Observations of a reversed hindsight effect have hardly ever been reported (see Christensen-Szalanski & Willham, 1991, for a compilation), and of course the prediction

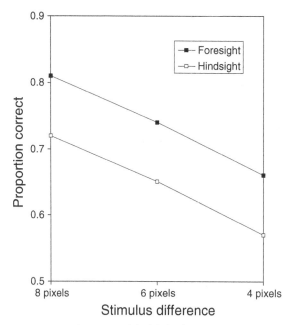

FIGURE 17.4. A reverse hindsight bias: True proportions correct and the proportions of retrospectively indicated correct answers in a hindsight task that involved sensory discrimination between pairs of stimuli with three different stimulus differences. Based on data from Winman, Juslin, & Björkman (1998).

of unusual events lends particular support to any model. Whereas underconfidence bias is uncommon in cognitive domains, as discussed in the previous section there is often underconfidence bias in sensory discriminations of a psychophysical nature (Björkman et al., 1993; Juslin & Olsson, 1997). In other words, if we are searching for tasks with underconfidence, by now we have a pretty good idea where to look. As far as we know, no other researchers have investigated hindsight bias in sensory discriminations.

In two studies using weight and line-length discriminations we have reported reversed hindsight effects of the magnitude expected. In Winman et al. (1998) participants decided which of two lines they would have thought to be the longer one after they had been told that of the two, the right one actually is longer. Figure 17.4 shows the data for the line-length discrimination experiment. Twenty-nine participants were exposed to 120 items each in foresight and hindsight. As expected a pervasive underconfidence bias was found, but more noteworthy a reversed hindsight bias effect was also found. This effect was remarkably strong and persistent (see also Figure 17.6B). Our interpretation of this result is that the inclination to reason that "this is so hard that I probably would have been wrong" is strong enough to override the participants' own sensory information that they could have relied on. The reversal of the hindsight phenomenon is problematic for several approaches, especially for motivational explanations. In Campbell & Tesser's (1983) model, for example, hindsight bias stems from a person's desire to control the environment, and the model will consequently never produce a reversed effect. In the task at hand, there is likewise apparently no need for participants to preserve a self-image, experience the world as predictable, or try to impress the experimenter. The result will also prove difficult to handle for cognitive models such as RAFT (Hoffrage, Hertwig, & Gigerenzer, 2000) in which the phenomenon can be understood as a byproduct of an adaptive knowledge-updating process rather than as a metacognitive response strategy. As discussed in the foregoing, persisting underconfidence that holds in entire cognitive domains is hard to find, and therefore the phenomenon is difficult to demonstrate as a general finding for cognitive domains. For particular items especially associated with underconfidence bias in foresight, however, it is possible to demonstrate a reliable reversed hindsight bias also with general knowledge items.

What Does Random Sampling of Items Do to the Hindsight Bias?

Given the relatedness of overconfidence and hindsight bias and the model discussed in the previous section it is natural to inquire about the role of item sampling in hindsight bias (Winman, 1997a,b). Can random sampling of items also make hindsight bias diminish or even go away? The model we present suggests that if random sampling eliminates overconfidence, the same thing will happen to hindsight bias. We found that with a random item sample in a hindsight task it is indeed hard to obtain a significant hindsight bias compared to a sample of selected items (see Table 17.1). Only three out of twenty participants showed hindsight bias for the representative sample. Inspection of the item samples revealed a tremendous variation in the hindsight bias associated with each item (Figure 17.5). For

TABLE 17.1. *Number of Items Associated with Hindsight Bias in the Two Samples. Data from Winman, 1997a. Table Reprinted with Permission from Blackwell Publishers.*

Stimulus Material	Hindsight Bias	No Bias	Reversed Bias
Selected set	69 (57.5%)	14 (11.65%)	37 (30.8%)
Random set	47 (39.15%)	37 (30.85%)	36 (30%)

Note: $\chi^2 = 14.6$, $p < .001$.

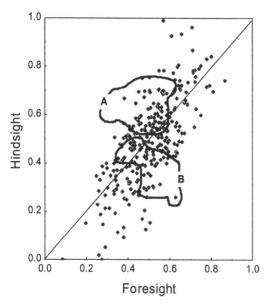

FIGURE 17.5. Proportion of correctly indicated answers in hindsight for each of 120 randomly selected general knowledge items plotted against the corresponding proportion correct. Points above the diagonal are items producing ordinary hindsight bias and points below the diagonal are items producing reverse hindsight bias. The two arbitrarily defined item sets A and B will produce entirely different conclusions about the extent and direction of the hindsight bias. Adapted from Winman (1997a).

a considerable number of items a reversed hindsight bias is observed. Remember that general knowledge is an area where hindsight bias has proven to be particularly robust.

Realization of the wide itemwise variability in hindsight bias is important in itself. It is not uncommon to use only one task and forget about the problem of item sampling altogether. In Figure 17.5 we illustrate two arbitrarily constructed item samples, A and B. The conclusion is that an experimenter could test participants with item sample A on Tuesday and sample B on Thursday and make hindsight bias appear, disappear, or even reverse within a week.

We can further demonstrate that items associated with more overconfidence also produce more hindsight bias. Figure 17.6A shows hindsight bias, measured by the difference between the proportion of (retrospectively indicated) correct answers and the true proportion of correct answers in foresight, as a function of overconfidence. Each over/underconfidence category consists of eighty items. It is obvious from the figure that, as overconfidence increases, so does hindsight bias. We have demonstrated this

A

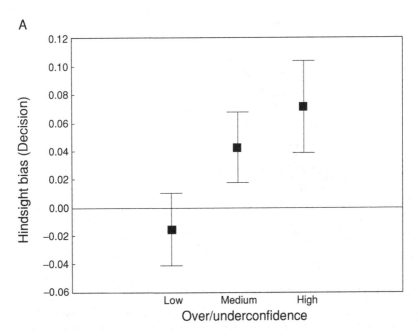

B

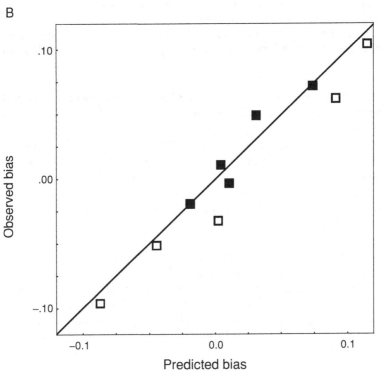

phenomenon both in within- and between-subjects designs with the only difference that participants in between-subjects designs tend to show a slight increase in average confidence between foresight and hindsight. Figure 17.6B plots observed hindsight bias as a function of observed over/underconfidence bias in foresight across ten experimental conditions in Winman et al. (1998). The confidence–hindsight mirror effect implied by the accuracy-assessment model allows very accurate quantitative prediction of the observed hindsight biases without the fitting of any free parameters.

Does the hindsight bias disappear with representative sampling of general knowledge items, as suggested by the data in Figure 17.1? Winman (1997a, Experiment 3) relied on large and independent random samples of up to 300 responses per participant and a systematic comparison of a within- and a between-subjects procedure. It is not bold to claim that this study is the most comprehensive attempt to address representative design in the area. Considering the number of items each participant received, these are also unusually reliable results at the level of individuals. Figure 17.7 presents data aggregated over two conditions. Each mean is based on 12,000 responses from 60 participants. As suggested by Figure 17.7, with these large and independent random samples of items, there are no statistically significant hindsight biases, although there is a tendency in this direction for the probability assigned by participants to the correct answer (Figure 17.7A). It seems fair to conclude that *there are* tasks in which humans have *accurate and undistorted* perceptions of their foresight performance.

In summary, we have shown that, as predicted, random sampling of general knowledge items can decrease or in some instances eliminate the hindsight bias, underlining the importance of item sampling and the role of representative design. Items associated with overconfidence bias in foresight are especially prone to cause hindsight bias, and items associated with underconfidence in foresight tend to produce *reverse hindsight bias*. A large body of research has shown that hindsight bias is robust to a number of debiasing procedures (Fischhoff, 1982b). No matter how much we pay, explain to, or inform the participants, they are still victims of the bias. However, as we have shown, merely depriving the psychologists from the privilege of creating the stimulus material can make the overall bias disappear.

←——————————————————————————————————

FIGURE 17.6. (A.) Observed hindsight bias defined as the difference between the proportion correct in the hindsight condition minus the (true) proportion correct in foresight for three item samples associated with different levels of overconfidence bias in foresight. Based on data from Winman (1997a). (B) Predicted versus observed hindsight bias across the conditions in three experiments. Adapted from Winman, Juslin, & Björkman (1998).

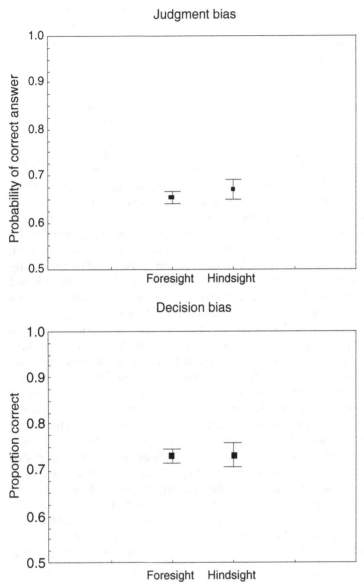

FIGURE 17.7. Data from a study with large, independent, and new item samples for each participant. (A) The mean probability assigned to the correct answer in foresight and hindsight with 95% confidence intervals. (B) The proportion of indicated correct answers in foresight and hindsight with 95% confidence intervals.

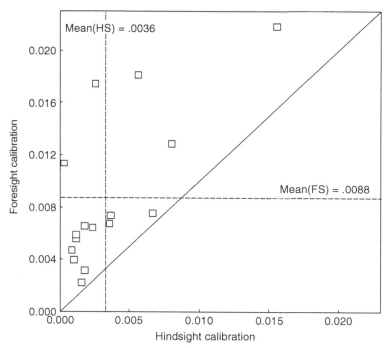

FIGURE 17.8. Calibration scores measuring agreement between confidence and proportion correct computed on data from foresight and hindsight data from 16 experimental conditions. Adapted from Winman (1997b).

"I Was Well Calibrated All Along!"

The data that appear in Figure 17.8 are taken from the conditions reported in three published articles (Winman, 1997a, 1999; Winman et al., 1998). All tasks were two-alternative items of the sort exemplified in the previous examples. Two tasks concerned sensory discrimination; the remaining fourteen tasks were general knowledge tasks. Calibration scores – measuring the mean-square deviation between confidence and proportion correct – from the hindsight conditions are plotted against the corresponding measures from the foresight conditions. When the calibration score is zero, confidence judgments are perfectly realistic and coincide with proportions correct. The higher a score is, the worse calibration becomes. Data points above the diagonal in Figure 17.8 are those where the hindsight calibration score is better (lower) than the foresight calibration score, as implied by Prediction 1. In *all sixteen conditions*, the calibration scores are considerably lower in hindsight than in foresight in that all points are located above the identity line. We conclude from the data illustrated in the figure that this prediction, the *"I was well calibrated all along"* effect, appears to be a robust phenomenon.

Are There Global Hindsight Effects?

We will first make a distinction between what we refer to as local versus global hindsight effects. A local hindsight effect is the false belief that one would have known the answer *to a particular question*. A global effect, in contrast, is the false belief that *for a set of questions*, one would have known the answers to all or most of these. In Experiment 4 of Winman et al. (1998), Predictions 3 and 5 were investigated. Sometimes, for different reasons even a randomly sampled stimulus material is associated with overconfidence bias. For death causes, for example, it has been assumed (e.g., Lichtenstein et al., 1978) that biased media coverage leads to biased ideas about frequencies, with overconfidence as a result. As expected we found a strong hindsight bias for this material. More interesting is the global hindsight effect operating over a set of items that has been implied by a number of researchers. Fischhoff, for example, hypothesized that this effect may have negative consequences for our desire to take part in education: "If we underestimate how much we are learning from the facts presented in a particular context, we should feel less reason to go on learning. If what we learn does not surprise us, then we overestimate how much we know already" (Fischhoff, 1977, p. 349). However, no investigations of this global effect had been undertaken. We undertook an experiment in which the task of participants was as follows:

The learning phase. Participants learn propositional facts, such as, for instance,
  1. Peru is larger than Belgium.
  2. London is situated farther south than New York.
  3. Nairobi has more inhabitants than Oslo.

The test phase. Participants are then asked for a global estimate:
  How many (what percentage) of these facts would you have known, had the correct answers not been indicated?

In this study there was no sign of a global hindsight effect at all. In fact, the experimental condition gave slightly lower estimates of what percentage of items participants would have answered correctly than members of the control group without prior learning of the facts, .67 compared to .69. Thus, for a stimulus material that produces a strong local hindsight effect of the traditional type there is no global effect. This dissociation closely mirrors the confidence–frequency effect for overconfidence described by Gigerenzer et al. (1991).

The last prediction is that the effect should be independent of the origin or source of uncertainty. It has been proposed that hindsight bias is a natural unavoidable consequence of reasoning processes per se and thus is linked to this particular source of uncertainty. This view is most distinctly communicated by Walster, who argued that the more cognitive processes that are involved in the task, the more probable the outcome would seem in retrospect: "When an outcome event is serious, and one spends very much time thinking about it, one probably becomes especially interested in seeing how the outcome and its antecedents fit together.... [T]he important outcome might, in fact, seem more probable than if one had not thought it through so completely" (Walster, (1967), p. 239).

Walster showed that the "knew-it-all-along effect" (although this term was not yet coined at the time) was stronger for outcomes with serious consequences and concluded that this was because the participants were spending more time thinking of these outcomes. Thus, according to this view hindsight bias emerges as a consequence of the intrinsic nature of higher cognitive processes: The more these processes are involved, the more bias is produced. In contrast, from an accuracy-assessment perspective, the uncertainty or difficulty experienced by the participants is the key aspect, independently of the degree to which higher reasoning processes or other sources underlie this subjective state of uncertainty.

To confront the two perspectives, an experimental manipulation of the cognitive processes was undertaken by stimulus material that differentially permitted reliance on higher order reasoning or inferential processes. An experimental condition with material that mainly allowed for associative retrieval was compared to a condition that only allows for inductive inference. In the first condition participants were required to learn meaningless syllables whereas participants in the second condition briefly read a text on which they later were tested. The short exposure and semantic representation of the text should force the participants to use inference, whereas the lack of semantic representations in the other condition would only allow for "lower" associative retrieval processes. Both of these conditions have earlier been found to be associated with reasonable calibration and comparable and low over/underconfidence (Juslin, Winman & Persson, 1995). Hence, over/underconfidence was constant whereas the degree of higher inferential processes involved was varied.

As can be seen in Figure 17.9 the degree of hindsight bias is small and virtually identical in both conditions. The size of bias is in line with what one can expect from the accuracy-assessment model, with modest and non-significant overconfidence in both conditions. These results are problematic for the view that hindsight bias is a natural cause of higher reasoning processes per se. However, these results are expected from an accuracy-assessment perspective, according to which it should not matter whether the origin of the experienced difficulty resides in less than perfectly reliable inferences or in weak strengths of associative retrieval processes, as long as the over/underconfidence bias in foresight is the same.

To summarize, we have described a tentative model of hindsight bias. We have provided empirical support of six predictions that all pose problems for existing approaches. We have previously shown (Winman et al., 1998) both that the presented approach can organize a number of

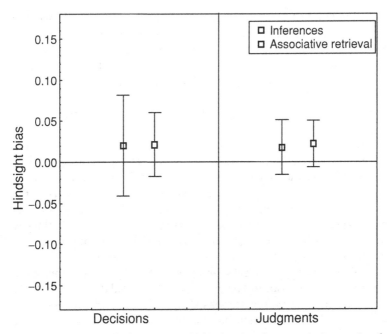

FIGURE 17.9. Hindsight bias in two tasks that involve associative retrieval and reconstructive inference. Adapted from Winman (1999).

regularities observed in the literature and that the accuracy-assessment model at least in some instances can deliver a good parameter-free quantitative fit to the data. This, however, does not mean that we consider hindsight bias an artifact. We ask of our participants to ignore the presented feedback. If our explanation is correct they do in fact, in a quite explicit way, *violate* this instruction and use the presented feedback. This violation is perhaps most conspicuous in sensory discriminations tasks. The metacognitive considerations made by participants (would I have been correct or wrong?) involved in accuracy assessment do, however, seem reasonable and as we have seen in many cases work just fine. Again, participants may simply just rely on sampling intuitions to make the best of a bad situation.

CONCLUSIONS

Behind the notion that subjective probability judgments derive from sampling probabilities is a conviction that the extensional properties – what is out there – are an essential part of reality that no biological forms of organisms can afford to ignore. Different habitats vary in degree of variability or predictability. Within some environments, resources are distributed evenly within time and space; in others the availability of, for example, food supply tomorrow is hard to predict. The inhabitants of these environments have to cope with the uncertainty and have often developed mechanisms for detecting and estimating cues to environmental variability. It has indeed been demonstrated that even "lower" organisms readily adapt their behavior to changes in the variability of an environment.

For example, experimentally altering the potential foraging period affects the food storage strategy of the marsh tit (*Parus palustris*); a more variable environment promotes a higher amount of food storage (Hurly, 1992). We propose that the backbone of the sensitivity to environmental variability in animals is analogous to that ability in humans to capitalize on knowledge of an environment to infer and assess sampling probabilities. As in the classical interpretation, this conception of probability makes little distinction between subjective and frequentistic aspects of probability. Therefore, subjective probability judgments become premised on the reference classes that pertain to people's natural environment.

An organism that functions by registering environmental frequencies needs to be equipped with a specialized fine-tuned and sophisticated mechanism taking care of this, at first sight, overwhelming task. Hasher & Zacks (1979) showed in a series of empirical studies that we indeed seem to have such required capacity of automatic, effortless, and accurate encoding of frequencies. Other studies have shown that frequency of occurrence of events can be learned effortlessly and implicitly and can be used correctly (see Sedlmeier & Betsch, 2002, for a recent overview). Likewise, many well-known biases in probabilistic reasoning apparently diminish or disappear when information is presented in a trial-by-trial manner derived from experience rather than presented as summary statistics, or when questions are phrased in terms of frequencies rather than probabilities. (e.g., Christensen-Szalanski & Bushyhead, 1981; Gigerenzer, 1991; Hasher & Zacks, 1984; Manis et al., 1980; Medin & Edelson, 1988).

Whereas there thus exists evidence that humans and animals are efficient frequency encoders, the link between frequency learning from sampling and subjective probability has not been investigated empirically. This remains an interesting future research area. We have recently obtained data from a yet unpublished experiment (Winman & Juslin, 2003) in which participants learn binary cues and give confidence ratings of these, which seems to lend further support for the m/n subjective probability view. Although more research supporting this contention is needed, these tentative data show that in an environment controlled experimentally subjective probability *can* match hit rates well when frequencies have been experienced on a trial-by-trial basis.

In this chapter we have presented empirical findings consistent with the idea that humans can pick up and use frequency information directly sampled from the environment, which sometimes, but not always, allows probability judgments, whether made in foresight or hindsight, to be in agreement with reality. This, of course, does not prove that our portrayal of reality is correct. But it does bring up the old wisdom that one cannot overstate the importance of developing a framework that incorporates the task presented to the participants with a model of how the participants solve the task and that any such attempt will highlight the need for a consideration of how representative the task is of the participants' natural environment (see Brunswik, 1952; 1956). In the end, there can be little doubt that all aspects of cognitive activity are constrained by the conditions under which they take place.

References

Baranski, J. V., & Petrusic, W. M. (1994). The calibration and resolution of confidence in perceptual judgments. *Perception and Psychophysics, 55,* 412–428.

Björkman, M. (1994). Internal cue theory: Calibration and resolution of confidence in general knowledge. *Organizational Behavior and Human Decision Processes, 57,* 386–405.

Björkman, M., Juslin, P., & Winman, A. (1993). Realism of confidence in sensory discrimination: The underconfidence phenomenon. *Perception & Psychophysics, 54,* 75–81.

Brunswik, E. (1952). *The conceptual framework of psychology.* Chicago: University of Chicago Press.

Brunswik, E. (1956). *Perception and the representative design of psychological experiments.* Berkeley: University of California Press.

Campbell, J. D., & Tesser, A. (1983). Motivational interpretations of hindsight bias: An individual difference analysis. *Journal of Personality, 51,* 605–620.

Christensen-Szalanski, J. J. J., & Bushyhead, J. B. (1981). Physicians' use of probabilistic information in a real clinical setting. *Journal of Experimental Psychology: Human and Performance, 7,* 928–935.

Christensen-Szalanski, J. J. J., & Willham, C. F. (1991). The hindsight bias: A meta-analysis. *Organizational Behavior and Human Decision Processes, 48,* 147–168.

Doherty, M. E., Brake, G., & Kleiter, G. D. (2001). The contribution of representative design to calibration research. In K. R. Hammond and T. R. Stewart, (Eds.), *The essential Brunswik.* New York: Oxford University Press.

Fischhoff, B. (1975). Hindsight ≠ Foresight: The effect of outcome knowledge on judgment under uncertainty. *Journal of Experimental Psychology: Human Perception and Performance, 1,* 288–299.

Fischhoff, B. (1977). Perceived informativeness of facts. *Journal of Experimental Psychology: Human Perception and Performance, 3,* 349–358.

Fischhoff, B. (1982a). For those condemned to study the past: Reflections on historical judgment. Reprinted in D. Kahneman, P. Slovic, & A. Tversky (Eds.), *Judgment under uncertainty: Heuristics and biases.* New York: Cambridge University Press.

Fischhoff, B. (1982b). Debiasing. In D. Kahneman, P. Slovic, & A. Tversky (Eds.), *Judgment under uncertainty: Heuristics and biases* (pp. 422–444). New York: Cambridge University Press.

Gigerenzer, G. (1991). How to make cognitive illusions disappear: Beyond "heuristics and biases." *European Review of Social Psychology, 2*, 83–115.

Gigerenzer, G., Hoffrage, U., & Kleinbölting, H. (1991). Probabilistic mental models: A Brunswikian theory of confidence. *Psychological Review, 98*, 506–528.

Gilovich, T., Griffin, D., & Kahneman, D. (Eds.) (2002). *Heuristics and biases: The psychology of intuitive judgment.* New York: Cambridge University Press.

Griffin, D., & Tversky, A. (1992). The weighing of evidence and the determinants of confidence. *Cognitive Psychology, 24*, 411–435.

Hacking, I. (1975). *The emergence of probability.* Cambridge, UK: Cambridge University Press.

Hammond, K. R., & Stewart, T. R. (Eds.) (2001). *The essential Brunswik: Beginnings, explications, applications.* New York: Oxford University Press.

Hasher, L., & Zacks, R. T. (1984). The automatic processing of fundamental information: The case of frequency of occurrence. *American Psychologist, 39*, 1372–1388.

Hasher, L., & Zacks, R. T. (1979). Automatic and effortful processes in memory. *Journal of Experimental Psychology: General, 108*, 356–388.

Hawkins, S. A., & Hastie, R. (1990). Hindsight: Biased judgments of past events after the outcomes are known. *Psychological Bulletin, 107*, 311–327.

Hoch, S. J., & Loewenstein, G. F. (1989). Outcome feedback: Hindsight and information. *Journal of Experimental Psychology: Learning, Memory, and Cognition, 15*, 605–619.

Hoffrage, U., & Pohl, R. F. (2003). Research on hindsight bias: A rich past, a productive present, and a challenging future. *Memory, 11*, 357–377.

Hoffrage, U., Hertwig, R., & Gigerenzer, G. (2000). Hindsight bias: A by-product of knowledge updating? *Journal of Experimental Psychology: Learning, Memory and Cognition, 26*, 566–581.

Hurly, T. A. (1992). Energetic reserves of marsh tits (Parus palustris): food and fat storage in response to variable food supply. *Behavioral Ecology, 3*, 181–188.

Juslin, P. (1994). The overconfidence phenomenon as a consequence of informal experimenter-guided selection of almanac items. *Organizational Behavior and Human Decision Processes, 57*, 226–246.

Juslin, P., & Olsson, H. (1997). Thurstonian and Brunswikian origins of uncertainty in judgment: A sensory sampling model of confidence in sensory discrimination, *Psychological Review, 104*, 344–366.

Juslin, P., Olsson, H., & Winman, A. (1998). The calibration issue: Theoretical comments on Suantak, Bolger, and Ferrell (1996). *Organizational Behavior and Human Decision Processes, 73*, 3–26.

Juslin, P., Wennerholm, P., & Olsson, H. (1999). Format dependence in subjective probability calibration. *Journal of Experimental Psychology: Learning, Memory and Cognition, 25*, 1038–1052.

Juslin, P., Winman, A., & Persson, T. (1995). Can overconfidence be used as an indicator of reconstructive rather than retrieval processes? *Cognition, 54*, 99–130.

Juslin, P., Winman, A., & Olsson, H. (2000). Naive empiricism and dogmatism in confidence research: A critical examination of the hard-easy effect. *Psychological Review, 107*, 384–396.

Juslin, P., Winman, A., & Olsson, H. (2003). Calibration, additivity, and source independence of probability judgments in general knowledge and sensory discrimination tasks. *Organizational Behavior and Human Decision Processes, 92,* 34–51.

Kahneman, D., Slovic, P., & Tversky, A. (1982). *Judgment under uncertainty: Heuristics and biases.* Cambridge, UK: Cambridge University Press.

Kahneman, D., & Tversky, A. (1982). Variants of uncertainty. *Cognition, 11,* 143–157.

Klayman, J., Soll, J. B., González-Vallejo, C., & Barlas, S. (1999). Overconfidence? It depends on how, what, and whom you ask. *Organizational Behavior and Human Decision Processes, 79,* 216–247.

Kleitman, S., & Stankov, L. (2001). Ecological and person-driven aspects of metacognitive processes in test-taking. *Applied Cognitive Psychology, 15,* 321–341.

Koriat, A., Lichtenstein, S., & Fischhoff, B. (1980). Reasons for confidence. *Journal of Experimental Psychology: Human Learning and Memory, 6,* 107–118.

Lichtenstein, S., & Fischhoff, B. (1977). Do those who know more also know more about how much they know? *Organizational Behavior and Human Performance, 20,* 159–183.

Lichtenstein, S., Slovic, P., Fischhoff, B., Layman, M., & Combs, B. (1978). Judged frequency of lethal events. *Journal of Experimental Psychology: Human Learning and Memory, 4,* 551–578.

Manis, M., Dovenalina, I., Avis, N. E., & Cardoze, S. (1980). Base rates can affect individual predictions. *Journal of Personality and Social Psychology, 38,* 287–298.

McCloskey, M. (1983). Naive theories of motion. In: D. Gentner & A. L. Stevens (Eds.), *Mental models* (pp. 299–324). London: Lawrence Erlbaum.

Medin, D. L., & Edelson, S. M. (1988). Problem structure and the use of base-rate information form experience. *Journal of Experimental Psychology: General, 117,* 68–85.

Oaks, M. (1986). *Statistical inference: A commentary for the social and behavioral sciences.* New York: Wiley.

Piaget. J. (1951). *The child's conception of physical causality.* New York: The Humanities Press.

Peirce, C. S., & Jastrow, J. (1884). On small differences of sensation. *Memoirs National Academy of Sciences, 3,* 73–83.

Ronis, D. L., & Yates, J. F. (1987). Components of probability judgment accuracy: Individual consistency and effects of subject matter and assessment method. *Organizational Behavior and Human Decision Processes, 40,* 193–218.

Savage, L. J. (1954). *The foundations of statistics.* New York: Dover.

Sedlmeier, P., & Betsch, T. (2002). *Etc. Frequency processing and cognition.* New York: Oxford University Press.

Soll, J. B. (1996). Determinants of overconfidence and miscalibration: The roles of random error and ecological structure. *Organizational Behavior and Human Decision Processes, 65,* 117–137.

Suantak, L., Bolger, F., & Ferrell, W. R. (1996). The "hard-easy effect" in subjective probability calibration. *Organizational Behavior and Human Decision Processes, 67,* 201–221.

Taylor, S. E., & Brown, J. D. (1988). Illusion of well-being: A social psychological perspective on mental health. *Psychological Bulletin, 103,* 193–210.

Tversky, A., & Kahneman, D. (1983). Extensional versus intuitive reasoning: The conjunction fallacy in probability judgment. *Psychological Review, 90,* 293–315.

Vickers, D., & Pietsch, A. (2001). Decision making and memory: A critique of Juslin and Olsson's (1997) sampling model of sensory discrimination. *Psychological Review, 108,* 789–804.

Walster, E. (1967). Second guessing important events. *Human Relations, 20,* 239–249.

Winman, A. (1997a). The importance of item selection in "knew-it-all-along" studies of general knowledge. *Scandinavian Journal of Psychology, 38,* 63–72.

Winman, A. (1997b). *Knowing if you would have known: A model of the hindsight bias.* Comprehensive summaries from the faculty of Social Sciences, 69. Uppsala University Library, Uppsala, Sweden.

Winman, A. (1999). Cognitive processes behind the "knew-it-all-along effect." *Scandinavian Journal of Psychology, 40,* 135–145.

Winman, A., & Juslin, P. (2003). *Learning of subjective probabilities.* Manuscript in preparation.

Winman, A., Juslin, P., & Björkman, M. (1998). The confidence-hindsight mirror effect in judgment: An accuracy-assessment model for the knew-it-all-along phenomenon. *Journal of Experimental Psychology: Learning, Memory, and Cognition, 24,* 415–431.

18

Natural Sampling of Stimuli in (Artificial) Grammar Learning

Fenna H. Poletiek

The capacity to learn implicitly complex structures from exemplars of this structure underlies many natural learning processes. This process is occasionally called grammar induction, whereby a grammar can be any structure or system generating exemplars. The most striking example of this process probably is natural language grammar induction: children learning (to use the rules of) the language of their caregivers, by exposure to utterances of this language. This case of grammar induction has been argued by linguists to be such a complex task that its occurrence cannot be explained without invoking a special inborn language device, containing prior information about these natural grammars. The contribution of experience, then, is to provide the learner with additional information needed to "set the parameters" of this device. This is the well-known linguistic position (Chomsky, 1977; Pinker, 1989).

The nativist view was put forward as an alternative to the empiricist position, explaining language learning as a process of imitation and conditioning on the basis of verbal stimuli (Reber, 1973). According to linguistics, the psychological position cannot fully explain the acquisition of the rules of natural grammar because the sample of exemplars to which a child is exposed during the language-acquisition period is demonstrably insufficient to master all these complex rules. This argument against the experience-based explanation of grammar acquisition is known as the "poverty of stimuli" argument (Chomsky, 1977; Haegeman, 1991; Marcus, 1993; Pinker, 1994). The poverty of stimuli argument is based on a theoretical and logical analysis of the natural language-acquisition task (Chater & Vitanyi, 2003).

This analysis is based on an idealization of the grammar-acquisition problem that is still the general basis of linguistics (Gold, 1967). Globally, the learner is represented as a computing agent operating on a sample of input sequences, with the goal to induce the correct grammar underlying this input. Given the complexity of the underlying (natural) input, and the goal that a learner should fully identify the correct grammar, the requirements this input should satisfy can be derived theoretically. Paradoxically, this idealization leads to the conclusion that natural grammar induction on the basis of the utterances a child is faced with is impossible, because the input set does not satisfy these requirements.

The input is poor in a number of ways. For example, the sample is too small. To induce a grammar of the complexity-class to which natural languages belong requires infinitely more exemplars. Another crucial shortcoming of the input sample is the absence of negative evidence: No information is given to the learner about the sequences *not* belonging to the language (Chater & Vitanyi, 2003; Hornstein & Lightfoot, 1981; Marcus, 1993). The solution of this problem, raised within the Gold idealization, is that language acquisition must be innate at least to some extent, since environmental input alone cannot explain that children actually do succeed in acquiring (using) language. The Gold idealization has uncovered a number of universal commonalities among languages and the universal stages learners go through when acquiring natural grammar (Brown & Hanlon, 1981; Wexler & Cullicover, 1980). The focus in this framework is mainly on complexity analyses of the grammatical structure and comparison among languages (Hornstein & Lightfoot, 1981).

In contrast to the formal approach, both empiricist and statistical approaches on language acquisition have focused on the actual richness of the input learners work with (Charniak, 1993). This richness, however, bears on those aspects of the stimulus set that are considered irrelevant to complexity-based analyses (Chater & Vitanyi, 2003; Redington, Chater & Finch, 1998; Wolff, 1982). For example, in the input sample, some types of sentences (e.g., short ones) are much more frequent than others (e.g., long ones) (Poletiek & Wolters, 2004). In addition, a learner will hear simple grammatical constructions not only more often than complex ones in a lifetime but also more often in the early stages of learning (Newport, 1990). In the formal idealization, repeated input is redundant and uninformative, nor are variations in frequencies useful, because they do not add logical information. In general, sample properties like frequency distribution and ordering of stimuli are alleged to affect learning according to the empiricist and the related statistical view.

The same difference applies to the assumed goal of a grammar learner. In the formal approach the language learner aims at identifying the true grammar underlying the linguistic stimuli. The achievement of this goal is tested by "grammaticality judgments," or "grammaticality intuitions," about new stimuli that every language user has (Haegeman, 1991). The idealized learner who has completed the task successfully can perfectly judge as either "correct" or "incorrect" every possible new sentence. In the empiricist view, however, the learner's goal is to use the language in the first place. That is, the learner tries to understand relevant speakers and to make him- or herself understandable to them. This pragmatic goal does not require perfect identification of the system, but only sufficient mastering for understanding and producing new language items. Because every new linguistic item, either perceived or produced, is an attempt to communicate, a language user will estimate whether an exemplar is useful enough to achieve a particular linguistic goal. If this estimation is successful, the task is well enough performed. Thus, the pragmatic language learner might rather be concerned with continuous intuitions about correctness of new items (the feeling that a sentence is "almost correct"), which might sufficiently predict that they work in verbal communication. In sum, the poverty of the stimulus sample on which natural learners operate is an implication of a theoretical idealization of the natural grammar induction problem. As an alternative, the empiricist view has stressed the richness of the natural linguistic sample to fulfill the goal of a pragmatic learner.

Nativist and empiricist views differ by even more aspects. Grammar induction in the empiricist framework is seen as a general cognitive skill, underlying, among others, natural grammar learning but also other forms of structured behavior, like social skills and motor skills learning. Natural language learning is not given a special position but is considered as a case of inductive learning from stimuli. Also, any kind of information in the stimulus set plays a role in these processes. In the case of natural grammar learning, the environmental input of verbal stimuli consists not only of syntactical correct utterances satisfying the grammatical structure but also of a large variety of other verbal and non verbal cues such as meaning, facial expressions, volume of voice, intonations, and context, all presumably contributing to learning. Distributional properties of the natural sample of stimuli to which a natural language learner is exposed comprise one of these extralinguistic sources of information.

The present chapter focuses on this latter property. I hypothesize that the distributional property of the environmental sample is highly effective for future verbal performance, when we assume that the learner's objective is to master the system sufficiently rather than perfectly (see also Wolff, 1988). In particular, it will be argued that the Artificial Grammar Learning (AGL) program, set up within the empiricist framework, has paid little attention to both kinds of ecological factors: the environment and the mind. In fact, AGL's experimental representation of cognitive grammar induction relies on the formal idealization. Finally, it is shown that AGL nonetheless provides a very good framework for studying the interface between environmental input and pragmatic learning. This is illustrated with an AGL experiment with simulated data.

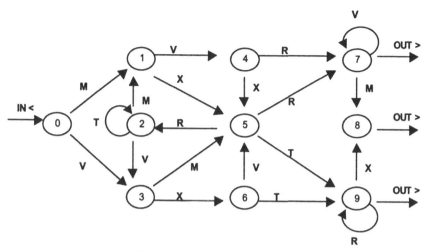

FIGURE 18.1. Grammar G (from Meulemans & Van der Linden, 1997).

AGL AS A REPRESENTATIVE DESIGN FOR NATURAL GRAMMAR LEARNING STUDY

The Artificial Grammar Learning program, started by Reber (1967), was developed to study experimentally this general human capability to induce implicitly structure knowledge underlying stimuli in the environment. The standard task developed by Reber consists of a learning phase and a test phase. During learning, participants are exposed to a sample of stimuli generated by a structure, typically, a finite state grammar. The grammar displayed in Figure 18.1 is an example of such a grammar used in this paradigm (Meulemans & Van der Linden, 1997). This "grammar" is a set of sequential rules and five elements, generating strings satisfying the rules of the grammar.

In a theoretical account of the AGL program, Reber (1993) argues that the AGL design aims at investigating grammar induction. Reber considers natural grammar induction to be an expression of the general skill of structure induction. The standard AGL task is designed to tap into this general skill. In other words, the AGL task domain is a sample of all possible structured stimulus domains. Hence, the AGL task is meant to represent all possible structured domains in the real world producing exemplars, including natural grammar. Thus, rather than being designed to simulate natural grammar learning only, AGL experiments are designed to simulate structure induction in general. Eventually, however, the results of AGL research are meant to contribute to explaining natural grammar induction, according to this logic.

Reber (1993) gives a thorough justification of the contrast between his paradigm and linguistic theories of grammar learning. In particular, he argues for the representativeness of his task design and stimulus domain for the purpose of explaining natural grammar learning on the one hand, but by studying general, language-independent structure learning behavior on the other (Reber, 1993). To achieve this, the experiments should satisfy a number of demands. First, the structured stimuli should be novel. They should not be something that is already within the subjects' "sphere of knowledge." Second, the rule system should be complex. The subjects should not be able to "crack the code" by simply testing explicit hypotheses. Third, the stimuli should be meaningless and emotionally neutral. Fourth, the stimuli should be synthetic and arbitrary, because "if our assumptions about implicit learning are correct, it should appear when learning about virtually any stimulus domain and the use of the synthetic and the arbitrary gives additional force to this argument" (Reber, 1993, p. 27). Thus, the AGL design is meant to be a representative design for various ecological human system induction tasks, among which is learning natural grammar. However, in contrast to the elaborate argumentation and the guidelines for representativeness of the AGL *design* and the stimulus *domain*, little is said about the selection and representativeness of the *sample* of stimuli that should be used to get a valid picture of general grammar induction processes.

Actually, in the history of the AGL paradigm, little attention is paid to how stimuli generated by the structure domain should be sampled. Researchers presumably sample on the basis of intuition (Redington & Chater, 1996). They sample a number of exemplars, varying in length and form, so as to get what is believed to be a heterogeneous and unbiased sample of stimuli. In fact, researchers proceed in a scientific way. They make sure that more-or-less all rules (paths) of the grammar are exemplified in the sample of strings, to ensure a representative and random example of what it possibly generates. Surprisingly, few AGL studies have varied quite straightforward characteristics of the input sample such as number of stimuli or the arrangement of the stimuli in the learning procedure (Meulemans & Van der Linden, 1997; Poletiek, 2003; Reber et al., 1980).

This disregard may have been strengthened by the "career" of the AGL paradigm, which shows similarities with other famous experimental tasks such as the selection task (Poletiek, 2001; Wason, 1968). As research with the task progresses and more publications appear, the focus shifts away from what the task tells about the ecological situations it originally intends to represent (Reber, 1973) to the observable and comparable effects in the experiment per se, that is, effects of manipulations of the AGL task. In other words, concerns with internal validity of the experiment increase at the cost of external validity (Gigerenzer, 2000). This development in the history of the AGL task may also explain why the AGL program seems to have gradually lost its initial mission to study natural grammar and structure learning. In recent studies with AGL, explicit claims about the generalizability of AGL findings to natural language learning are mainly avoided. It seems that the program, of which the contribution to explaining natural grammar learning had been its major rationale in the beginning, ended up renouncing to any claim of generalizability beyond the task itself. In the next section, I will argue that AGL, in spite of its history, is a very good paradigm to study natural factors affecting cognitive grammar induction.

HOW NATURAL SAMPLING OF EXPERIMENTAL STIMULI CAN BRING BACK THE OLD AGENDA OF AGL

A survey of ecological situations involving learning of an abstract structure of stimuli clearly intimates that the distributions of the samples are not random (Newport, 1990; Poletiek & Wolters, 2004; Redington et al., 1998). Consider, for example, an environment in which two types of plants contain a toxic substance: large plants with blue flowers, which are very frequent in the environment, and small plants with white flowers, which are very rare. The stimuli from which the rules underlying the toxicity of plants must be induced are all observed cases of people eating these plants and getting sick. This set of exemplars is not a flat distribution with each possible case occurring equally often: The sick-and-blue-flower cases will be observed more often than the sick-and-white-flower cases. Also, the sick-and-small-white-flowers even might not be observed during some years, because these cases are extremely rare. Hence, the sample generated by the structure that nature produces will lead to more knowledge about some plant–sickness relations than others, depending on the frequency of the plants being eaten. Similarly, in no linguistic natural situations, the input sample of items on the basis of which crucial rules have to be abstracted, will be distributed equally, leading to better knowledge about some rules and less about others.

Linguistic utterances do not come in equal distributions. The sample of utterances a child is faced with is strongly biased. First, short and easy sentences are used and repeated much more often than long and complex ones. In fact, very complex grammatical constructions might not be heard in the first years of a life, or may never be heard in a lifetime. Also, the stimuli do not come in random order over time. The easy and short ones come mainly in the beginning and more often; the long and complex ones come rarely and later. A mother will not start her verbal communication with her child by using sentences with relative clauses, but by using two- or three-word sentences containing simple words she endlessly repeats (Newport, 1990). Hence, natural grammar induction starts with a strongly biased learning set, which was sampled with (rather than without) replacement, containing exemplars with extremely varying frequencies, and finally, being ordered from most "easy" ones in the beginning, to complex ones over time. In contrast, sampling in AGL seems concerned with avoiding all these natural biases.

These statistical properties of natural nonrandom stimulus sampling have been argued to be extremely informative in theories of statistical language learning (Charniak, 1993). This information is unrelated to the content of the exemplars themselves. Thus, the hypothesized effect of natural sampling on the learnability of a (grammatical) structure is comparable to extralinguistic effects such as nonverbal information, intonation, and context, none of which play a role in formal theories of language acquisition. Interestingly, however, the contribution of these sampling effects can be calculated and accounted for more precisely and formally than qualitative effects such as context, intonation, and meaning. Hence, exploring these natural sample properties within the AGL framework might enhance the relevance of AGL for studying natural grammar induction, while at the same time allowing for precise quantification. This project, currently undertaken in our lab, has revealed preliminary empirical effects of natural frequency distribution and sample size on categorization performance in AGL (Poletiek, 2003). In the following AGL study, the effect of natural input on a pragmatic learner is illustrated with simulated data.

THE IMPACT OF THE FREQUENCY DISTRIBUTION OF LEARNING EXEMPLARS ON THE LEARNABILITY OF THE GRAMMAR: A SIMULATION

Assume there are two learners in an AGL task. Both learners are exposed to 32 unique strings of a grammar G that are presented 10,000 times in total. Both learners use the same learning strategy. In Learner A's training set, the 32 strings are equally distributed (the traditional AGL sampling condition). Thus each string is seen about 312 times. In Learner B's training set, the frequencies of the 32 strings are not equal; the 32 strings are distributed in accordance with their probability to be generated by the grammar (the natural sampling condition). The probability of a string being generated by G is the product of the path probabilities. In Figure 18.2, the standard finite state grammar displayed in Figure 18.1 is shown with path probabilities added.

In principle, paths can have any probability, but all paths starting from one state must have probabilities adding up to one. In the present simulation, we assume that all paths starting from one state have the same probabilities. On the basis of the path probabilities of G, the probabilities [p(string|G)] of the 32 unique training strings to be generated by G can be derived. These are displayed in Table 18.1.

The representative frequencies in the natural sampling condition are proportional to the p(string|G) values of the strings. The frequency

TABLE 18.1. *Set of 32 Unique Exemplars of G, and Their Probability of Being Generated by G (String|G)*

0.0416667	MVR	0.00308642	MXTRR
0.00154321	MVRVVV	0.00102881	MXTRRRX
0.000771605	MVXRMXR	0.00231481	VMRMVR
0.000771605	MVXRMXT	0.000514403	VMRTMXR
0.000771605	MVXRVMR	0.000514403	VMRVMRV
0.00154321	MVXRVV	0.000514403	VMRVMTR
0.00115741	MVXRVXT	0.00308642	VMRVVM
0.00154321	MVXTRR	0.00102881	VMRVVVM
0.00462963	MVXTRX	0.00231481	VMRVXT
0.00231481	MXRMVR	0.000771605	VMRVXVT
0.00231481	MXRMVRM	0.0277778	VMTX
0.000771605	MXRMVRV	0.000514403	VXTRRRR
0.00154321	MXRVMRM	0.000385802	VXVRMVXRM
0.000342936	MXRVVVV	0.000771605	VXVRVMT
0.000771605	MXRVXTR	0.00154321	VXVTRR
0.000771605	MXRVXVR	0.00154321	VXVTRRX

distribution in the traditional sampling condition is flat. The difference between these two distributions is visualized in the histograms of Figure 18.3.

Furthermore, we must assume a learning strategy. We assume that both learners use the same chunk-based learning strategy (Knowlton & Squire, 1994, 1996; Meulemans & Van der Linden, 1997; Perruchet & Pacteau, 1990; Poletiek, 2002). This strategy can be formalized as follows. From every unique pair of letters (bigram) displayed at training, the learner

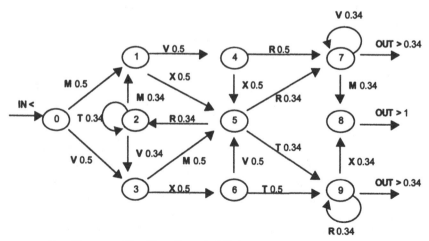

FIGURE 18.2. Grammar G with path probabilities.

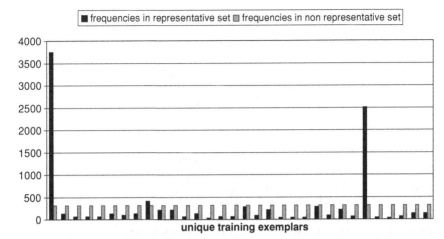

FIGURE 18.3. Frequency distribution of representative and nonrepresentative learning sets of exemplars.

learns how often the first one is followed by the second one. Thus, the probability is learned of any letter X figuring in the learning set to be followed by Y, that is, $p(Y|X)$ (Poletiek & Wolters, 2004). Notice that, in this chunk-based learning model, the first and last letter positions in a string are taken into account. The first position is represented by a dummy letter $<$, and the last one by the dummy $>$. For example, the bigram "$<M$" is "M in the first position." However, except for the first and last positions, the fragment-based strategy does not result in any knowledge about the position of the fragments (X,Y) in the strings of the training set. It can be shown that this strategy will result in different knowledge for Learner A and Learner B because of the stimulus distributions they are exposed to. Indeed, owing to the different frequency distributions of the unique strings, learners will see a given bigram of letters more or less often.

The fragment-based knowledge induced during training is projected on a test set containing novel strings generated by the grammar G and novel strings that violate grammar G but do not contain illegal bigrams. The criterion task is to judge grammaticality of the strings. Grammaticality is defined as the probability of that string being generated by that grammar. In the formal linguistic tradition, grammaticality is two-valued. A string can be judged either grammatical (value 1) or ungrammatical (value 0). In the present study, grammaticality is represented continuously as well: Ungrammatical strings always take value 0, but grammatical strings can take values varying from 0 to 1, in accordance with the strings' probability in grammar G, $p(\text{string}|G)$ (Poletiek & Wolters, 2004). Moreover, we assume that both learners learn in the same way. They will base their grammaticality judgements on fragment knowledge gathered at training. That is, the "chunk associativeness" of a new string to the learning set determines its perceived familiarity (Knowlton & Squire, 1994; Meulemans & Van der Linden, 1997). Familiarity, in turn, determines the grammaticality judgement.

This projection of fragment knowledge on test strings can be formalized as follows: The learner "calculates" the familiarity of a test string by retrieving the $p(Y|X)$ values of all bigrams (X_i, Y_j) it contains and then multiplies them, according to

$$\prod_{i,j} p(Y_j | X_i) \tag{18.1}$$

in which X and Y are elements figuring in the test string. By applying Eq. 18.1, a test string will have a chunk associativeness value of 0 if it contains bigrams never seen in the learning phase. Such a string would thus be seen as completely unfamiliar to a learner. In the present simulation, the ungrammatical test strings were selected in such a way that they did not contain illegal bigrams; that is, they all consisted of bigrams seen at least once at training. Thus, no ungrammatical test string has value 0: Ungrammatical strings only contain legal bigrams but at wrong locations.[1] In Table 18.2, the familiarity values of the test strings are displayed, as calculated by Learner A and Learner B, on the basis of their chunk knowledge induced at training. In addition, the grammaticality of the test string is indicated by 1 (grammatical) or 0 (ungrammatical).

Now, we can evaluate Learner A and Learner B's performance. To test the predictive accuracy of the estimators calculated by both learners, the correlation between dichotomous grammaticality (taking value 0 if the string is ungrammatical and value 1 if it is grammatical; see Table 18.2) and the familiarity values is calculated for all test strings. This correlation is higher for Learner B ($df = 18; r = .31; p > .05$), who has been trained with a representative distribution of exemplars, than for Learner A ($df = 18; r = -.06; p > .05$), but neither correlation reaches significance.

Correlations were also calculated for the pragmatic grammaticality criterion. This criterion variable is the (continuous) probability of a string being generated by the grammar [$p(\text{string}|G)$] taking any value between 0 and 1. This criterion takes a higher value as the string is more common in G, more typical for G, and shorter. Learner B's familiarity estimates correlated very high with these "G-probabilities" ($df = 18; r = .95; p < .01$). Learner A's estimates also correlated with the strings probabilities, but to a lesser extent ($df = 18; r = .50; p < .05$).

In Figures 18.4a and 18.4b, the familiarity estimates of both simulated learners are displayed in histograms.

---

[1] This was done to sharpen the contrast between the knowledge induced by the two learners. Strings having illegal bigrams would amount to zero for both learners, thereby decreasing the distinguishability between the two learners.

TABLE 18.2. *Familiarity Values of Test String, Calculated by Learner A (Trained with an Equally Distributed Set of Exemplars) and Learner B (Trained with a Representatively Distributed Set of Exemplars), and Probability Values in G*

	Grammaticality (0/1)	p(string$\mid$G)	Familiarity Values by Learner A	Familiarity Values by Learner B
VMT	1	0.0277778	0.0024306	0.0034672
MVXTX	1	0.0138889	0.0001022	0.0027651
MVXRVM	1	0.00462963	0.0002201	0.0000369
MXTRRX	1	0.00308642	0.0000253	0.0000160
VMRVXTX	1	0.00231481	0.0000053	0.0000517
MXRVMTX	1	0.00154321	0.0000028	0.0000539
MVXRMVR	1	0.00115741	0.0000500	0.0001206
VXVRVXTX	1	0.00115741	0.0000007	0.0000030
VXVRMVRM	1	0.00115741	0.0000006	0.0000004
VMTRRR	1	0.00102881	0.0000309	0.0000517
MRM	0	0	0.0028826	0.0005424
MXRRVM	0	0	0.0001068	0.0000059
MVXVXT	0	0	0.0002299	0.0000149
VMVXRVV	0	0	0.0000193	0.0000009
VXRVMR	0	0	0.0002970	0.0000429
VVMRRXTR	0	0	0.0000015	0.0000003
MRRXT	0	0	0.0001093	0.0000116
MVXTRVM	0	0	0.0000743	0.0000116
VMXRVXR	0	0	0.0001429	0.0000100
MVXRVXTM	0	0	0.0000041	0.0000000

The twenty test strings are numbered. In Figure 18.4c, the grammaticality values of the test strings are displayed by means of their probability in G. The strings numbered 11 to 20 take probability value p(string$\mid$G) = 0: Their probability of being generated by G is 0. The dichotomous grammaticality values are 0 as well for the ungrammatical strings, and 1 for all grammatical strings. It can be seen from the histograms that the familiarity distribution of Learner B (Figure 18.4b) gives a better approximation of the actual grammaticality distribution of the strings than Learner A's distribution (Figure 18.4a).

In summary, the present simulation suggests that learning a structure by induction from a sample of repeatedly presented exemplars distributed in accordance with their natural frequencies can lead to better grammaticality estimates of new strings than learning from an evenly distributed sample of exemplars. In addition, this facilitation is especially effective if the criterion for performance is not to approximate perfect categorization, but to approximate the probabilities of new strings in grammar G.

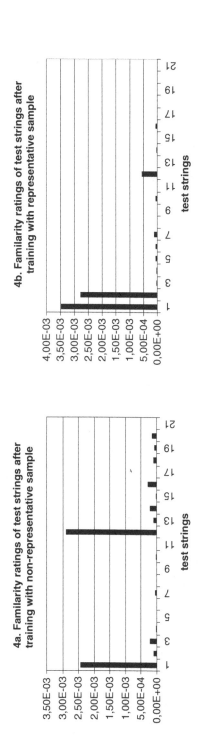

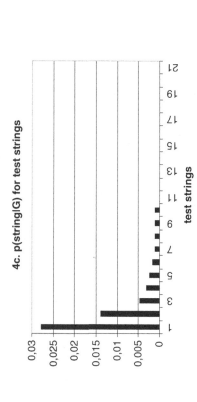

FIGURE 18.4. Familiarity rating distributions of test strings, after training with a nonrepresentatively distributed set of exemplars (a) and after training with a representatively distributed set of exemplars (b), and probability values in G (c).

DISCUSSION

As our simulation illustrates, AGL studies may be improved with regard to their validity for natural grammar induction and, in this manner, generate new explanations for the easiness or difficulty with which cognitive grammar induction proceeds. In real life, stimuli generated by a structure are not sampled randomly and without replacement, in any scientific way. In contrast, the stimuli come in a certain order and with a certain frequency distribution. These sampling biases that are very striking in natural language learning may not be mere accidental variations but may contain precisely the information a pragmatic language user motivated to understand and produced language well enough will need (Newport, 1990). In other words, the "text" on which a grammar learner operates in real life is not unconstrained as assumed in the formal idealization of grammar acquisition (Chater & Vitanyi, 2003; Gold, 1967). Its shape mostly enables one to make reasonable estimations of the chance that an exemplar will *work* rather than allow for perfect identification of the abstract structure.

The old mission of AGL, that is, to simulate experimentally natural grammar induction, can be rehabilitated by doing justice to the natural sampling properties of the learning stimuli. Indeed, one of the psychological arguments for grammar acquisition as an expression of a general inductive learning skill, and not as an expression of a special grammar module, is that natural language stimuli contain extralinguistic information, such as nonverbal and contextual information and intonation. To this list, we can add sampling information. Natural characteristics of the learning sample like frequency distribution are particularly easy to simulate in experimental AGL studies; they can be quantified and manipulated, and their effect can be observed very precisely, as was illustrated in the present simulation.

The other interesting suggestion about natural grammar induction emerging from this simulation involves the validity of the grammaticality judgements criterion. In the traditional AGL methodology, and in formal linguistics as well, the criterion for knowing the rules of language are two-valued grammaticality judgements. Grammar knowledge is assessed by means of a learner's capability to categorize sentences as correct or incorrect. However, in everyday life, this is not how we cope with language utterances. Rather than rating them as "correct" or "incorrect" we try to make sense of them. Part of this process may be parsing. However, in making sense of new utterances, either understanding or producing them, we may be more or less successful. That is, we may be able to parse an exemplar for 90%, but not understand it fully, precisely because the string is very unlikely in G. Or it may take us more or less time to understand a sentence because it is long, or because it makes use of exceptional syntactical rules. Consequently, assessing the grammaticality (i.e., understanding) of a new exemplar in everyday life situations may be a continuous rather than a dichotomous process, and in this manner, we may be very adequate in estimating the likeliness of the utterance to belong to the language. Hence, to represent this judgmental process, a continuous measure of grammaticality might be more adequate than a two-valued measure. Moreover, the criterion to test it might be to fit these judgements against true probabilities. Such a grammaticality measure was proposed here with the p(string|G) measure (Johnstone & Shanks, 1999; Poletiek & Wolters, 2004). Surprisingly, if the fit between the judgments and the true probabilities in G are taken as a measure of performance, performance in AGL increased dramatically, suggesting that learning is quite good when the validity of grammaticality estimations of new items is evaluated on their approximation of the probabilities of occurrence in the grammar.

Why has the interface between natural samples in the environment and pragmatic learning goals been unattended to in the AGL literature, in spite of its commitment to the empiricist view? An explanation might be that the AGL experiments, though carefully designed to test general induction skills, essentially follow the linguistic assumptions about natural grammar induction, from the very beginning. The stimulus set was sampled in a scientific way. It was assumed that a learner processes the exemplars as cumulative logical information. More or less repetitions, in this view, are useless because once an exemplar is seen, the information it provides is added only once to the stack of knowledge. In a formal linguistic framework, knowledge is not strengthened. There is no associative learning going on, requiring repeated exposure, and there is no forgetting.

Similarly, the categorization task in AGL is directly inspired from the linguistic idea that the grammatical knowledge is evidenced by means of grammaticality categorizations of new utterances. In sum, the typical AGL procedure relies highly on a linguistic view of language learning: seeing unique exemplars, submitting them to a rule abstraction process in which each exemplar contributes linearly, and building up a body of rule knowledge that enables the learner to categorize new utterances without error. Thus, although the AGL paradigm originally was designed to study general inductive learning skills, and not language learning in particular, it seems to have been influenced implicitly by the formal idealization of language learning. Also, since the paradigm drifted away from its original commitment to address problems of natural language acquisition, and became an enterprise for its own sake, questions about its resemblance to any natural situation disappeared.

As our analysis suggests, it may pay off to look closely both at how nature samples the structured language stimuli and what the language learning mind is really concerned with. The AGL paradigm might, in a second round, contribute to uncover how we proceed to perform the grammar induction task.

References

Brown, R., & Hanlon, C. (1970). Derivational complexity and order of acquisition in child speech. In J. Hayes (Ed.), *Cognition and the development of language* (pp. 11–53). New York: Wiley

Charniak, E. (1993). *Statistical language learning*. Cambridge. MIT Press.

Chater, N., & Vitanyi, P. (2003). *A simplicity principle for language learning: Reevaluating what can be learned from positive evidence*. Unpublished manuscript.

Chomsky, N. (1977). *Language and responsibility*. New York: Pantheon Books.

Gigerenzer, G. (2000). *Adaptive thinking*. Oxford: Oxford University Press.

Gold, E. M. (1967). Language identification in the limit. *Information and Control, 16*, 447–474.

Haegeman, L. (1991). *Introduction to government and binding theory*. Oxford: Blackwell Science.

Hornstein, N. & Lightfoot, D. W. (1981). *Explanation in linguistics: the logical problem of language acquisition*. London: Longman

Johnstone, T. & Shanks, D. R. (1999). Two mechanisms in artificial grammar learning? Comment on Meulemans and van der Linden (1997). *Journal of Experimental Psychology: Learning, Memory and Cognition, 25*, 524–531.

Knowlton, B. J., & Squire, L. R. (1994). The information acquired during artificial grammar learning. *Journal of Experimental Psychology: Learning, Memory and Cognition, 20*, 79–91.

Knowlton, B. J., & Squire, L. R. (1996). Artificial grammar learning depends on implicit acquisition of both abstract and exemplar specific information. *Journal of Experimental Psychology: Learning, Memory and Cognition, 22*, 168–181.

Marcus, G. F. (1993). Negative evidence in language acquisition. *Cognition, 46*, 53–85.

Meulemans, T., & Van der Linden, M. (1997). associative chunk strength in artificial grammar learning. *Journal of Experimental Psychology: Learning, Memory and Cognition, 23*(4), 1007–1028.

Newport, E. L. (1990). Maturational constraints on language learning. *Cognitive Science, 14*, 11–28.

Perruchet, P., & Pacteau, C. (1990). Synthetic grammar learning: Implicit rule abstraction of explicit fragmentary knowledge? *Journal of Experimental Psychology: General, 119*, 264–275.

Pinker, S. (1989). Language acquisition. In M. I. Posner (Ed.), *Foundations of cognitive science*. Cambridge, MA: MIT Press.

Pinker, S. (1994). *The language instinct*. Harmondsworth, UK: Penguin.

Poletiek, F. H. (2001). *Hypothesis testing behaviour*. Hove, UK: Psychology Press.

Poletiek, F. H. (2002). Learning recursion in an artifical grammar learning task, *Acta Psychologica, 111*, 323–335.

Poletiek, F. H. (2003). *The influence of stimulus set size on performance in Artifcial Grammar Learning*. Paper presented at the 44th annual meeting of the Psychonomic Society, Vancouver, Canada, November 6–9, 2003.

Poletiek, F. H., & Wolters, G. (2004). *One probabilistic measure for grammaticality and chunk associativeness in Artificial Grammar Learning*. Manuscript submitted for publication.

Reber, A. S. (1967). Implicit learning of artificial grammars. *Journal of Verbal Learning and Verbal Behavior, 6*, 317–327.

Reber, A. S. (1973). On psycho-linguistic paradigms. *Journal of psycholinguistic research, 2*, 289–318.

Reber, A. S. (1993). *Implicit learning and tacit knowledge: An essay on the cognitive unconscious.* New York: Oxford University Press.

Reber, A., Kassin, S., Lewis, S., & Cantor, G. (1980). On the relationship between implicit and explicit modes in the learning of a complex rule structure. *Journal of Experimental Psychology: Human learning and memory, 6*, 492–502.

Redington, M., & Chater, N. (1996). Transfer in artificial grammar learning: A reevaluation. *Journal of Experimental Psychology: General, 125*, 123–138.

Redington, M., Chater, N., & Finch, S. (1998). Distributional information: A powerful cue for acquiring syntactic categories. *Cognitive Science, 22*, 425–469.

Wason, P. C. (1968). Reasoning about a rule. *Quarterly Journal of Experimental Psychology, 20*, 273–281.

Wexler, K., & Cullicover, P. (1980). *Formal principles of language acquisition.* Cambrdige, MA: MIT Press.

Wolff, J. G. (1982). Language acquisition, data compression and generalization. *Language and Communication, 2*, 57–89.

Wolff, J. G. (1988). Learning syntax and meanings through optimisation and distributional analysis. In Y. Levy, I. M. Schlesinger, & M. D. S. Braine (Eds.), *Categories and processes in language acquisition* (pp. 179–215). Hillsdale, NJ: Lawrence Erlbaum.

19

Is Confidence in Decisions Related to Feedback?
Evidence from Random Samples of
Real-World Behavior*

Robin M. Hogarth[†]

A central theme of this book is how people learn from samples. In this chapter, we look at this issue from two viewpoints. First, we question the manner in which we – as scientists – sample the environments of the experimental participants who engage in the judgment and decision-making tasks that we study. To what extent are these samples representative of the natural decision-making ecology that our participants face in their everyday lives? Second, by actually sampling people's activities in their natural ecologies, we seek to characterize how they experience these environments. In particular, we investigate how one feature of environments (the presence or absence of feedback) affects inferences (the confidence people express in their decisions).

The first question is, of course, not new. For many years, psychologists have been concerned about how to generalize behavior from experimental evidence (see, e.g., Brunswik, 1956; Chapanis, 1961, 1967; Cronbach, 1975; Ebbesen & Konečni, 1980; Hammond, 1978; Hogarth, 1986; Lipshitz et al., 2001). I do not propose to add to this debate. Instead, the contribution is to demonstrate how the use of readily available technology can greatly facilitate random sampling of decision behavior outside the psychological laboratory. Indeed, as I shall argue in the following, obtaining appropriate samples of human decision behavior is not as difficult as might be imagined by researchers trained within experimental paradigms.

The intellectual stimulus for this work was a study reported by Egon Brunswik in 1944 and the development of the *experience sampling*

* This research was financed partially by grants from the Spanish Ministerio de Ciencia y Tecnología, CREI at Universitat Pompeu Fabra, and CEDEP (Fontainebleau, France).
† The author is ICREA Research Professor at Universitat Pompeu Fabra, Barcelona. He is most grateful to Irina Cojuharenco and Carlos Trujillo for their excellent research assistance and to illuminating comments from Antonio Cabrales, Irina Cojuharenco, Klaus Fiedler, Peter Juslin, Jochen Reb, and an anonymous reviewer. Contact address: robin.hogarth@upf.edu

method or ESM by Csikszentmihalyi and others (see Brunswik, 1944; Csikszentmihalyi & Larson, 1987; Hurlburt, 1997). Briefly stated, this chapter reports a study of the decision behavior of two groups of people, business executives and students. For several days, participants received four or five text messages on their mobile telephones at random moments, at which point they completed brief questionnaires about their current decision-making activities. With this methodology, therefore, it was possible to achieve random samples of each participant's decision behavior although, of course, no claim can be made that the participants themselves were anything other than "convenience" samples of their respective populations, that is, business executives and students.

The samples of decisions obtained by the ESM can be used to infer the characteristics of people's decision environments. This, in turn, can be used to address our second issue that deals with how such characteristics affect the inferences that people make. To do this, we focus on the issue of confidence and ask how one aspect of environments – namely, the presence or absence of feedback – affects the confidence that people express in their decisions.

The issue of whether people exhibit appropriate confidence in their judgments has attracted much attention in the decision-making literature. Here, the key concept has been calibration; that is, do people's assessments of uncertainty match empirical relative frequencies? For example, do events that are assessed subjectively as occurring, say, 65% of the time actually occur 65% of the time, and so on? Early findings suggested that people are overconfident; that is, their assessments of probabilities of target events are systematically higher than empirical relative frequencies (see, e.g., Lichtenstein, Fischhoff, & Phillips, 1982). However, these experimental findings have been challenged as researchers have shown, inter alia, differential effects of (a) some realistic task environments (Murphy & Winkler, 1984), (b) framing questions in terms of frequencies as opposed to probabilities (Gigerenzer, Hoffrage, & Kleinbölting, 1991), (c) "hard" and "easy" questions (Harvey, 1997; Juslin, 1993), and (d) whether confidence is expressed in the form of confidence intervals or beliefs about binary choices (Klayman et al., 1999).

More importantly, Gigerenzer et al. (1991) adopted a Bruswikian perspective and demonstrated how the representative sampling of questions from people's natural environments led to judgments of confidence that were well calibrated. In addition, Juslin and his colleagues have made a careful examination of how many other experimental tasks have been sampled and have demonstrated that nonrepresentative sampling of items could account for much overconfidence (Juslin, 1994; Juslin, Winman, & Olsson, 2000). In other words, if research on this topic had sought to sample tasks in the ecologically valid manner advocated by Brunswik (1956), "overconfidence" might not be considered such an important issue in behavioral decision making as it is today (see, e.g., Russo & Schoemaker, 2002).

The calibration paradigm tests people's ability to express their confidence in probabilistic form for particular events or classes of events. However, it does not address the more general issue of whether people feel confident in their everyday decision-making activities and, if so, whether these feelings are justified. Nonetheless, in terms of behavior, such feelings are important. They affect how individuals feel and act; and they can also influence others. In the workplace, for example, the confidence expressed by one party (e.g., a boss or a salesman) may determine whether a decision is implemented (e.g., whether to undertake an action or buy a product).

Some twenty-five years ago, Einhorn and I (Einhorn & Hogarth, 1978) addressed the issue of confidence in judgment from a theoretical perspective. We noted that the structure of many decision-making tasks in the real world is such that either people do not receive feedback on their decisions or, if feedback is received, it can be misleading. For example, consider hiring decisions made by managers. By observing how well new employees perform in their jobs, managers clearly receive feedback on their hiring decisions. However, this feedback is incomplete because, by hiring some applicants but not others, there is no feedback concerning the performance of rejected candidates. In addition, the very act of hiring may engender self-fulfilling prophecies with respect to "successful" candidates. The theoretical implication is that people's general confidence in their decision-making abilities can be continually reinforced by the fact that many of their actions involve positively biased or even no feedback (with the latter being interpreted as "no news is good news"). Subsequent experiments by Camerer (1981) and Schwartz (1982) supported this analysis. [For related ideas and evidence, see Fiedler (2000).]

Einhorn and I subsequently refined these ideas by characterizing environments in which erroneous beliefs might be created and maintained by distorted or even missing feedback. Einhorn (1980) referred to outcome irrelevant learning structures. More recently (Hogarth, 2001), in reviewing how people develop "good" and "bad" intuitions, I introduced the notion that the environments in which people acquire such knowledge can be classified as "kind" or "wicked" according to whether the feedback they receive is or is not veridical (i.e., accurate).[1]

This chapter addresses the relation between decision making and confidence in people's natural decision-making environments. Specifically, is there a relation between the confidence that people express in their decisions and the feedback that they either receive or expect to receive? The data reported are generated by a novel methodology for research on

---

[1] See also the discussion in Goldstein et al. (2001, pp. 186–187).

decision making inspired by Brunswik's (1944) study of "size constancy."[2] (See also Hoffrage & Hertwig, this volume.) In this study, Brunswik arranged for a person (a student) to be followed by an associate for extensive periods of time in her natural environment (the University of California at Berkeley) over a period of four weeks. The participant was instructed to behave in her normal fashion but was asked – at irregular moments – by the associate to estimate the sizes of objects in her visual field as well as the distance from the objects. The associate then measured the objects and distances. Note that in this study the "experimenter" did not manipulate or choose the objects about which the participant made judgments; that is, the "experimental tasks" were selected in a manner such that the objects of the participant's judgments constituted a random sample in her natural environment. Thus, although this study involved but a single participant, valid inferences could be made about the "size constancy" that she exhibited.

Briefly stated, Brunswik's point was that the environments of psychological experiments distort reality because participants are asked to act or make judgments in situations that have been artificially created by manipulating variables in some orthogonal, factorial manner. Instead, Brunswik argued, many variables in the real world are correlated to some extent and people learn to deal with a variety of situations as opposed to isolated incidents that have been created by an experimenter. Thus, if you want to know how people "do something" in the real world, you need to sample situations from the environments (or real worlds) in which they actually live.

This chapter exploits the ESM to explore issues of confidence in decisions as well as the kind of feedback people receive and expect on the decisions they make. Specifically, studies were conducted with two distinct populations, business executives and students. However, to gain more insight into estimates of behavior generated by the ESM, a retrospective questionnaire was also administered in which another group of student respondents was asked to summarize aspects of their decision-making activities over the two preceding weeks.[3] This questionnaire requested aggregate-level estimates of characteristics of decision-making activities that corresponded to the questions that participants in the ESM studies answered for all the specific decisions on which they reported.

The retrospective questionnaire was motivated by methodological and substantive concerns. First, there is evidence that retrospective reports of events differ from data concerning the same events collected at the time

---

[2] "Size constancy" is the ability of the perceptual system to see objects as having approximately constant size despite the fact that their projection onto the retina varies as a function of distance and other conditions.

[3] I am grateful to an anonymous referee for suggesting that this study be conducted to complement results from the ESM.

the events actually occurred (see, e.g., Ericsson & Simon, 1984; McFarlane, Martin, & Williams, 1988). To what extent, therefore, would estimates of behavior obtained by ESM differ from retrospective reports? Moreover, in what way would they differ? Second, within the calibration paradigm (see the foregoing discussion), people have been shown to exhibit less confidence in judgments when these are viewed within the framework of a series of judgments as opposed to considering each judgment separately (Gigerenzer et al., 1991). Does this finding generalize to feelings that people have about confidence in the decisions they make (i.e., retrospective reports versus confidence expressed when deciding)?

The chapter is organized as follows. I first describe the participants in the study and the methodology. Specific questions centered on the kinds of decisions made on an ongoing basis, the extent to which these decisions are repeated frequently or are unique, confidence expressed that decisions are "right," whether people receive or expect to receive feedback on the appropriateness of their decisions, the kind of feedback and its timing relative to the moment of decision, and confidence as to whether the feedback is or will be appropriate.

Next, I present results of the study. To summarize, people – both students and executives – express considerable confidence in the decisions they make on a daily basis even though they do not receive or expect to receive feedback for some 40% of these decisions. They are also quite confident in the quality of feedback (when received) and are well aware that this can lead to disconfirming the validity of their actions. As could be expected, executives and students differ in the kinds of decisions they face, both in terms of content and structure. For student populations, the data collected retrospectively differ significantly from those collected by the ESM. In particular, reported confidence in decisions as well as relative presence of feedback is larger when estimated in retrospect than from data gathered concurrently.

Finally, I discuss the results of the study in terms of both methodology and substance. The ESM is clearly a useful adjunct to a decision researcher's toolbox and will become even more powerful when used in conjunction with more traditional methodologies. In terms of substance, the study emphasizes the importance of characterizing habitual behavior or routines that affect the cumulative consequences of the many decisions we make each day.

METHOD

Participants

Three groups of participants took part in the ESM studies. Two groups were composed of business executives. The third group consisted of students at Universitat Pompeu Fabra in Barcelona, Spain. A further group of students at the same university completed the retrospective questionnaire concerning decisions made in the two preceding weeks.

The Executives. The first group of executives were recruited by e-mail notices sent by the president of the local alumni club of an executive MBA program in Barcelona that requested volunteers to participate in a study organized by the author (who actually knew many of the potential participants). However, not many details of the study were announced except that participants would be required to respond to text messages on their mobile telephones during a period of two weeks. In return for participation, potential respondents were promised feedback on the study. Twenty managers volunteered to participate. However, data were only received from thirteen. Two executives explicitly replied that they had been unable to participate (one because of technical problems with his telephone); no data – or even any responses – were received from five other executives and it can only be presumed that they experienced difficulty in complying with the study's requirements and so gave up on the task. Of the thirteen executives who responded, ten were male and three were female. Their ages varied between thirty and forty-five. Most of the executives worked in the Barcelona area except for two who were in Madrid, one in Mallorca, and one in another European country.

Following a brief presentation by the author, the second group of executives was recruited at an executive education program at CEDEP in Fontainebleau, France. These executives were similar in age and other demographic characteristics to the first group except that they worked in France, Belgium, the Netherlands, and the United Kingdom. Responses were subsequently received from eleven of the sixteen executives who had expressed an interest in the study. A difference between the first and second groups of executives was that, in the second group, the executives were only asked to participate in the study for one week, that is, five working days. The decision to reduce the length of participation reflected the difficulties experienced by the first group of executives. As with the first group, participants were promised feedback on the study.[4]

The Students. Eleven students (ten undergraduates and one graduate student) responded to advertisements placed on notice boards on campus. These stated that the task would take approximately thirty minutes each day during two weeks and that participants would be paid forty euros for participating. They were required to own or have access to mobile telephones capable of receiving text messages. On completion of the experiment, the participants were paid and were debriefed as to the purposes

---

[4] Both groups of executives received feedback in the form of a report summarizing results of the study. The author also volunteered to hold a meeting at which the results could be discussed. At the time of writing, this had not yet occurred.

of the study. The median age of the undergraduates was twenty (with a range of nineteen to twenty-one); the graduate student was twenty-eight years old; there were two males and nine females. The responses of one participant (a nineteen-year-old male) were excluded from analysis because it was deemed that they had not been completed with sufficient care.

Respondents for the retrospective questionnaire were recruited after an undergraduate class at Universitat Pompeu Fabra. The twenty-five volunteer respondents had a median age of twenty-one. There were eleven females and fourteen males. They were paid five euros each for completing the questionnaire, a task that took between twenty and thirty minutes. Since the responses of males and females did not differ significantly, they were pooled for the purpose of analysis.

Procedure

The student participants were asked to keep their mobile telephones on from 9 AM to 9 PM each day, Monday through Saturday, for a period of two weeks (twelve days). They were told that, each day, they would receive approximately five text messages at random moments between 9 AM and 9 PM. Each time they received a message, they were asked to complete a short questionnaire in respect of the most recent "decision action" they had taken. Messages followed a standard format, for example, "Message #2 – Robin" would signal that this was the second message being sent on that day. The procedure for the executive participants was the same except that messages were only sent on Mondays through Fridays and between 9 AM and 7 PM.

To send the messages each day, time (9 AM to 9 PM or 9 AM to 7 PM) was divided into ten-minute segments and random numbers were assigned to the segments. The segments corresponding to the five largest numbers were the time intervals during which the messages were sent. In a few cases, I failed to send messages and thus participants did not receive exactly five messages each day. The messages were sent over the Internet using commercially available messaging services provided by mobile telephone operators. Forty-eight messages were sent to the first group of executives; and between twenty and twenty-four messages were sent to the second group. Sixty or sixty-one messages were sent to the students (not all of whom started the study on the same day).

The first group of executives received all instructions and questionnaires in written form through e-mail and returned either hard or electronic versions of the questionnaires. A package of all materials was given to the second group of executives including an addressed envelope to facilitate returns. English was the sole language used with the executives. (At least seven respondents were native-English speakers and all had studied or were studying management courses in English.) The student participants were given verbal instructions (in Spanish) and written instructions (in English) and were provided with questionnaire forms to complete and return to the author. The questionnaires were in English but almost all student participants responded in Spanish.

Instructions asked participants to focus on what were called "decision actions" (or DAs).[5] These were defined as "any *decision* or *judgment* that you make." Moreover, it was emphasized that these could vary from important to trivial and that, from the viewpoint of the study, it did not matter whether the DAs involved "large" or "small" consequences. Participants were further instructed to focus on the DA that was "closest in time to the moment that you receive the text message." Since it was known beforehand that executives would not keep their mobile telephones on all day, instructions to this group specified "If on reconnecting your mobile telephone you find a message, please act as though the message had just been sent, that is, assume that the message was sent just before you reconnected your telephone." At one level, this procedure could have produced a small bias in random sampling of decisions; however, its design helped to reduce distortions caused by faulty memory.

The object of the questionnaire administered to the second group of students was to obtain retrospective estimates of the aggregate features of the data that had been collected from the first group of students on a concurrent basis. Thus, the questionnaire explicitly asked respondents to limit their estimates to the two preceding weeks, specifically, Mondays through Saturdays, from 9 AM to 9 PM. The questionnaire also emphasized that all decisions, including the trivial, should be considered.

To illustrate the manner in which the questions were asked, consider the question in the ESM studies about confidence in decisions. Here, participants were asked to state for each decision "How confident are you that you took the 'right' decision?" Responses were made by checking one of five levels: "Very confident," "Confident," "Somewhat confident," "Not confident," and "Not at all confident." In the retrospective questionnaire, respondents were asked "Of all the decisions that you took, what percentages of times could you express the following levels of confidence that you had taken the 'right' decision?" Responses were made by assigning percentage estimates to each of the aforementioned five levels (using exactly the same words). In the following we refer to the responses of the questionnaire study as the "student controls" to indicate that these responses represent a baseline against which the ESM student data can be assessed.[6]

---

[5] Copies of the instructions and the questionnaires as well as the coding scheme for qualitative data (see the discussion to follow) may be obtained from the author.

[6] The second group of students is, of course, not a control group in the accepted use of this term (i.e., allocation of participants to the first and second groups was not made at random).

RESULTS

Responses

As previously noted, valid responses for the ESM studies were received from twenty-four executives and ten students. The executives provided analyzable responses for 613 of 876 messages sent, that is, 70%. However, there was considerable variation in individual response rates from 15% to 100%. The median was 85%. One could exclude the responses of those executives who provided few responses; however, there is no reason to believe that their responses were any more or less valid than those who gave many responses. Executives who submitted few responses simply stated that they had difficulty completing all requests for data.

The response rate from the students was much higher. In total, the students gave responses to 585 of the 605 messages sent to them, that is, 97%. The median response rate per student was 98%. Given that the methodology interrupted their daily life at unpredictable moments, this is a highly satisfactory response rate.

Across both populations, there were approximately 1,200 decisions to analyze. In the questionnaire study, all twenty-five participants provided usable data.

Checks on Data

Participants were asked to record both the time at which they received the text messages and the time at which the decisions they reported took place. Deviations between the former and the time at which the messages were sent provide a check on whether messages were being received at the appropriate moments. Checks between the time messages were received and the time reported decisions took place are important because the smaller the gap between these two times, the less likely it is that the decisions reported suffer from selection biases or distortions in memory. (Recall that participants were asked to report on decisions that immediately preceded the receipt of messages.)

For the twenty-four executives, the median times between sending messages and the recording of these varied between 1 and 46 minutes with an overall median (by participant) of 5 minutes. Thus, the executives did not always report receiving the messages when they were sent, mainly because several were unable to maintain their mobile telephones in a ready state

---

What the second group does represent is simply a number of students drawn from the same population as those who participated in the ESM study. In a future study, it would be of interest to replicate the present experiments by allocating respondents at random to the two groups.

but only consulted them from time to time. (This was a result, for example, of air travel or attending important meetings.) For all but one of the student participants, the analogous median times were only 1 or 2 minutes. (The outlier was 10 minutes.) The students clearly received the messages when they were sent.

Given the goals of the study, the more important deviation is between the moments that messages were received and when reported decisions were made. For the executives, individual median times varied between 0 and 35 minutes with an overall median of 14 minutes. For the individual student participants, the median times varied from 4 to 35 minutes with an overall median (across participants) of 9 minutes.

ESM Questionnaire Results

Each ESM questionnaires posed eleven questions. Two simply asked for the date and time. Six required quantitative responses (e.g., estimates of time or checking one of several possible answers). For three of the questions, participants had to write descriptions. These involved what the participants were doing when they received the messages, a description of the most recent "decision action" that they had taken, and an explanation of the feedback (if any) that they would receive on their decisions (see also the discussion in the following).

The strategy for dealing with the three latter questions involved (1) accepting what participants had written at face value (i.e., if a participant wrote about a decision in a particular way, it was assumed that was how she or he had actually perceived the situation) and (2) developing a coding scheme to classify responses in a consistent manner. Coding involved three tasks. One was to classify what respondents were currently doing and the domain of activity to which specific decision actions applied. Categories for classification were created after reading all questionnaire responses and examining categories used in previous ESM studies (in particular, Sjöberg & Magneberg, 1990). The second task concerned the structure of each decision. This was analyzed in terms of the number of alternatives stated or implied and whether the decision was positively or negatively framed. The third task, the explanation of feedback, was classified as to whether it could be considered confirming, disconfirming, or possibly both. The schema used for analysis is discussed in greater detail in the following (see results concerning feedback).

All data were coded independently by two research assistants who were ignorant of the author's expectations for the study. For each qualitative question, the assistants recorded their level of agreement and then discussed all disagreements until they reached consensus. (Where appropriate, I report the initial level of agreement between the two coders.) In what follows, the data are primarily presented in aggregate form, contrasting the total responses of the executives, on the one hand, and the students, on the other. Although there are variations in responses by individuals, the mean aggregate responses (of both executives and students) are almost identical to the mean responses of individuals in each group.[7]

Finally, in the tables presenting the data, I have included responses of the "student controls." These are the responses of the twenty-five participants in the retrospective questionnaire study who estimated aggregate characteristics of their decision-making activities over the two preceding weeks (9 AM to 9 PM, Mondays through Saturdays).

Current Activities and Domains of Decisions

When they received messages, participants were asked to record what they were doing (their "current activity") as well as to describe their most recent decision or DA. Table 19.1 presents classifications of both current activities and the domains of most recent decisions[8] – for executives and students, separately, with the data of the "student controls" appearing on the right-hand side of the student data.

The data show, first, that the domains in which decisions are made are closely related to current activities (as one would suspect). Second, there are both similarities and differences in what the executives and students are doing and the domains of their decisions. For both, about one-third of their activities involve their lives as business people or students ("basic occupation") with the executives making somewhat more decisions than students in this category (36% versus 30%). However, whereas 19% of executives' decisions involve "professional communication" (the second most important category), the analogous figure for students is a mere 2%. Indeed, if the categories of basic occupation and professional communication are summed, it is clear that the executives are involved in many more work-related activities than the students (i.e., 55% versus 32%).

As to further differences, "sleep, rest, and recreation" are quite high on the students' list but low on that of the executives. In addition, if we add to this category the activities and decisions devoted to "eating and drinking," it is clear that these occupy more space in the lives of the students than in those of the executives (27% of decisions as opposed to 13%). Interestingly, between 10% and 15% of the decisions of both groups lie in the domain "housework, personal time, and funds management." [For other studies that describe people's activities across time using the ESM, see Sjöberg & Magneberg (1990).]

---

[7] This does not, of course, have to be the case because participants (and particularly the executives) varied in the number of responses they provided.

[8] Here, as in other cases where qualitative responses were provided, I report the initial intercoder agreement rate on classification (i.e., prior to the reaching of a consensus).

TABLE 19.1. *Current Activities and Domains of Decisions*

				Executives	
	Domain of Decisions[a]	Current Activity	Domain of Decisions (%)	Current Activity (%)	
1. Basic occupation[c]	217	208	36	34	
2. Professional communication	117	169	19	27	
3. Eating and drinking	61	52	10	8	
4. Housework, personal time, and funds management	93	49	15	8	
5. Transportation	25	48	4	8	
6. Personal communication	27	27	4	4	
7. Acquiring information	11	22	2	4	
8. Entertainment and sports	14	16	2	3	
9. Sleep, rest, recreation	20	15	3	2	
10. Developing additional skills	7	11	1	2	
11. Personal hygiene, beautification, dressing	13	2	2	0	
12. Ethics	6	0	1	0	
	611	619			

(*continued*)

TABLE 19.1 (continued)

	Students				
	Domain of Decisions[b]	Current Activity	Domain of Decisions (%)	Current Activity (%)	Activities of Controls (%)
1. Basic occupation[c]	173	243	30	41	30
2. Eating and drinking	107	95	18	16	12
3. Sleep, rest, recreation	50	56	9	10	14
4. Housework, personal time, and funds management	61	39	10	7	4
5. Transportation	26	38	4	6	7
6. Personal communication	60	36	10	6	8
7. Entertainment and sports	26	22	4	4	7
8. Developing additional skills	19	32	3	5	3
9. Personal hygiene, beautification, dressing	29	10	5	2	4
10. Acquiring information	17	11	3	2	6
11. Professional communication	13	6	2	1	2
12. Ethics	1	0	0	0	3
	582	588			

[a] Intercoder agreement: 76%.
[b] Intercoder agreement: 81%.
[c] The data are ordered by frequency of current activity.

Pretesting of the retrospective questionnaires revealed that respondents would probably not be able to distinguish differences in estimates of the time they had spent in different activities and the domains in which they took decisions. Thus, they were only asked to estimate percentages of time spent on different activities. The distribution across activities shown in Table 19.1 under "Activities of controls" (lower right-hand column) differs from the distribution of current activities reported by the ESM students in that the former is less skewed than the latter.[9]

Action Types

Decisions can be described in many different ways. In the foregoing, I reported the content or domain of decisions. Here, I consider their structure. Specifically, each decision was defined as belonging to one of three different "action types." The first is a straightforward situation where someone decides "to do" as opposed "not to do" something. What is not done is unspecified and the description of the decision focuses only on what is to be done. As an example, consider a decision described as "to have a cup of coffee." Note that the formulation of this decision has a *positive* focus. In the second action type, we consider the decision "not" to do something, such as "not to have a cup of coffee." This is a *negative* focus. The third action type reflects more complex decisions with multiple alternatives. An example is "to have a cup of coffee or to have a cup of tea." More complicated examples could include "deciding on a schedule of activities" or "making a list of priorities." These all involve several different possible actions where the alternative is not just the negation of a single action.

Table 19.2 reports some characteristics of the data including "action type," distinguishing between responses of students and executives. Actions with a positive focus dominate those with a negative focus for both groups of participants. Indeed, the relative lack of decisions with a negative focus is striking. For the third action type, executives clearly see more specific alternatives than students, 36% versus 16%. However, of particular interest here is the individual variability among executives. Whereas for thirteen (out of twenty-four) respondents, this figure is less than 15%, for six it is greater than 50% (including three with more than 95%).

Orientation

Respondents were asked whether their decisions were professional or private. As might be expected, the percentage of private responses was higher

---

[9] All comparisons between the distributions of responses of the student controls and the data collected concurrently from the students in the ESM study are significantly different (using χ^2 tests, $p < .01$).

TABLE 19.2. *Characteristics of Decisions*

	Executives	Students	Total	Executives (%)	Students (%)	Students Controls (%)
Action type						
Positive focus	358	453	811	58	80	28
Negative focus	36	27	63	6	5	56
Multiple alternatives	218	89	307	36	16	16
	612[a]	569[b]	1,181			
Orientation						
Professional	375	153	528	61	26	
Private	222	397	619	36	69	
Both	15	28	43	2	5	
	612	578	1,190			
Frequency: Was this something that you do						
Frequently and without really thinking?	99	134	233	16	23	34
Frequently but you do think about it?	161	213	374	26	37	45
From time to time?	197	122	319	32	21	12
Rarely?	96	80	176	16	14	6
This was the first time!	58	34	92	9	6	3
	611	583	1,194			
Confidence in the "right" decision						
Very confident	189	203	392	31	35	42
Confident	232	181	413	38	31	32
Somewhat confident	150	77	227	25	13	16
Not confident	35	103	138	6	18	7
Not at all confident	6	20	26	1	3	3
	612	584	1,196			

for the students than for the executives. Indeed, it is surprising that the students should have such a low proportion of professional responses (24%) when so many of their activities were centered on their studies (see Table 19.1). Similarly, although the executives were mainly at work, they perceived 36% of their decisions as being private in nature.

Frequency

To what extent does decision making involve frequently occurring or new activities? Here students and executives had somewhat different responses, with executives reporting somewhat less frequently occurring events. (Transforming the questionnaire responses to a 1–5 scale and testing the difference in means leads to a t statistic of 2.04, $p < .05$.) Students classified 23% of decisions as being made "frequently and without really thinking" and for executives this percentage was 16%. There must, of course, be some doubt as to whether these latter figures represent an accurate assessment of decisions that are made "automatically" since, almost by definition, people may not be aware of taking such decisions (see, e.g., Hogarth, 2001). Nevertheless, roughly 20–25% of all decisions fell into the categories of being done "rarely" or for "the first time."

Confidence in the "Right" Decision

Overall, confidence that the "right" decision was taken was high. For both students and executives, the categories of "very confident" and "confident" were checked in almost 70% of cases and relatively few responses indicated lack of confidence. Interestingly, one respondent alone accounted for almost 50% of the students' responses to the last two categories ("not confident" and "not at all confident") and when this participant's responses are omitted, the data for the students and executives are more similar.[10] In the following, I analyze the correlates of confidence and comment further on these findings.

Finally, the retrospective data of the student controls indicate more confidence that the "right" decisions had been taken than the responses of the students in the ESM study (76% versus 66% when summing the "very confident" and "confident" categories). In other words, the data suggest that overall assessments, based on memory, involve a greater sense of confidence than the aggregation of estimates expressed at the time events occurred. This finding runs contrary to what would be predicted from

---

[10] This "outlier" was a twenty-eight-year-old graduate student who was quite different from the undergraduate population. Incidentally, statistical tests of any differences between the distributions of responses of students and executives on this question revealed no significant differences irrespective of whether the outlier was included.

evidence in the calibration literature where overall confidence expressed for a series of events has been found to be less than that obtained by aggregating the confidence levels expressed for each of the events (Gigerenzer et al., 1991).

Feedback

Table 19.3 reports data concerning feedback. In light of a pilot study, the ESM questionnaire did not explicitly use the word "feedback." Instead, respondents were asked "Will you ever know whether you took the 'right' decision or do you already know?" with the response categories being "Yes" or "No." Following "Yes" responses, participants were requested to answer three questions: (1) "When will you know (please express as the length of time between taking the DA and the moment of knowing)?" (2) "Please explain how you know or will know." (3) "How confident can you be in this explanation?" (This latter question had five possible responses – see Table 19.3.)

Overall, students and executives report that some 52% to 65% of decisions, respectively, are accompanied by feedback or expected feedback. (There was considerable variation in student responses but differences are not statistically significant. In particular, two students reported lack of feedback in 85% and more of cases; the largest comparable figure for any executive was 71%.) In addition, both executives and students express high levels of confidence in the accuracy of the feedback they receive or expect to receive (see bottom of Table 19.3).

Where the executives and students differ is in the timing of feedback following action. Both groups do receive some feedback immediately after taking actions (32% and 20%, respectively, in less than 2 minutes). Here, however, the similarity ends. Whereas within 45 minutes students receive feedback on 55% of occasions, the analogous figure for executives is 29%. For feedback received after a week, the figures are 20% and 35% for students and executives, respectively. In addition, in 10% of the cases the timing of feedback for executives is indeterminate; that is, it is unclear from their descriptions when this feedback will be received, if at all. At the individual level, fourteen of the twenty-four executives reported that at least 40% of their feedback would only be received after one week or was indeterminate, whereas this was true of only one of the ten students.

The student controls differ from their ESM counterparts in that they claimed that they received much more feedback (70% versus 52%) and the distribution of this feedback was more spread out in time. (For eample, the median time for the student controls to receive feedback after taking decisions was between "90 minutes to 3 hours," whereas for the ESM student participants, this median was between "15 and 45 minutes.") The ESM participants also express greater confidence in the feedback than do the student controls (74% versus 67%).

	Executives	Students	Total	Executives (%)	Students (%)	Student Controls (%)
Feedback						
Yes	395	305	700	65	52	70
No	217	279	496	35	48	30
	612	584	1,196			
Time until feedback						
Less than 2 minutes	80	96	176	20	32	17
2 to 5 minutes	3	3	6	1	1	8
5 to 15 minutes	16	31	47	4	10	4
15 to 45 minutes	17	35	52	4	12	11
45 to 90 minutes	18	13	31	5	4	8
90 minutes to 3 hours	8	21	29	2	7	7
3 hours to 15 hours	24	21	45	6	7	5
15 hours to 2 days	21	12	33	5	4	15
2 days to 1 week	30	9	39	8	3	7
1 week to 1 month	56	21	77	14	7	8
More than 1 month	81	40	121	21	13	11
Indeterminate	41	2	43	10	1	0
	395	304	699			
Confidence in feedback						
Very confident	179	119	298	46	40	35
Confident	130	102	232	33	34	32
Somewhat confident	77	26	103	20	9	23
Not confident	6	43	49	2	15	7
Not at all confident	0	6	6	0	2	3
	392	296	688			

473

Feedback

	Confirming	Disconfirming
Action taken	Cell a	Cell b
*Action not taken**	Cell c	Cell d

* This could mean that another action was taken as opposed to no action having been taken.

FIGURE 19.1. Types of feedback

What type of feedback do people receive and expect? Participants were asked to "explain how you know or will know" that the "right" decision had been taken. For both cases, participants' responses were classified according to a 2×2 table that has often been used to describe the relations between actions or beliefs and types of feedback. This is reproduced here as Figure 19.1.

The rows of Figure 19.1 represent actions taken and their alternatives; the columns distinguish between feedback that can be thought of as, alternatively, confirming and disconfirming the "correctness" or "appropriateness" of actions taken. Cell a represents the conjunction of taking the action and observing/expecting confirming evidence (that the "right" decision was taken); cell b is the conjunction of taking the action and observing evidence that it was not the right decision (i.e., disconfirming evidence); cell c is the conjunction of not taking the action but seeing evidence that would have confirmed that the action should have been taken; and cell d is the conjunction of not taking the action and evidence that would have shown it to be inappropriate. Of course, "not taking the action" also includes taking another explicit action when alternatives have been made explicit. As seen previously in Table 19.2, most decisions had a simple, positive focus even though there were differences between executives and students as to the extent to which alternative actions were specified (i.e., simple negations versus multiple alternatives).

Table 19.4 provides a classification of the types of feedback received or expected. The analysis distinguishes between feedback that has been or is being received (labeled "current") and feedback that had not yet been received (labeled "expected"). Consistent with the fact that the executives received or expected to receive feedback later than the students (relative to when decisions were taken), the students report more current feedback than the executives (32% versus 15%) and, correspondingly, less expected feedback.

TABLE 19.4. *Type of Feedback*

Classification		Executives[a]	Students[b]	Total	Executives (%)	Students (%)	Student Controls[c] (%)
Unobservable		36	9	45	9	3	0
Current	Cell a	60	88	148	14	29	22
	Cell b	1	8	9	0	3	9
	Cell c	1	0	1	0	0	0
	Cell d	4	1	5	1	0	0
Expected	Cell a	61	67	128	15	22	35
	Cell b	9	12	21	2	4	17
	Cell c	2	1	3	0	0	1
	Cell d	4	4	8	1	1	1
	Cell a or b	233	110	343	56	36	12
	Cell c or d	7	6	13	2	2	1
		418	306	724			

[a] Intercoder agreement: 78%.
[b] Intercoder agreement: 79%.
[c] Responses normalized to facilitate comparisons.

475

The predominant type of feedback for both students and executives – and for both time periods – involves cells *a* and *b*, either separately or as a conjunction. Indeed, there are relatively few instances that involve other cells. The classification "unobservable" covers cases where participants stated that they would receive or had received feedback but, as far as could be determined from what they had written, these simply represented beliefs for which no evidence was forthcoming. Of particular interest for expected feedback is that participants were often acutely aware that their decisions could be wrong and expressed this by describing feedback in terms of both cells *a* and *b* or cells *c* and *d*. Indeed, 56% of expected feedback of the executives refers to both cells *a* and *b* (for students this figure is 36%).

To seek insight into when participants are more likely to expect "*a* or *b* cell feedback," the corresponding percentage responses for all ESM participants were regressed on different combinations of percentage responses to other questions as well as a dummy variable to distinguish executives and students. This analysis revealed positive significant effects for executives versus students ($t = 2.41$), decisions taken in the domain of "basic occupation" ($t = 2.69$), and multiple alternative action types ($t = 1.86$), and a negative effect for feedback received within 2 minutes of making decisions ($t = -2.15$). The R^2 from this regression equation was 0.54 ($F = 8.35$, $p < .001$) These results can be interpreted by stating that respondents were more open to the possibility that feedback could be either favorable (cell *a*) or unfavorable (cell *b*) in cases involving more complex, work-related decisions for which feedback was not expected within a short time. Moreover, this was truer for executives than students.

Finally, an attempt was made in the retrospective questionnaire to elicit estimates of the kind of feedback that respondents thought they had received or would receive. These student control data are similar to the ESM results concerning current feedback but quite different for expected feedback (see right-hand side of Table 19.4).

Confidence, Feedback, and Time

What are the correlates of feedback? First, recall that whereas participants received or expected to receive feedback for some 60% of their decisions, they were either "confident" or "very confident" for almost 70% of their decisions. However, there was no relation between receiving or not receiving feedback and confidence in decisions. At the individual level, the relation between feedback and confidence in decisions was statistically significant ($\chi^2 < .05$) for only three of the ten students, and for one of these respondents, the relation was negative. For the twenty-four executives, there was only one statistically significant relation ($\chi^2 < .05$). At the group level, the correlation between confidence in decisions and receiving or expecting to receive feedback is 0.07 (ns). In addition, levels of confidence in decisions were unrelated to and did not vary as a function of whether feedback was received shortly after making decisions ($r = 0.08$, ns), or much later ($r = -0.07$, ns).

Attempts to find relations in the data that would "explain" confidence in decisions taken proved unsuccessful. Nonetheless, there is a strong positive relation between the confidence participants have that their decisions are "right" and the confidence they express in the appropriateness of the feedback they receive or expect to receive ($r = 0.77$, $p < .01$). This therefore raises the possibility of "explaining" confidence in feedback. To what extent could this be stimulated by prior feelings of confidence in decisions taken and other variables? To explore this issue, I regressed confidence-in-feedback on confidence-in-decisions as well as other variables. As well as a positive significant effect for confidence-in-decisions ($t = 7.59$), this analysis revealed a significant effect for frequency of decisions, that is, their "habitual" nature ($t = 2.23$). The R^2 from this regression equation was 0.65 ($F = 29.1$, $p < .001$). Whereas extreme care should be exercised in interpreting such a regression, it does suggest a link between confidence and the frequency of certain kinds of decisions; in other words, the more habitual a decision, the more confident a person feels.

DISCUSSION

The study illuminates four different but related issues. The first concerns differences between the decision environments of executives and students; the second is whether people receive feedback with respect to their decisions, and if so, how much; the third is expectation of feedback on decisions taken; and the fourth is the general topic of confidence in decision making. A fifth issue involves differences between data collected concurrently (by the ESM) and retrospectively (the student controls).

The decision environment of the executives and students varied significantly in both content and structure. In terms of content, students perceived the majority of their decisions to be private in nature (69%) whereas executives classified most of their decisions as professional (61%).[11] Second,

---

[11] An anonymous referee correctly noted that comparisons between executives and students might be biased because the times for which the two groups were sampled were not identical (9 AM to 7 PM, Mondays through Fridays, for the executives; and 9 AM to 9 PM, Mondays through Saturdays, for the students). To assess this bias, the data from all messages sent to students after 7 PM and on Saturdays were eliminated. Across all questions, the differences between the original data (i.e., all responses) and responses left after eliminating the noncomparable subset were minimal. The only differences that might merit attention were that the subset of data indicated that more time was spent on "basic occupation" (46% versus 41%) and, correspondingly, more decisions were professional in nature (31% versus 26%). If anything, these results are the opposite of what one might expect (presumably

even though about one-third of the decisions of both executives and students concerned their "basic occupation," executives had a further 19% in the area of professional communication compared with 2% for the students.[12] In addition, about one-quarter (27%) of the executives' activities when they received the text messages occurred while they were engaged in some form of professional communication. Indeed, communication has often beeen quoted as a key managerial skill (see, e.g., Goleman, 1998) and one that has to be developed in younger executives. To the extent that these data are representative, it suggests an important gap between demands in the decision-making environments of executives and students and thus possibly why training in communication is so important for younger executives.

Executives clearly saw the structure of decisions as more complex, as evidenced by the fact that whereas 80% of students' decisions were perceived to have a positive focus, the comparable figure for executives was 58%. Students and executives also varied on reported frequency of the kinds of decisions they took. For students, 60% of decisions were described as occurring frequently (whether they "really" thought about them or not), whereas for executives the comparable figure was 42% (see Table 19.2). Both executives and students received or expected to receive feedback for about 60% of their decisions. However, for executives, feedback was more delayed in time than it was for students.

An anonymous referee suggested that one possible explanation for differences between the inferred decision environments of executives and students could be a reporting bias; that is, the two groups differed in what they considered appropriate DAs to report. In particular, it was argued, executives might be reluctant to report "trivial" decisions (such as when to have coffee). Although it is hard to reject this explanation definitively, a reading of the ESM questionnaire responses does not suggest that the executives refrained from reporting "trivial" decisions. Indeed, these form a large part of their data.

As previously noted, some 40% of decisions involved no actual or expected feedback. Whether this figure is high or low is unclear because, to the best of my knowledge, no other studies have attempted to make such an estimate on the basis of random samples of people's behavior. In addition, evidence from tacit learning would suggest that people may not always

---

the subset contained less leisure time for the students). To conclude, there might well be important differences between how the executives and students responded to the text messages. However, such differences cannot be attributed to the different times during which responses were elicited.

[12] Note too that even if one sums the categories of professional and personal communication, the executives both spend more time involved in this activity than the students (31.7% versus 6.7%) and take relatively more decisions (23.5% versus 11.6%).

be aware of feedback and its effects (see, e.g., Reber, 1993). However, the significance of the 40% estimate is that it emphasizes the informational poverty of environments in which we learn about our decision-making effectiveness. In many cases, I suspect, people simply rely on internal feelings to assess whether their decisions are correct – feelings that are probably based on having received no negative feedback in similar situations in the past. However, it should be clear that this strategy is liable to lead to self-fulfilling beliefs and actions. This, in turn, raises the important issue of seeking means to improve the level of feedback that can be obtained following actions and of making people aware of this necessity (Hogarth, 2001).

What kind of feedback do people expect to test the validity of their actions? Is this only confirmatory in nature? For decisions for which feedback has already been received (the "current" category in Table 19.4), note that the vast majority of actions were accompanied by confirmatory evidence. However, given the manner in which the study was conducted, this feedback must have been received shortly after the actions were taken.

For expected feedback, students and executives do expect confirmation (cell *a*) for 22% and 15% of their decisions, respectively. However, they indicate significant awareness that their actions could also result in negative or disconfirmatory feedback. Indeed, more than half (56%) of all the executive feedback falls in the expected *a* and *b* category. (The figure for students is 36%.) There is also occasional, albeit minimal, reference to potential evidence from cells *c* and *d*. Curiously, some 9% of the feedback executives claimed they would receive was "unobservable."

The overall picture that emerges from these data does not match a stereotype where people *only* think of confirming evidence. Both executives and students are clearly aware that not all actions will result in positive feedback. However, participants do not always indicate that they know what feedback is appropriate to assessing the validity of the actions they have taken.

Of course, in the present work there are no data that can test whether the levels of confidence exhibited by the participants are justified. A priori, the lack of a direct relation between feedback and confidence is some cause for concern. Do participants really discriminate between situations where they are or are not "justified" in expressing confidence? As speculated by Einhorn and Hogarth (1978), the mere fact of making judgments and decisions without receiving direct disconfirming evidence may be sufficient to both create and sustain feelings of confidence. In addition, illusions of confidence may sometimes have beneficial effects in that they encourage taking positive actions across time (cf., Taylor & Brown, 1988, 1994). A critical issue raised, therefore, by this research is how to determine the bases on which people establish feelings of confidence in the decisions they make.

Two important sources of data not considered here are people's individual histories for certain types of decisions[13] and possible individual differences in general levels of confidence. The task given to the students who completed the retrospective questionnaire was not easy and thus the fact that responses were different from the concurrent ESM data should not be surprising. The most interesting differences concerned confidence and feedback. In the retrospective study, students reported being more confident in their decisions than their ESM counterparts, receiving more feedback, and, on average, receiving feedback with greater delays. Whereas here I interpreted the difference in confidence as contradicting research in the calibration paradigm that would lead to expecting the opposite finding (i.e., that retrospective would be less confident than concurrent; cf. Gigerenzer et al., 1991), I caution against emphasizing this result. First, it should be replicated in a study where both retrospective and concurrent data are collected from the same participants. Second, the real interest in the difference between concurrent and retrospective data in the present research is to demonstrate that they do not lead to the same results (cf. Ericsson & Simon, 1984; McFarlane et al., 1988).

CONCLUSIONS

To the best of my knowledge, no other studies have used the ESM methodology for examining decision behavior. Thus, there are no benchmarks for considering the results reported in this study. For example, should we be surprised that participants receive or expect to receive feedback on about 60% of their decisions or that they are confident about 70%? What do these estimates imply in terms of how people acquire decision-making skills? Given that this is the first ESM study on decision making, I limit myself here to its promise for studying further aspects of decision behavior. I first consider some methodological issues, and then turn to substantive ones.

In terms of methodology, the present study raises a number of concerns. The major one can be summarized under the heading of selection biases. It was clear that not all executive participants were able to complete the task in the manner requested. This therefore raises the possibility that only executives with certain kinds of decision environments could complete the demands of the study and raises the question of how to overcome this problem in future studies. Jobs undoubtedly differ in the extent to which they allow executives to be more or less available to respond to questionnaires when they receive text messages. In addition, the effort needed to

[13] It was, of course, possible to categorize the data collected by domains of applications and comparative frequencies of decisions (see Tables 19.1 and 19.2). However, neither of these categories is sufficient to characterize a person's decision-making history for kinds of decisions.

respond to each message can be a barrier. Some suggestions to overcome these difficulties include reducing the number of days over which studies are conducted and/or the number of messages sent per day. Simplification of the questionnaires is also a possibility. For example, questions could possibly be designed in more of a check-list format. In addition, it may be feasible to enlist technology to help facilitate the process. For example, a couple of executives in the present study transformed the questionnaire into a spreadsheet and used this to record responses. Specially programmed pocket computers have been used to collect data in other ESM studies (see, e.g., Teuchmann, Totterdell, & Parker, 1999). The use of such technology also increases the feasibility of being able to ask people different questions on different occasions and/or being able to tailor questions to prior responses on a real-time basis.

A second form of selection bias centers on which particular decisions participants chose to report. In this study, participants were requested to focus on the *decision action* that was closest in time to the moment the message was received. However, this instruction still left much leeway to the participants to select or avoid specific types of decisions. For example, although both groups reported that some 16% to 23% of their decisions were taken "without really thinking," my guess is that the real percentage could be much higher. One way of assessing the seriousness of this bias could be to have participants "shadowed" by an investigator as in Brunswik's (1944) study. It would be the investigator's task to "audit" the decisions reported by the participant. Although expensive, this could possibly be done on a small-sample basis.

It would, of course, be naïve to believe that the ESM is *the* solution to methodological problems in research on decision making. No methodology owns the truth. Clearly, the ESM has great potential but this potential is most likely to be realized when it is used in conjunction with other approaches. For example, ESM could be used as an adjunct measurement tool in studies of so-called naturalistic decision making (cf. Lipshitz et al., 2001). It is also easy to imagine it as a supplement to questionnaires such as used in studies of consumer behavior or social psychology. Its greatest value, however, would seem to be its ability to calibrate the frequency of behaviors that have been identified by other means and to assess the kinds of samples of situations that people experience in their natural environments. We all know, for example, that we do not receive feedback on all of our decisions. However, how often does this occur? How does this vary according to different conditions, that is, by persons and/or environments? And how important is this?

In terms of substance, what this study has achieved, inter alia, is a glimpse into the decision environments of groups of executives and students. It will therefore be important to replicate the present study with different populations and to develop specific hypotheses about the types of decisions they encounter. For example, it would be intriguing to investigate differences between people with varying levels of experience in specific occupations or people with different functional responsibilities.

For future studies, several questions would seem well suited to the ESM. One is the relation between mood or emotions and types of decision taken. To what extent do people use different ways of making decisions when they are in "good" or "bad" moods? Mood has often been a dependent variable in ESM work and can be measured quite easily (see, e.g., Csikszentmihalyi, 1990). However, is it related to confidence in decisions made or the kind of information that people might expect as feedback? Can one detect a relation between mood and risk-taking in people's naturally occurring environments (cf. Isen, 1993)? A further topic concerns how people make decisions, for example, analytically, using specific heuristics, through feelings and intuitions, or by some combination of the preceding. Assuming that one could teach people how to classify decisions just taken, it would be intriguing to learn more about the frequencies of different "methods" as well as their correlates, for example, confidence, perceived risk, and so on. One complication, of course, with the present approach is that it is difficult to assess whether people are or are not making "good" decisions. Finding ways to assess this represents a daunting challenge for future research.

A criticism of the present study could be that most of the approximately 1,200 decisions examined had trivial consequences and could thus be dismissed as irrelevant. It is important to recognize, however, that life's consequences do not depend solely on how people make important decisions but may be more significantly affected by the *cumulative* effects of small and seemingly irrelevant decisions. There are two reasons for this. First, the sheer size of the cumulative consequences of trivial decisions can be huge. Second, even though people may use somewhat different processes when making important as opposed to trivial decisions, the former are undoubtedly affected by the latter. It is difficult to suspend habits or routines that have built up over years of experience. Indeed, a case could be made that even incremental improvements in the manner in which people make small, everyday decisions could have huge effects on the outcomes they experience in life (see also Hogarth, 2001).

In conclusion, the present study demonstrates the feasibility of using the ESM to study decision behavior. Clearly, it cannot examine all issues that have been studied by other research methods. However, it can illuminate the kinds of decision environments that people really experience and thereby clarify possible misconceptions of what is involved in ongoing, everyday decision making. In other words, what are the samples of tasks that constrain responses of "intuitive statisticians"? Moreover, the ESM can help researchers sample decision-making behaviors in ways that can differentiate what is and what is not important. For example, it has been argued that judgmental "biases" can be the result of "heuristic" strategies that are generally useful (cf. Tversky & Kahneman, 1974). However, how often do people encounter situations in which heuristics are dysfunctional? As this chapter demonstrates, we have the means to sample people, behavior, and circumstances and thus to answer these kinds of questions by achieving a more accurate understanding of the natural ecology of decision making.

References

Brunswik, E. (1944). Distal focusing of perception: Size constancy in a representative sample of situations. *Psychological Monographs, 56* (254), 1–49.

Brunswik, E. (1956). *Perception and the representative design of psychological experiments.* Berkeley: University of California Press.

Camerer, C. F. (1981). *The validity and utility of expert judgment.* PhD dissertation, University of Chicago.

Chapanis, A. (1961). Men, machines, and models. *American Psychologist, 16,* 113–131.

Chapanis, A. (1967). The relevance of laboratory studies to practical situations. *Ergonomics, 10,* 557–577.

Cronbach, L. J. (1975). Beyond the two disciplines of scientific psychology. *American Psychologist, 30,* 116–127.

Csikszentmihalyi, M. (1990). *Flow: The psychology of optimal experience.* New York: Harper-Collins.

Csikszentmihalyi, M., & Larson, R. (1987). Validity and reliability of the experience-sampling method. *Journal of Nervous and Mental Disease, 175,* 526–536.

Ebbesen, E. B., & Konecni, V. J. (1980). On the external validity of decision-making research: What do we know about decisions in the real world? In T. S. Wallsten (Ed.), *Cognitive processes in choice and decision behavior.* Hillside, NJ: Lawrence Erlbaum.

Einhorn, H. J. (1980). Learning from experience and suboptimal rules in decision making. In T. S. Wallsten (Ed.), *Cognitive processes in choice and decision behavior.* Hillside, NJ: Lawrence Erlbaum.

Einhorn, H. J., & Hogarth, R. M. (1978). Confidence in judgment: Persistence of the illusion of validity. *Psychological Review, 85,* 395–416.

Ericsson, A., & Simon, H. A. (1984). *Protocol analysis: Verbal reports as data.* Cambridge, MA: MIT Press.

Fiedler, K. (2000). Beware of samples! A cognitive-ecological sampling approach to judgment biases. *Psychological Review, 107,* 659–676.

Gigerenzer, G., Hoffrage, U., & Kleinbölting, H. (1991). Probabilistic mental models: A Brunswikian theory of confidence. *Psychological Review, 98,* 506–528.

Goldstein, D. G., et al. (2001). Group report: Why and when do simple heuristics work? In G. Gigerenzer & R. Selten (Eds.), *Bounded rationality: The adaptive toolbox.* Cambridge, MA: MIT Press.

Goleman, D. (1998). *Working with emotional intelligence.* New York: Bantam.

Hammond, K. R. (1978). *Psychology's scientific revolution: Is it in danger?* Reprint 211, Center for Research on Judgment and Policy, University of Colorado, Boulder, CO.

Harvey, N. (1997). Confidence in judgment. *Trends in Cognitive Science, 1,* 78–82.

Hogarth, R. M. (1986). Generalization in decision research: The role of formal models. *IEEE Transactions in Systems, Man, and Cybernetics, SMC-16*(3), 439– 449.

Hogarth, R. M. (2001). *Educating intuition.* Chicago: University of Chicago Press.

Hurlburt, R. T. (1997). Randomly sampling thinking in the natural environment. *Journal of Consulting and Clinical Psychology, 67,* 941–949.

Isen, A. (1993). Positive affect and decision making. In M. Lewis & J. M. Haviland (Eds.), *Handbook of emotions* (pp. 261–277). New York: Guilford.

Juslin, P. (1993). An explanation of the hard-easy effect in studies of realism of confidence in one's general knowledge. *European Journal of Cognitive Psychology, 5,* 55–71.

Juslin, P. (1994). The overconfidence phenomenon as a consequence of informal experimenter-guided selection of almanac items. *Organizational Behavior and Human Decision Processes, 57,* 226–246.

Juslin, P., Winman, A., & Olssen, H. (2000). Naïve empiricism and dogmatism in confidence research: A critical examination of the hard-easy effect. *Psychological Review, 107,* 384–396.

Klayman, J., Soll, J. B., González-Vallejo, C., & Barlas, S. (1999). Overconfidence: It depends on how, what, and whom you ask. *Organizational Behavior and Human Decision Processes, 79,* 216–247.

Lichtenstein, S., Fischhoff, B., & Phillips, L. D. (1982). Calibration of probabilities: The state of the art to 1980. In D. Kahneman, P. Slovic, & A. Tversky (Eds.), *Judgment under uncertainty: Heuristics and biases* (pp. 306–334). Cambridge, UK: Cambridge University Press.

Lipshitz, R., Klein, G., Orasanu, J., & Salas, E. (2001). Taking stock of naturalistic decision making. *Journal of Behavioral Decision Making, 14,* 331–352.

McFarlane, J., Martin, C. L., & Williams, T. M. (1988). Mood fluctuations: Women versus men and menstrual versus other cycles. *Psychology of Women Quarterly, 12,* 201–223.

Murphy, A., & Winkler, R. L. (1984). Probability forecasting in meteorology. *Journal of the American Statistical Association, 79,* 489–500.

Reber, A. S. (1993). *Implicit learning and tacit knowledge: An essay on the cognitive unconscious.* New York: Oxford University Press.

Russo, J. E., & Schoemaker, P. J. H. (2002). *Winning decisions.* New York: Doubleday.

Schwartz, B. (1982). Reinforcement-induced behavioral stereotypy: How not to teach people to discover rules. *Journal of Experimental Psychology: General, 111,* 23–59.

Sjöberg, L., & Magneberg, R. (1990). Action and emotion in everyday life. *Scandinavian Journal of Psychology, 31,* 9–27.

Taylor, S. E., & Brown, J. D. (1988). Illusion and well-being: A social psychological perspective on mental health. *Psychological Bulletin, 103,* 193–210.

Taylor, S. E., & Brown, J. D. (1994). Positive illusions and well-being revisited: Separating fact from fiction. *Psychological Bulletin, 116,* 21–27.

Teuchmann, K., Totterdell, P., & Parker, S. K. (1999). Rushed, unhappy, and drained: An experience sampling study of relations between time pressure, perceived control, mood, and emotional exhaustion in a group of accountants. *Journal of Occupational Health Psychology, 4,* 37–54.

Tversky, A., & Kahneman, D. (1974). Judgment under uncertainty: Heuristics and biases. *Science, 211,* 453–458.

Index

Printed in the United States
By Bookmasters